Springer Texts in Business and Economics

For further volumes:
http://www.springer.com/series/10099

Fuad Aleskerov • Hasan Ersel
Dmitri Piontkovski

Linear Algebra for Economists

Prof. Dr. Fuad Aleskerov
National Research University
Higher School of Economics
Mathematics for Economics
Myasnitskaya Street 20
101000 Moscow
Russia
alesk@hse.ru

Dr. Dmitri Piontkovski
National Research University
Higher School of Economics
Mathematics for Economics
Myasnitskaya Street 20
101000 Moscow
Russia
dpiontkovski@hse.ru

Dr. Hasan Ersel
Sabanci University
Faculty of Arts and Social Sciences
Orhanli-Tuzla, Istanbul
Turkey
hasanersel@yahoo.com

ISBN 978-3-642-27002-4   ISBN 978-3-642-20570-5 (eBook)
DOI 10.1007/978-3-642-20570-5
Springer Heidelberg Dordrecht London New York

© Springer-Verlag Berlin Heidelberg 2011
Softcover reprint of the hardcover 1st edition 2011
This work is subject to copyright. All rights are reserved, whether the whole or part of the material is concerned, specifically the rights of translation, reprinting, reuse of illustrations, recitation, broadcasting, reproduction on microfilm or in any other way, and storage in data banks. Duplication of this publication or parts thereof is permitted only under the provisions of the German Copyright Law of September 9, 1965, in its current version, and permission for use must always be obtained from Springer. Violations are liable to prosecution under the German Copyright Law.
The use of general descriptive names, registered names, trademarks, etc. in this publication does not imply, even in the absence of a specific statement, that such names are exempt from the relevant protective laws and regulations and therefore free for general use.

Cover design: eStudio Calamar S.L.

Printed on acid-free paper

Springer is part of Springer Science+Business Media (www.springer.com)

Preface

The structure of the book was proposed in the lectures given by Aleskerov in 1998–2000 in Boğaziçi University (Istanbul, Turkey), however, different parts of the course were given by Ersel in 1975–1983 in the Ankara University (Turkey) and by Piontkovski in 2004–2010 in the National Research University Higher School of Economics (Moscow, Russia).

The main aim of the book is, naturally, to give students the fundamental notions and instruments in linear algebra. Linearity is the main assumption used in all fields of science. It gives a first approximation to any problem under study and is widely used in economics and other social sciences. One may wonder why we decided to write a book in linear algebra despite the fact that there are many excellent books such as [10, 11, 19, 27, 34]? Our reasons can be summarized as follows. First, we try to fit the course to the needs of the students in economics and the students in mathematics and informatics who would like to get more knowledge in economics. Second, we constructed all expositions in the book in such a way to help economics students to learn mathematics and the proof making in mathematics in a convenient and simple manner. Third, since the hours given to this course in economics departments are rather limited, we propose a slightly different way of teaching this course. Namely, we do not try to give all proofs of all theorems presented in the course. Those theorems which are not proved are illustrated via figures and examples, and we illustrated all notions appealing to geometric intuition. Those theorems which are proved are proved in a most accurate way as it is done for the students in mathematics. The main notions are always supported with economic examples. The book provides many exercises referring to pure mathematics and economics.

The book consists of eleven chapters and five appendices. Chapter 1 contains the introduction to the course and basic concepts of vector and scalar. Chapter 2 introduces the notions of vectors and matrices, and discusses some core economic examples used throughout the book. Here we begin with the notion of scalar product of two vectors, define matrices and their ranks, consider elementary operations over matrices. Chapter 3 deals with special important matrices – square matrices and their determinants. Chapter 4 introduces inverse matrices. In Chap. 5 we analyze the systems of linear equations, give methods how to solve these systems. Chapter ends with the discussion of homogeneous equations. Chapter 6 discusses more general type of algebraic objects – linear spaces. Here the notion of linear independence of vectors is introduced, which is very important from economic point of view for it defines how diverse is the obtained information. We consider here the isomorphism of linear spaces and the notion of subspace. Chapter 7 deals with important case of linear spaces – the Euclidean ones. We consider the notion of orthogonal bases and use it to construct the idea of projection and, particularly, the least square method widely used in social sciences. In Chapter 8 we consider linear transformations, and all related notions such as an image and kernel of transformation. We also consider linear transformations with respect to different bases. Chapter 9 discusses eigenvalues and eigenvectors. Here we consider self-adjoint transformations, orthogonal transformations, quadratic forms and their geometric representation. Chapter 10 applies the concepts developed before to the linear production model in economics. To this end we use, particularly, Perron–Frobenius Theorem. Chapter 11 deals with the notion of convexity, and so-called separation theorems. We use this instrument to analyse the linear programming problem.

We observe during the years of our teaching experience that induction argument creates some difficulties among students. So, we explain this argument in Appendix A. In Appendix B we discuss how to evaluate the determinants. In Appendix C we give a brief introduction to complex numbers, which are important for better understanding the eigenvalues of linear operators. In Appendix D we consider the notion of the pseudoinverse, or generalized inverse matrix, widely used in different economic applications.

Each chapter ends with the number of problems which allow better understanding the issues considered. In Appendix E the answers and hints to solutions to the problems from previous chapters and appendices are given.

Fuad Aleskerov
Hasan Ersel
Dmitri Piontkovski
Moscow–Istanbul

Acknowledgements

We are very thankful to our students for many helpful comments and suggestions. We specially thank Andrei Bodrov, Svetlana Suchkova and other students from National Research University Higher School of Economics (HSE) for their comments and solutions of exercises. Among them, we are particularly grateful to Maria Lyovkina who, in addition, helped us a lot with computer graphics.

Fuad Aleskerov thanks the Laboratory DeCAn of HSE, Scientific Fund of HSE, the Russian Foundation for Basic Research (grant #09-01-91224-CT-a) for partial financial support.

He would like to thank colleagues from HSE and Institute of Control Sciences of Russian Academy of Sciences for permanent help and support.

Fuad Aleskerov is grateful to the Magdalene College of the University of Cambridge where he could complete the final draft of this book. He very much appreciates the hospitality and support of Dr. Patel, Fellow of Magdalene College.

Hasan Ersel would like to thank Tuncer Bulutay, his first mathematical economics teacher, for creating an excellent research environment under quite unfavourable conditions of 1970s at the Faculty of Political Sciences (FPS), Ankara University and for his continuous support and encouragement that span more than 40 years. Hasan Ersel also thanks his colleagues Yilmaz Akyuz, Ercan Uygur and Nuri Yildirim, all from the FPS during 1970s, for their kind help, comments and suggestions.

Dmitri Piontkovski thanks Scientific Foundation of HSE (project 10-01-0037) for partial financial support.

We are very thankful to Boğazici University, Bilgi University, Sabanci University and Turkish colleagues, who organized visits of Fuad Aleskerov and Dmitri Piontkovski to these universities. These visits were also helpful for us to work jointly on the book.

Finally, we thank our families for their patience and support.

Contents

1 Some Basic Concepts ... 1
   1.1 Introduction ... 1
      1.1.1 Linearity ... 1
      1.1.2 System of Coordinates on the Plane $\mathbb{R}^2$ ... 2
   1.2 Microeconomics: Market Equilibrium ... 7
      1.2.1 Equilibrium in a Single Market ... 8
      1.2.2 Multi-Market Equilibrium ... 9
   1.3 Macroeconomic Policy Problem ... 10
      1.3.1 A Simple Macroeconomic Policy Model with One Target ... 11
      1.3.2 A Macroeconomic Policy Model with Multiple Targets and Multiple Instruments ... 13
   1.4 Problems ... 15

2 Vectors and Matrices ... 17
   2.1 Vectors ... 17
      2.1.1 Algebraic Properties of Vectors ... 18
      2.1.2 Geometric Interpretation of Vectors and Operations on Them ... 19
      2.1.3 Geometric Interpretation in $\mathbb{R}^2$ ... 22
   2.2 Dot Product of Two Vectors ... 23
      2.2.1 The Length of a Vector, and the Angle Between Two Vectors ... 24
   2.3 An Economic Example: Two Plants ... 27
   2.4 Another Economic Application: Index Numbers ... 29
   2.5 Matrices ... 30
      2.5.1 Operations on Matrices ... 31
      2.5.2 Matrix Multiplication ... 32
      2.5.3 Trace of a Matrix ... 35
   2.6 Transpose of a Matrix ... 35
   2.7 Rank of a Matrix ... 36
   2.8 Elementary Operations and Elementary Matrices ... 38
   2.9 Problems ... 44

3 Square Matrices and Determinants ... 49
- 3.1 Transformation of Coordinates ... 49
  - 3.1.1 Translation ... 49
  - 3.1.2 Rotation ... 50
- 3.2 Square Matrices ... 51
  - 3.2.1 Identity Matrix ... 51
  - 3.2.2 Power of a Matrix and Polynomial of a Matrix ... 52
- 3.3 Systems of Linear Equations: The Case of Two Variables ... 52
- 3.4 Determinant of a Matrix ... 53
  - 3.4.1 The Basic Properties of Determinants ... 57
  - 3.4.2 Determinant and Elementary Operations ... 60
- 3.5 Problems ... 61

4 Inverse Matrix ... 65
- 4.1 Inverse Matrix and Matrix Division ... 65
- 4.2 Rank and Determinants ... 70
- 4.3 Problems ... 71

5 Systems of Linear Equations ... 75
- 5.1 The Case of Unique Solution: Cramer's Rule ... 78
- 5.2 Gauss Method: Sequential Elimination of Unknown Variables ... 80
- 5.3 Homogeneous Equations ... 85
- 5.4 Problems ... 87
  - 5.4.1 Mathematical Problems ... 87
  - 5.4.2 Economic Problems ... 89

6 Linear Spaces ... 91
- 6.1 Linear Independence of Vectors ... 92
  - 6.1.1 Addition of Vectors and Multiplication of a Vector by a Real Number ... 96
- 6.2 Isomorphism of Linear Spaces ... 97
- 6.3 Subspaces ... 98
  - 6.3.1 Examples of Subspaces ... 98
  - 6.3.2 A Method of Constructing Subspaces ... 99
  - 6.3.3 One-Dimensional Subspaces ... 99
  - 6.3.4 Hyperplane ... 99
- 6.4 Coordinate Change ... 100
- 6.5 Economic Example: Production Technology Set ... 101
- 6.6 Problems ... 104

7 Euclidean Spaces ... 107
- 7.1 General Definitions ... 107
- 7.2 Orthogonal Bases ... 109
- 7.3 Least Squares Method ... 117
- 7.4 Isomorphism of Euclidean Spaces ... 119
- 7.5 Problems ... 120

8	**Linear Transformations**	123
	8.1 Addition and Multiplication of Linear Operators	130
	8.2 Inverse Transformation, Image and Kernel under a Transformation	132
	8.3 Linear Transformation Matrices with Respect to Different Bases	135
	8.4 Problems	137
9	**Eigenvectors and Eigenvalues**	141
	9.1 Macroeconomic Example: Growth and Consumption	148
	9.1.1 The Model	148
	9.1.2 Numerical Example	149
	9.2 Self-Adjoint Operators	150
	9.3 Orthogonal Operators	153
	9.4 Quadratic Forms	156
	9.5 Problems	161
10	**Linear Model of Production in a Classical Setting**	165
	10.1 Introduction	165
	10.2 The Leontief Model	169
	10.3 Existence of a Unique Non-Negative Solution to the Leontief System	172
	10.4 Conditions for Getting a Positive (Economically Meaningful) Solution to the Leontief Model	176
	10.5 Prices of Production in the Linear Production Model	179
	10.6 Perron–Frobenius Theorem	184
	10.7 Linear Production Model (continued)	187
	10.7.1 Sraffa System: The Case of Basic Commodities	189
	10.7.2 Sraffa System: Non-Basic Commodities Added	191
	10.8 Problems	191
11	**Linear Programming**	195
	11.1 Diet Problem	195
	11.2 Linear Production Model	197
	11.3 Convexity	200
	11.4 Transportation Problem	206
	11.5 Dual Problem	207
	11.6 Economic Interpretation of Dual Variables	209
	11.7 A Generalization of the Leontief Model: Multiple Production Techniques and Linear Programming	211
	11.8 Problems	212
A	**Natural Numbers and Induction**	217
	A.1 Natural Numbers: Axiomatic Definition	217
	A.2 Induction Principle	219
	A.3 Problems	223

B	**Methods of Evaluating Determinants**	225
	B.1 Transformation of Determinants	225
	B.2 Methods of Evaluating Determinants of High Order	226
	B.2.1 Reducing to Triangular Form	226
	B.2.2 Method of Multipliers	227
	B.2.3 Recursive Definition of Determinant	228
	B.2.4 Representation of a Determinant as a Sum of Two Determinants	230
	B.2.5 Changing the Elements of Determinant	230
	B.2.6 Two Classical Determinants	232
	B.3 Problems	233
C	**Complex Numbers**	237
	C.1 Operations with Complex Numbers	238
	C.1.1 Conjugation	238
	C.1.2 Modulus	239
	C.1.3 Inverse and Division	239
	C.1.4 Argument	239
	C.1.5 Exponent	241
	C.2 Algebraic Equations	241
	C.3 Linear Spaces Over Complex Numbers	244
	C.4 Problems	245
D	**Pseudoinverse**	249
	D.1 Definition and Basic Properties	250
	D.1.1 The Basic Properties of Pseudoinverse	251
	D.2 Full Rank Factorization and a Formula for Pseudoinverse	252
	D.3 Pseudoinverse and Approximations	255
	D.4 Problems	259
E	**Answers and Solutions**	263
	References	275
	Index	277

Some Basic Concepts

1.1 Introduction

Suppose we study a number of firms by analyzing the following parameters: a_1 – number of workers, a_2 – capital stock and a_3 – annual profit. Then each firm can be represented as a 3-tuple $\mathbf{a} = (a_1, a_2, a_3)$.

The set of all n-tuples $(a_1, \ldots, a_n)$ of real numbers is denoted by $\mathbb{R}^n$.

For $n = 1$, we have the real line $\mathbb{R}$. A point a is represented by its value x_a on the real line $\mathbb{R}$.

Example 1.1. Let a, b, c denote three different automobiles with the respective prices $p_a > p_b > p_c > 0$. One can order these three cars on the price axis, p, as shown in Fig. 1.1.

Note that since prices are non-negative, p-axis coincides with nonnegative real line $\mathbb{R}_+$ in the above example.

1.1.1 Linearity

Price of the two cars a and b is equal to the sum of their prices:

$$p(a \oplus b) = p_a + p_b.$$

The addition is a *linear* operation.

But it can also be the case that

$$p(a \oplus b) = p_a^2 + p_b^2.$$

This, however, is not a linear relation, and we will not study such nonlinear relations in this book.

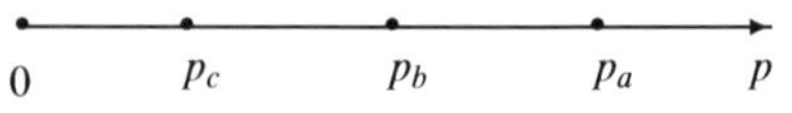

Fig. 1.1 Price axis

Fig. 1.2 Coordinates in the plane $\mathbb{R}^2$

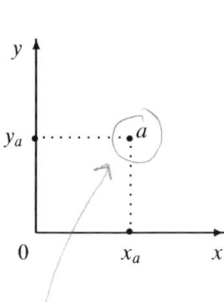

1.1.2 System of Coordinates on the Plane $\mathbb{R}^2$

A point a can be represented in $\mathbb{R}^2$ by its coordinates, i.e. the ordered pair (x_a, y_a), see Fig. 1.2.

Distance on the plane between two points a and b is denoted by $d(a, b)$, where

$$d(a,b)^2 = (x_b - x_a)^2 + (y_b - y_a)^2,$$

SUM OF SQUARE DIFFERENCES

and

$$d(a,b) = \sqrt{(x_b - x_a)^2 + (y_b - y_a)^2}.$$

Example 1.2. Let a be a firm, x_a the number of workers, and y_a the cost of the output (say, in $) of the firm a.

Consider now an equation which relates the variable x to y as

$$y = kx + b,$$

WORKERS
$ OUTPUT

where k, b are constants. The above equation represents an linear function.

Let us first study the case when $k = 2, b = -1$, so that y is an increasing function of x. Since $y = 2x - 1$, for $x = 0$ we have $y = -1$ (Fig. 1.3). This means, for example, that even in the absence of workers the factory should be kept in some condition which, in turn, costs $1. This is called *fixed cost*.

Example 1.3. Consider a firm whose profit function is

$$y = -2x + 3,$$

where x is the output level (Fig. 1.4). It indicates that at some output level the firm's profit declines to zero, and beyond this point it incurs loss.

1.1 Introduction

Fig. 1.3 The linear cost function: fixed cost case

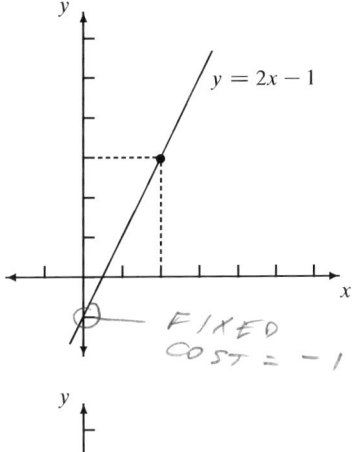

Fig. 1.4 The linear cost function: the case of loss at some output level

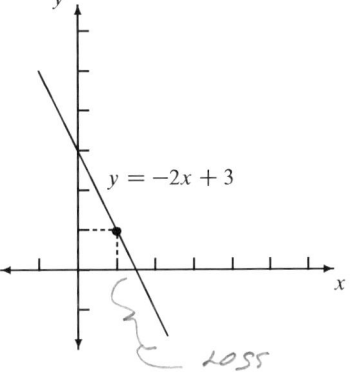

Fig. 1.5 The linear demand function

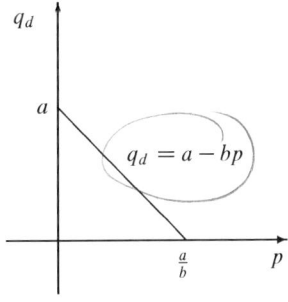

Let us switch to a more realistic model. Consider a market with a single good, and assume that the quantity demanded q_d is a decreasing function of the price p of the good, i.e.,

$$q_d = a - bp,$$

where $a, b > 0$ (Fig. 1.5).

Fig. 1.6 The linear supply function

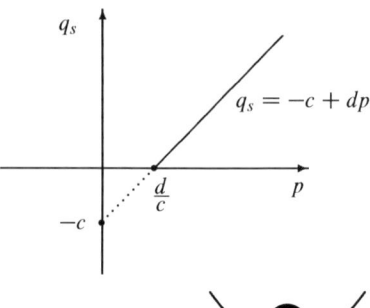

Fig. 1.7 An equilibrium

Assume next that the supplied quantity q_s of the good is an increasing, linear function of p, i.e.,
$$q_s = -c + dp,$$
where $c, d > 0$, see Fig. 1.6.

Consider now the following general definition:

Definition 1.1. Equilibrium is a state of system in which there is no tendency to change.

Example 1.4. Ball inside a bowl (see Fig. 1.7).

Let us define an "excess demand" as the difference between demand and supply. A market is said to be in equilibrium at the price p if the excess demand for the good is zero, i.e.,
$$q_d(p) - q_s(p) = 0.$$

For the market example, the equilibrium price can be found by simultaneously solving the two equations

and
$$\begin{cases} q_d(p) = q = a - bp, \\ q_s(p) = q = -c + dp. \end{cases}$$

A typical equilibrium is illustrated in Fig. 1.8.

A unique solution for the equilibrium as in the above figure can arise only for certain values of parameters a, b, c and d.

In general, two lines in a plane are either intersected or parallel or coincident as depicted in Fig. 1.9.

Solving a system of three equations in two unknowns corresponds to intersecting three lines on the plane. There may arise a number of possibilities as shown in the figures below, see Fig. 1.10.

1.1 Introduction

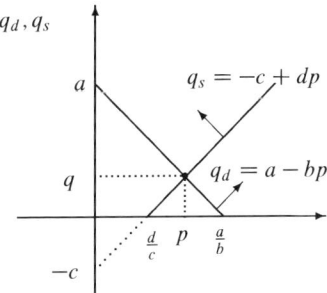

Fig. 1.8 The market equilibrium

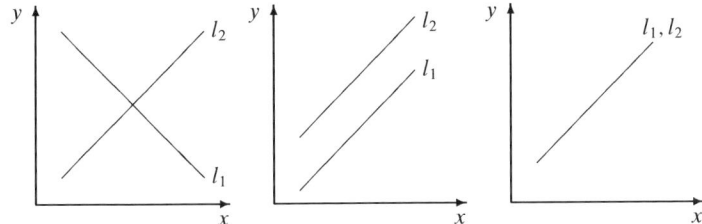

Fig. 1.9 Two lines on a plane with different locational characteristics

In the first of the graphs in Fig. 1.10 three lines intersect at a unique point. In the second graph every pair of the three lines intersects at a unique point, giving three distinct intersection points. In the third graph, the lines l_2 and l_3 coincide while they intersect with l_1 at a unique point. In the fourth graph all lines coincide. In the fifth graph lines l_2 and l_3 coincide which are parallel to the separate line l_1. In the next graph, the separate lines l_2 and l_3 are parallel to each other, intersecting with line l_1 at two distinct points. Finally, in the last graph the three separate lines are parallel.

Solving the system of three equations in three unknowns

$$\begin{cases} a_{11}x_1 + a_{12}x_2 + a_{13}x_3 = b_1, \\ a_{21}x_1 + a_{22}x_2 + a_{23}x_3 = b_2, \\ a_{31}x_1 + a_{32}x_2 + a_{33}x_3 = b_3 \end{cases}$$

is much more difficult than in the case of two equations with two unknowns.

It is obvious that in the case of higher dimensions the number of possibilities is enormously increasing, so that graphical tools then become almost inapplicable. Thus we need some analytical tools to solve and analyze arbitrarily large finite systems of equations. To this end, in the following chapters we will deal with vectors and matrices.

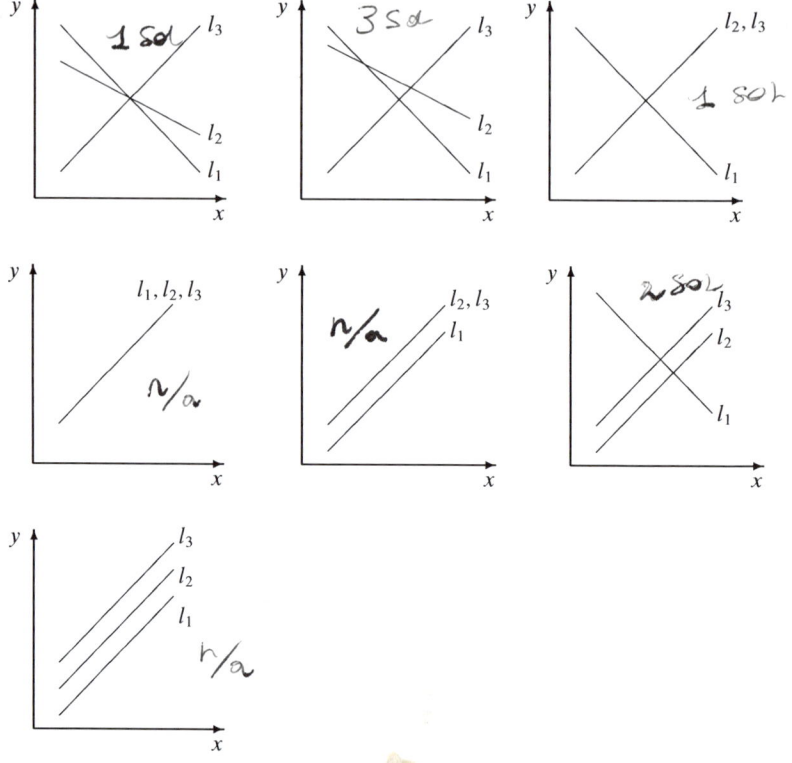

Fig. 1.10 Three lines on a plane with different locational characteristics

Example 1.5. Consider the following three good economy. Suppose that all of these goods are used in the production of others. Let x_{ij} denote the amount of the i-th good used to produce good j, and let x_i denotes the total amount of the produced good i.

Let the amount of good i used to produce one unit of good j is given by

$$a_{ij} = x_{ij}/x_j.$$

These are called input/output coefficients. Suppose also that each of these goods demanded for final use (i.e., for the consumption of households).

Suppose we have the following statistical information.

	x_1	x_2	x_3
x_1	$a_{11} = 0.3$	$a_{12} = 0.2$	$a_{13} = 0.3$
x_2	$a_{21} = 0.2$	$a_{22} = 0.3$	$a_{23} = 0.2$
x_3	$a_{31} = 0.4$	$a_{32} = 0.6$	$a_{33} = 0.4$

The final demand for good 1 is $y_1 = 20$, for good 2 is $y_2 = 30$ and for good 3 is $y_3 = 40$. Then the total output level which satisfies the input requirements and final demand can be found by solving the following system of linear equations

$$\begin{cases} x_1 = 0.3x_1 + 0.2x_2 + 0.3x_3 + 20, \\ x_2 = 0.2x_1 + 0.3x_2 + 0.2x_3 + 30, \\ x_3 = 0.4x_1 + 0.6x_2 + 0.4x_3 + 40. \end{cases}$$

1.2 Microeconomics: Market Equilibrium

In Sect. 1.1 we have raised the problem about intersection of lines in a space. In this and the next sections we give some micro- and macroeconomic reasons for this problem.

In contemporary economies a wide variety of goods are produced to satisfy the needs of people. These goods are produced by many different producers and demanded by many different consumers (individuals or institutions). Throughout history, societies developed various methods to find an answer to satisfy both consumers and producers. One such method that is a widely used is price mechanism. In this framework, both demand and supply are assumed to be influenced by prices of goods plus other factors. When other factors are taken as given, a change in the price of a good affects both its demand and its supply.

Economists therefore developed the partial-equilibrium approach to examine a market for a particular good in isolation of the other goods. In this framework, by taking factors other than price as given, quantity demanded and supplied for a good can be considered as a function of its price. Using the supply and demand apparatus, one can find the price and corresponding quantity at which both consumers and producers are satisfied, i.e. the point at which demand equals supply. This problem is discussed below.

Partial-equilibrium approach, however, fails to take into account the fact that markets interact with each other. A change in the price of a good affects the demand and/or the supply of another. Therefore in a more realistic framework, where many goods simultaneously produced, demanded and exchanged, a more general framework is needed. Multi-market equilibrium approach is a first step in this direction.[1]

---

[1] Partial equilibrium analysis has a long history in economics. It was elaborated and widely used by French economist Antoine Augustin Cournot (1801–1877) and English economist Alfred Marshall (1842–1924). General equilibrium analysis in economics, on the other hand, seeks to explain the behavior of supply, demand and prices in an economy with many markets. French economist Marie-Esprit-Léon Walras (1834–1910) is considered as the father of this approach.

1.2.1 Equilibrium in a Single Market

Let us consider an isolated market for good i. Suppose that both the demand (q_i^d) and supply (q_i^s) of this good is a function of its price (p_i), only.[2]

$$q_i^d = \alpha_0 - \alpha_1 p_i \quad (1.1)$$

$$q_i^s = -\beta_0 + \beta_1 p_i \quad (1.2)$$

These functions assume that there is a linear relation between the quantity demanded (or supplied) and the price of the good i.

Remark 1.1. Although, in economics texts (1.1) and (1.2) are usually referred to as *linear demand and supply functions,* from a strictly mathematical point of view, they are not. They have an extra constant slope term. In mathematics linear functions with a constant slope are called *affine functions*. Notice that in the case of affine functions, when the explanatory variable (in this example price) in the equation is zero, explained variable (in this example quantity demanded or supplied, respectively) can take non-zero values. Linear function formulation, on the other hand, does not allow it.

Question 1.1. Why in economics affine functions are used in formulating demand and supply functions?

All of the coefficients in (1.1) and (1.2) are assumed to be positive. The negative sign in front of the slope coefficient in the demand function indicates that, as price of the good increases, its demand declines. The reverse holds for the supply. The negative sign in front of the intercept term in the supply equation implies that supply will become positive only after the price of the good in question is positive and sufficiently high.

The market for good i is said to be in 'equilibrium' when the demand for good i is equal to its supply, i.e.

$$q_i^d = q_i^s \quad (1.3)$$

Recall that the difference between demand and supply is called *excess demand*. Using this concept, market equilibrium can also be characterized as the point at which excess demand is zero, i.e.

$$E(p_i) = q_i^d - q_i^s = 0 \quad (1.4)$$

---

[2] This basic demand and supply model is, obviously, an oversimplification. In economics, both demand and supply of a good is treated as functions of many variables, including prices of other goods, income etc. In the next section on multi-market equilibrium, prices of other goods will be allowed to influence supply and demand.

1.2 Microeconomics: Market Equilibrium

Suppose we want to find the equilibrium of the market described by (1.1), (1.3) or (1.1), (1.4). Such an equilibrium point can be characterized by two variables, equilibrium price and the quantity. An easy way to find this point, is to substitute (1.1) and (1.2) into (1.4), i.e.

$$E(p_i) = (\alpha_0 - \alpha_1 p_i) - (-\beta_0 + \beta_1 p_i) = 0,$$

from which the equilibrium price $\hat{p}$ can be derived as

$$\hat{p}_i = \frac{\alpha_0 + \beta_0}{\alpha_1 + \beta_1} \qquad (1.5)$$

Equilibrium quantity level, on the other hand, can be obtained by substituting (1.5) either in (1.1) (or (1.2)), which gives

$$\hat{q}_i = \frac{\alpha_0 \beta_1 - \alpha_1 \beta_0}{\alpha_1 + \beta_1} \qquad (1.6)$$

Obviously, (1.6) gives an economically meaningful result, i.e. $\hat{q}_i > 0$ only if $\alpha_0 \beta_1 - \alpha_1 \beta_0 > 0$.

Example 1.6. Let

$$\begin{cases} q^d = 38 - 2p \\ q^s = -6 + 9p \end{cases}$$

be the demand function and be the supply function for some good. Find the equilibrium price and corresponding quantity level.

Answer: $\hat{p} = 4, \hat{q} = 30$.

1.2.2 Multi-Market Equilibrium

Consider an economy with two goods. Suppose that the supply and demand of each good are functions of its own price as well as the price of other good. Such a system can be represented by the following four equations:

$$\begin{cases} q_1^d = \alpha_{01} + \alpha_{11} p_1 + \alpha_{12} p_2 & (1.7) \\ q_1^s = \beta_{01} + \beta_{11} p_1 + \beta_{12} p_2 & (1.8) \end{cases}$$

$$\begin{cases} q_2^d = \alpha_{02} + \alpha_{12} p_1 + \alpha_{22} p_2 & (1.9) \\ q_2^s = \beta_{02} + \beta_{12} p_1 + \beta_{22} p_2 & (1.10) \end{cases}$$

For such economy, multi-market equilibrium is the quadruple $(\hat{p}_1, \hat{p}_2, \hat{q}_1, \hat{q}_2)$ at which

$$E_1(\hat{p}_1, \hat{p}_2) = q_1^d(\hat{p}_1, \hat{p}_2) - q_1^s(\hat{p}_1, \hat{p}_2) = 0 \qquad (1.11)$$

and

$$E_2(\hat{p}_1, \hat{p}_2) = q_2^d(\hat{p}_1, \hat{p}_2) - q_2^s(\hat{p}_1, \hat{p}_2) = 0 \qquad (1.12)$$

are simultaneously satisfied.

Exercise 1.1. (i) Using (1.7)–(1.12) find the equilibrium prices for goods 1 and 2; (ii) what is the equilibrium quantity for good 1?

Hint. Use (1.11) and (1.12) to derive two equations with two unknowns (prices). Using one of the equations express one price in terms of the other. Substituting this relation in the other equation one gets an expression for one of the prices in terms of the parameters of the model. The expressions for other variables can be obtained in a similar fashion through substitution.

Example 1.7. Consider an economy with two goods (q_1, q_2). Let supply and demand functions for these goods be as follows. For good 1:

$$q_1^d = 3 - 2p_1 + 2p_2,$$
$$q_1^s = -4 + 3p_1.$$

For good 2:

$$q_2^d = 22 + 2p_1 - p_2,$$
$$q_2^s = 20 + p_2.$$

Find the equilibrium prices for goods 1 and 2.

Answer: $p_1 = 3, p_2 = 4$.

1.3 Macroeconomic Policy Problem

After the Second World War, governments' role in management of the economy was widely accepted. The mode of government intervention varied from extensive planning both at macro and micro levels in socialist economies to the more market oriented monetary and fiscal policies adopted in advanced economies. Many developing countries (such as India and Turkey) chose to implement development planning which was less comprehensive than the socialist planning but still requires much more intensive government intervention that the economic policies used in advanced market economies.

As the commitment of the government became increasingly more extensive, it became necessary to have a framework to deal with them simultaneously and in a consistent manner. The seminal contribution in this field was made by Jan

1.3 Macroeconomic Policy Problem

Tinbergen[3]'s celebrated book, [31], which became the cornerstone of the theory of economic policy since then.

Tinbergen's framework consists of three basic ingredients:
1. A set of *instruments* which are controlled by the policy maker.
2. A set of *targets*. They are not controlled by the policy maker, but they are of interest to the policy maker due to their contribution to the welfare of the society.
3. A quantitative model describing the relationships between the targets, instruments and other variables (i.e. endogenous variable that are not important from economic policy purpose and purely exogenous variables, i.e., those variables, whose values are externally given and can not be controlled by the policy maker).

1.3.1 A Simple Macroeconomic Policy Model with One Target

Consider the following simple one sector macroeconomic model,

$$Y = C + I + G + X - M, \tag{1.13}$$

$$C = cY_d, \quad 0 < c < 1 \tag{1.14}$$

$$Y_d = Y - T \tag{1.15}$$

$$T = tY, \quad 0 < t < 1 \tag{1.16}$$

$$G = T + D, \tag{1.17}$$

$$M = mY, \quad 0 < m < 1, \tag{1.18}$$

where Y – Gross Domestic Product (GDP), C – Private Consumption, T – Tax Revenues, I – Private Investment, G – Government Expenditure, X – Exports, M – Imports, D – Budget Deficit.

The first equation (1.13) of the model, is *definitional*. It defines the GDP from expenditure side. Equation (1.14) is a consumption function, which is a behavioral equation. It relates private consumption to disposable income, which is defined in (1.15) as the income that households can spend after paying their taxes. In this equation c is *marginal propensity to consume*. It gives the increase in aggregate private consumption, when GDP increased one unit. Equation (1.16) is an *institutional equation*, reflects the tax code. A certain percentage (t) of the GDP is collected as taxes. Equation (1.17) is another *definitional equation* which indicates that government expenditures can be financed either through collecting taxes or through borrowing. Equation (1.18) is import function, which connects imports to GDP. It is a behavioral equation, which asserts that as GDP increases

---

[3] Jan Tinbergen (1903–1994) was a distinguished Dutch economist. He awarded the First Nobel Prize in economics (1969), which he shared for having developed and applied dynamic models for the analysis of economic processes.

demand for foreign goods increase. In this equation m denotes *marginal propensity to import*.

The model has six equations. Each one of these equations describes a different variable. These are *endogenous variables* of the model. Their values are determined by the model. Notice that the model has three *exogenous variables*, I, X and D. Here the private investment I is treated as exogenous for the sake of simplicity.[4] These are referred to as *data variables*. Let us assume that their exogenously given values are denoted by I^* and X^*, respectively. Third exogenous variable is budget deficit (D). It requires special attention. The magnitude of this variable is determined by the parliament when it approves the national budget submitted by the government. In other words, in contrast to I and X, D can be controlled by the policy makers. Therefore it is called a policy *instrument*. Let us distinguish it from other exogenous variables by denoting its value as $\tilde{D}$, i.e. $D = \tilde{D}$.

The following informal statement is called Tinbergen theorem.

In a Tinbergen type economic policy framework, the number of targets should be equal to number of instruments.

Explanation. Tinbergen in [31, Chap. 4] discusses the meaning of the equality of the number of instruments and targets is discussed. When this condition is satisfied, the unique values of the instruments to achieve given targets can be determined. The problems that arise when the number of instruments are not equal to the number of targets is addressed in the following chapter of the Tinbergen's book [31, Chap. 5]. The problem will be discussed from mathematical point of view in Chap. 5 of this book. □

In the light of this fundamental theorem, the macro model given above allows the policy maker to choose one target variable. Suppose the policy maker chose GDP as the target variable. Then the question at hand can be formulated as follows: How much the government should borrow in order to achieve the targeted GDP level?

Substituting (1.14) to (1.18) in (1.13), and rearranging, we have

$$Y = \frac{1}{1 - c(1-t) - t + m} \left(I^* + X^* + \tilde{D} \right) \qquad (1.19)$$

When the target value of the GDP, say $\hat{Y}$ is given, the required amount of budget deficit (i.e. the value of the instrument) is obtained from (1.19) as

$$\tilde{D} = (1 - c(1-t) - t + m)\hat{Y} - \left(I^* + X^* \right) \qquad (1.20)$$

Example 1.8. Consider the model given by (1.13)–(1.18). Suppose that the following coefficients are estimated

---

[4] In fact, private investment is a very important component of the GDP and therefore any macroeconomic model that claims to be characterizing of the working economy should be in a position of explaining private investment activity.

$$c = \text{marginal propensity to consume} = 0.9,$$
$$t = \text{tax/GDP ratio} = 0.15,$$
$$m = \text{marginal propensity to import} = 0.3.$$

Assume that $I = \$50$ and $X = \$50$ (billion). Suppose the authorities targeted the GDP level as $320 billion.

How much the government should borrow?

Answer: $23.2 billion.

1.3.2 A Macroeconomic Policy Model with Multiple Targets and Multiple Instruments

In most instances, governments have more that one target. For example, they find themselves both attaining a satisfactory employment level and constraining the current account deficit at a reasonable level. Suppose that it is the case for an economy which is represented by the following model.

$$Y = C + I + G + X - M, \quad (1.21)$$
$$G = G_C + G_I, \quad (1.22)$$
$$C = cY, \quad 0 < c < 1 \quad (1.23)$$
$$I = k_1(Y - Y_{-1}) + k_2 G_I, \quad k_1, k_2 > 0 \quad (1.24)$$
$$M = m_C C + m_I I + m_G G_I + m_X X, \quad 0 < m_C, m_I, m_G, m_X < 1, \quad (1.25)$$
$$N = nY, \quad n > 0 \quad (1.26)$$
$$B = p_x X - p_m M. \quad (1.27)$$

In this system of equations there are three groups of variables:
1. *Endogenous variables*
   Y – GDP, C – Private Consumption, I – Private Investment, M – Imports, N – Employment, B – Current Account of the Balance of Payments.
2. *Exogenous Variables*
   (a) Data variables: exports X and last years GDP Y_{-1}.
   (b) Instruments: G_C – Public Consumption Expenditures, G_I – Public Investment Expenditures.
3. *Target Variables* The target variables and their values determined by the policy maker are as follows
$$B = \hat{B},$$
$$N = \hat{N}.$$

The parameters are the following: c – share of consumption in income, t – average tax rate, m – import per unit of output, n – employment per unit of output (the reverse of productivity), p_x and p_m – export and import prices, the coefficient k_1 shows how much investment will be undertaken by the private agents in response to an increase in GDP from its previous period level by unit, and k_2 shows how much extra investment will be undertaken if government increases its investment by unit.

Example 1.9. Consider the model given by (1.21)–(1.27). Let

$$c = 0.8, k_1 = 0.2, k_2 = 0.05, m_C = 0.1, m_I = 0.4, m_G = 0.3, m_X = 0.2, n = 0.4.$$

Suppose
$$Y_{-1} = 100, X = 30.$$

Suppose also that the government wants to achieve the following targets

$$B = 0, N = 60.$$

Find the amounts of G_C and G_I.

Solution. We have the system of linear equations:

$$\begin{cases} Y = C + I + G + 30 - M, \\ G = G_C + G_I, \\ C = 0.8Y, \\ I = 0.2(Y - 100) + 0.05G_I, \\ M = 0.1C + 0.4I + 0.3G_I + 0.2 \cdot 30, \\ 60 = 0.4Y, \\ 0 = p_x 30 - p_m M. \end{cases}$$

After the elimination of all variables but G_I and G_C, we obtain

$$G_I = 93.75r - 68.75 \text{ and } G_C = 62.1875 - 68.4375r,$$

where $r = p_X/p_M$.

Question 1.2. Explain and categorize equations (1.21)–(1.27) (definitional, technical etc.)

Exercise 1.2. Find the values of the instruments that enable the system to achieve the targeted levels of current account balance and employment.

1.4 Problems

1. Plot each of the following pair of points on $\mathbb{R}^2$ and draw (and calculate the length of) the line segment connecting them
   (a) $(4, -10), (0, 1)$; (b) $(0, 0), (-7, -8)$; (c) $(\sqrt{2}, \sqrt{5}), (\sqrt{2}, -\sqrt{5})$

2. Consider two points on x-axis. Show that the distance between them is equal to the absolute value of the difference of their coordinates.

3. Draw the following lines in $\mathbb{R}^2$
   (a) $3x - 4y = 12$; (b) $x + y = 10$; (c) $2x - 5y = 10$; (d) $x = 5$.

4. Write the equation of the lines determined by the two points in each part of problem 1.

5. Draw a line having (a) an x-intercept but no y-intercept, (b) a y-intercept but no x-intercept, (c) x-intercept and y-intercept as coincident.

6. Show that if $a \neq 0, b \neq 0$, then the intercepts of the line

$$ax + by + c = 0$$

are $(0, -c/b)$ and $(-c/a, 0)$.

7. Show that

$$\frac{x}{a} + \frac{y}{b} = 1$$

is the equation of a straight line with the intercepts $(0, b)$ and $(a, 0)$.

8. Solve $7x - 10 = 0$ graphically by considering $y = 7x - 10$.

9. Draw the lines for each of the following equations.
   (a) $|x| + |y| = 1$; (b) $|x + y| = 1$.

10. Solve the following systems of equations.

    (a) $\begin{cases} 3x - 5y = 15, \\ 2x + y = 5; \end{cases}$ (b) $\begin{cases} 3x - 5y = 15, \\ 6x - 10y = 30; \end{cases}$ (c) $\begin{cases} 6x - 10y = 30, \\ 3x - 5y = 10. \end{cases}$

11. Let

$$\begin{cases} z_1 = a_{11}y_1 + a_{12}y_2 \\ z_2 = a_{21}y_1 + a_{22}y_2 \end{cases}$$

and

$$\begin{cases} y_1 = b_{11}x_1 + b_{12}x_2 \\ y_2 = b_{21}x_1 + b_{22}x_2. \end{cases}$$

Express z_1 and z_2 as functions of x_1 and x_2.

Vectors and Matrices

2.1 Vectors

Ordered n-tuple of objects is called a vector

$$\mathbf{y} = (y_1, y_2, \ldots, y_n).$$

Throughout the text we confine ourselves to vectors the elements y_i of which are real numbers.

In contrast, a variable the value of which is a single number, not a vector, is called *scalar*.

Example 2.1. We can describe some economic unit **EU** by the vector

$$\mathbf{EU} = (\text{output, \# of employees, capital stock, profit})$$

Given a vector $\mathbf{y} = (y_1, \ldots, y_n)$, elements y_i, $i = 1, \ldots, n$ are called *components* of the vector. We will usually denote vectors by bold letters.[1] The number n of components is called the *dimension* of the vector $\mathbf{y}$. The set of all n-dimensional vectors is denoted by $\mathbb{R}^n$ and called n-dimensional real space[2].

Two vectors $\mathbf{x}, \mathbf{y} \in \mathbb{R}^n$ are equal if $x_i = y_i$ for all $i = 1, 2, \ldots, n$.

Let $\mathbf{x} = (x_1, \ldots, x_n)$ and $\mathbf{y} = (y_1, \ldots, y_n)$ be two vectors. We compare these two vectors element by element and say that $\mathbf{x}$ is greater than $\mathbf{y}$ if for all i $x_i > y_i$, and denote this statement by $\mathbf{x} > \mathbf{y}$. Analogously, we can define $\mathbf{x} \geq \mathbf{y}$.

Note that, unlike in the case of real numbers, for vectors when $\mathbf{x} > \mathbf{y}$ does not hold, this does not imply $\mathbf{y} \geq \mathbf{x}$. Indeed, consider the vectors $\mathbf{x} = (1, 0)$ and $\mathbf{y} = (0, 1)$. It can be easily seen that neither $\mathbf{x} \geq \mathbf{y}$ nor $\mathbf{y} \geq \mathbf{x}$ is true.

---

[1] Some other notations for vectors are $\bar{y}$ and $\vec{y}$.
[2] The terms *arithmetic space, number space* and *coordinate space* are also used.

A vector $\mathbf{0} = (0, 0, \ldots, 0)$ (also denoted by $\bar{0}$) is called a *null vector*.[3]
A vector $\mathbf{x} = (x_1, x_2, \ldots, x_n)$ is called non-negative (which is denoted by $\mathbf{x} \geq \mathbf{0}$) if $x_i \geq 0$ for all i.

A vector $\mathbf{x}$ is called positive if $x_i > 0$ for all i. We denote this case by $\mathbf{x} > \mathbf{0}$.

2.1.1 Algebraic Properties of Vectors

One can define the following natural arithmetic operations with vectors.
Addition of two n-vectors

$$\mathbf{x} + \mathbf{y} = (x_1 + y_1, x_2 + y_2, \ldots, x_n + y_n)$$

Subtraction of two n-vectors

$$\mathbf{x} - \mathbf{y} = (x_1 - y_1, x_2 - y_2, \ldots, x_n - y_n)$$

Multiplication of a vector by a real number λ

$$\lambda \mathbf{y} = (\lambda y_1, \lambda y_2, \ldots, \lambda y_n)$$

Example 2.2. Let $\mathbf{EU}_1 = (Y_1, L_1, K_1, P_1)$ be a vector representing an economic unit, say, a firm, see Example 2.1 (where, as usually, Y is its output, L is the number of employees, K is the capital stock, and P is the profit). Let us assume that it is merged with another firm represented by a vector $\mathbf{EU}_2 = (Y_2, L_2, K_2, P_2)$ (that is, we should consider two separate units as a single one). The resulting unit will be represented by a sum of two vectors

$$\mathbf{EU}_3 = (Y_1 + Y_2, L_1 + L_2, K_1 + K_2, P_1 + P_2) = \mathbf{EU}_1 + \mathbf{EU}_2.$$

In this situation, we have also $\mathbf{EU}_2 = \mathbf{EU}_3 - \mathbf{EU}_1$. Moreover, if the second firm is similar to the first one, we can assume that $\mathbf{EU}_1 = \mathbf{EU}_2$, hence the unit

$$\mathbf{EU}_3 = (2Y_1, 2L_1, 2K_1, 2P_1) = 2 \cdot \mathbf{EU}_1$$

gives also an example of the multiplication by a number 2.

This example, as well as other 'economic' examples in this book has an illustrative nature. Notice, however, that the profit of the merged firm might be higher or lower than the sum of two profits $P_1 + P_2$.

---

[3]The null vector is also called *zero vector*.

2.1 Vectors

The following properties of the vector operations above follow from the definitions:

1a. $\mathbf{x} + \mathbf{y} = \mathbf{y} + \mathbf{x}$ (commutativity).
1b. $(\mathbf{x} + \mathbf{y}) + \mathbf{z} = \mathbf{x} + (\mathbf{y} + \mathbf{z})$ (associativity).
1c. $\mathbf{x} + \mathbf{0} = \mathbf{x}$.
1d. $\mathbf{x} + (-\mathbf{x}) = \mathbf{0}$.
2a. $1\mathbf{x} = \mathbf{x}$.
2b. $\lambda(\mu \mathbf{x}) = \lambda\mu(\mathbf{x})$.
3a. $(\lambda + \mu)\mathbf{x} = \lambda\mathbf{x} + \mu\mathbf{x}$.
3b. $\lambda(\mathbf{x} + \mathbf{y}) = \lambda\mathbf{x} + \lambda\mathbf{y}$.

Exercise 2.1. Try to prove these properties yourself.

2.1.2 Geometric Interpretation of Vectors and Operations on Them

Consider $\mathbb{R}^2$ plane. Vector $\mathbf{z} = (\alpha_1, \alpha_2)$ is represented by a directed line segment from the origin $(0, 0)$ to (α_1, α_2), see Fig. 2.1.

The sum of the two vectors $\mathbf{z}_1 = (\alpha_1, \beta_1)$ and $\mathbf{z}_2 = (\alpha_2, \beta_2)$ is obtained by adding up their coordinates, see Fig. 2.2.

In this figure, the sum $\mathbf{z}_1 + \mathbf{z}_2 = (\alpha_1 + \alpha_2, \beta_1 + \beta_2)$ is represented by a diagonal of a parallelogram sides of which being formed by the vectors $\mathbf{z}_1$ and $\mathbf{z}_2$.

Multiplication of a vector by a scalar has a contractionary (respectively, expansionary) effect if the scalar in absolute value is less (respectively, greater) than unity. The direction of the vector does not change if the scalar is positive, and it changes by 180 degrees if the scalar is negative. Figure 2.3 plots scalar multiplication for a vector $\mathbf{x}$, two scalars $\lambda_1 > 1$ and $1 < \lambda_2 < 0$.

The difference of the two vectors $\mathbf{z}_2$ and $\mathbf{z}_1$ is shown on Fig. 2.4.

The projection of the vector $\mathbf{a}$ on x–axis is denoted by $pr_x \mathbf{a}$, and is shown in Fig. 2.5 below.

Let $\mathbf{z}_1, \ldots, \mathbf{z}_s$ be a set of vectors in $\mathbb{R}^n$. If there exist real numbers $\lambda_1, \ldots, \lambda_s$ not all being equal to 0 and

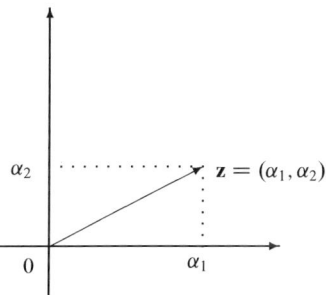

Fig. 2.1 A vector on the plane $\mathbb{R}^2$

Fig. 2.2 The sum of two vectors

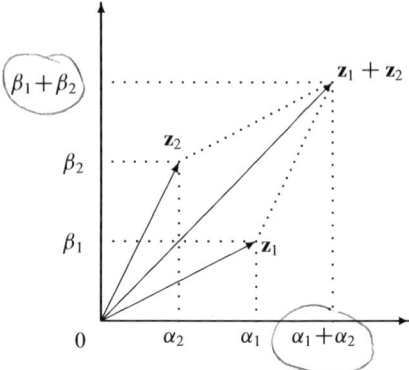

Fig. 2.3 The multiplication of a vector by a scalar

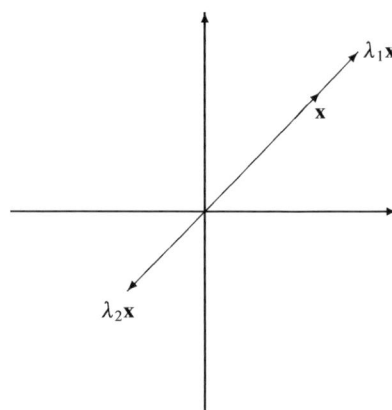

Fig. 2.4 The difference of vectors

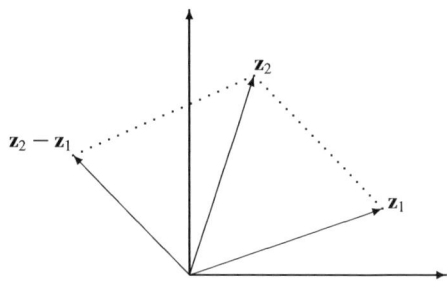

$$\lambda_1 \mathbf{z}_1 + \lambda_2 \mathbf{z}_2 + \cdots + \lambda_s \mathbf{z}_s = \mathbf{0},$$

then these vectors are called *linearly dependent*.

Example 2.3. Three vectors $\mathbf{a} = (1, 2, 3)$, $\mathbf{b} = (4, 5, 6)$ and $\mathbf{c} = (7, 8, 9)$ are linearly dependent because

2.1 Vectors

Fig. 2.5 The projection of a vector a on the x-axis

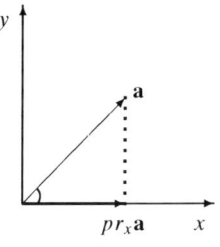

Fig. 2.6 Unit vectors in $\mathbb{R}^3$

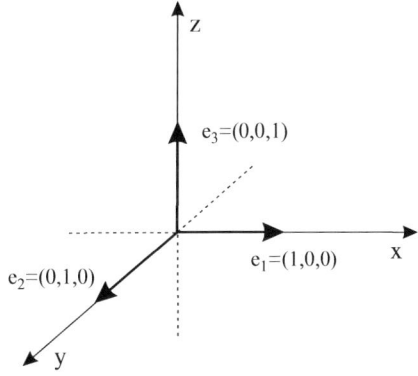

$$1\mathbf{a} - 2\mathbf{b} + 1\mathbf{c} = \mathbf{0}.$$

The vectors $\mathbf{z}_1, \ldots, \mathbf{z}_s$ are called *linearly independent* if

$$\lambda_1 \mathbf{z}_1 + \cdots + \lambda_s \mathbf{z}_s = \mathbf{0}$$

holds only whenever $\lambda_1 = \lambda_2 = \cdots = \lambda_s = 0$.

Note that the n vectors $\mathbf{e}_1 = (1, 0, \ldots, 0)$, $\mathbf{e}_2 = (0, 1, \ldots, 0)$, $\ldots$, $\mathbf{e}_n = (0, 0, \ldots, 1)$ (see Fig. 2.6 for the case $n = 3$) are linearly independent in $\mathbb{R}^n$.

Assume that vectors $\mathbf{z}_1, \ldots, \mathbf{z}_s$ are linearly dependent, i.e., there exists at least one λ_i, where $1 \leq i \leq s$, such that $\lambda_i \neq 0$ and

$$\lambda_1 \mathbf{z}_1 + \lambda_2 \mathbf{z}_2 + \cdots + \lambda_i \mathbf{z}_i + \cdots + \lambda_s \mathbf{z}_s = \mathbf{0}.$$

Then

$$\lambda_i \mathbf{z}_i = -\lambda_1 \mathbf{z}_1 - \lambda_2 \mathbf{z}_2 - \cdots - \lambda_{i-1} \mathbf{z}_{i-1} - \lambda_{i+1} \mathbf{z}_{i+1} - \cdots - \lambda_s \mathbf{z}_s,$$

and

$$\mathbf{z}_i = \mu_1 \mathbf{z}_1 + \cdots + \mu_{i-1} \mathbf{z}_{i-1} + \mu_{i+1} \mathbf{z}_{i+1} + \cdots + \mu_s \mathbf{z}_s, \qquad (2.1)$$

where $\mu_j = -\lambda_j / \lambda_i$, for all $j \neq i$ and $j \in \{1, \ldots, s\}$.

A vector **a** is called a *linear combination* of the vectors $\mathbf{b}_1, \ldots, \mathbf{b}_n$ if it can be represented as

$$\mathbf{a} = \alpha_1 \mathbf{b}_1 + \cdots + \alpha_n \mathbf{b}_n,$$

where $\alpha_1, \ldots, \alpha_n$ are real numbers. In particular, (2.1) shows that the vector $\mathbf{z}_i$ is a linear combination of the vectors $\mathbf{z}_1, \ldots, \mathbf{z}_{i-1}, \mathbf{z}_{i+1}, \ldots, \mathbf{z}_s$.

These results can be formulated as

Theorem 2.1. *If vectors $\mathbf{z}_1, \ldots, \mathbf{z}_s$ are linearly dependent, then at least one of them is a linear combination of other vectors. Vectors one of which is a linear combination of others are linearly dependent.*

2.1.3 Geometric Interpretation in $\mathbb{R}^2$

Are the vectors $\mathbf{z}$ and $\lambda \mathbf{z}$ (see Fig. 2.7) linearly dependent?

Note from Fig. 2.8 that the vector $\lambda_1 \mathbf{z}_1 + \lambda_2 \mathbf{z}_2$ is a linear combination of the vectors $\mathbf{z}_1$ and $\mathbf{z}_2$. Any three vectors in $\mathbb{R}^2$ are linearly dependent!

Remark 2.1. Consider the following n vectors in $\mathbb{R}^n$.

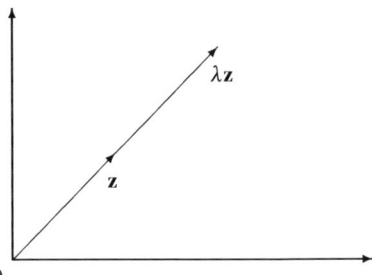

Fig. 2.7 Are these vectors linearly dependent?

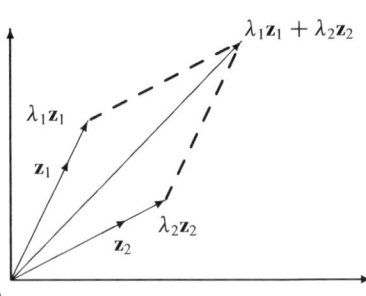

Fig. 2.8 A linear combination of two vectors

2.2 Dot Product of Two Vectors

$$\mathbf{a}_1 = (1, -2, 0, 0, \ldots, 0)$$
$$\mathbf{a}_2 = (0, 1, -2, 0, \ldots, 0)$$

$$\mathbf{a}_{n-1} = (0, 0, \ldots, 0, 1, -2)$$
$$\mathbf{a}_n = (-2^{-(n-1)}, 0, \ldots, 0, 0, 1)$$

These vectors are linearly dependent since

$$2^{-n}\mathbf{a}_1 + 2^{-(n-1)}\mathbf{a}_2 + \cdots + 2^{-1}\mathbf{a}_n = \mathbf{0}.$$

If $n > 40$ then $2^{-(n-1)} < 10^{-12}$, a very small number. Moreover, if $n > 64$, then $2^{-n} = 0$ for computers. So, for $n > 64$, we can assume that in our system $\mathbf{a}_n$ is given by $\mathbf{a}_n = (0, \ldots, 0, 1)$. Thus, the system is written as

$$\begin{cases} \mathbf{a}_1 &= (1, -2, 0, 0, \ldots, 0) \\ \mathbf{a}_2 &= (0, 1, -2, 0, \ldots, 0) \\ & \vdots \\ \mathbf{a}_{n-1} &= (0, 0, \ldots, 0, 1, -2) \\ \mathbf{a}_n &= (0, 0, \ldots, 0, 0, 1) \end{cases}$$

But this system is linearly independent. (Check it!)

This example shows how sensitive might be linear dependency of vectors to rounding.

Exercise 2.2. Check if the following three vectors are linearly dependent:
(a) $\mathbf{a} = (1, 2, 1), \mathbf{b} = (-2, 3, -2), \mathbf{c} = (7, 4, 7)$;
(b) $\mathbf{a} = (1, 2, 3), \mathbf{b} = (0, -1, 3), \mathbf{c} = (2, -1, 2)$.

2.2 Dot Product of Two Vectors

Definition 2.1. For any two vectors $\mathbf{x} = (x_1, \ldots, x_n)$ and $\mathbf{y} = (y_1, \ldots, y_n)$, the dot product[4] of $\mathbf{x}$ and $\mathbf{y}$ is denoted by $(\mathbf{x}, \mathbf{y})$, and is defined as

$$(\mathbf{x}, \mathbf{y}) = x_1 y_1 + x_2 y_2 + \cdots + x_n y_n = \sum_{i=1}^{n} x_i y_i. \tag{2.2}$$

---

[4] Other terms for dot product are *scalar product* and *inner product*.

Example 2.4. Let $\mathbf{a}_1 = (1, -2, 0, \ldots, 0)$ and $\mathbf{a}_2 = (0, 1, -2, 0, \ldots, 0)$. Then

$$(\mathbf{a}_1, \mathbf{a}_2) = 1 \cdot 0 + (-2) \cdot 1 + 0 \cdot (-2) + 0 \cdot 0 + \ldots + 0 \cdot 0 = -2.$$

Example 2.5 (Household expenditures). Suppose the family consumes n goods. Let $\mathbf{p}$ be the vector of prices of these commodities (we assume competitive economy and take them as given), and $\mathbf{q}$ be the vector of the amounts of commodities consumed by this household. Then the total expenditure of the household can be obtained by dot product of these two vectors

$$E = (\mathbf{p}, \mathbf{q}).$$

Dot product $(\mathbf{x}, \mathbf{y})$ of two vectors $\mathbf{x}$ and $\mathbf{y}$ is a real number and has the following properties, which can be checked directly:
1. $(\mathbf{x}, \mathbf{y}) = (\mathbf{y}, \mathbf{x})$ (symmetry or commutativity)
2. $(\lambda \mathbf{x}, \mathbf{y}) = \lambda (\mathbf{x}, \mathbf{y})$ for all $\lambda \in \mathbb{R}$ (associativity with respect to multiplication by a scalar)
3. $(\mathbf{x}_1 + \mathbf{x}_2, \mathbf{y}) = (\mathbf{x}_1, \mathbf{y}) + (\mathbf{x}_2, \mathbf{y})$ (distributivity)
4. $(\mathbf{x}, \mathbf{x}) \geq 0$ and $(\mathbf{x}, \mathbf{x}) = 0$ iff $\mathbf{x} = \mathbf{0}$ (non-negativity and non-degeneracy).

2.2.1 The Length of a Vector, and the Angle Between Two Vectors

Definition 2.2. The length of a vector $\mathbf{x}$ in $\mathbb{R}^n$ is defined as $\sqrt{(\mathbf{x}, \mathbf{x})}$ and denoted by $|\mathbf{x}|$. If $\mathbf{x} = (x_1, \ldots, x_n)$ then $|\mathbf{x}| = \sqrt{x_1^2 + \cdots + x_n^2}$. The angle φ between any two nonzero vectors $\mathbf{x}$ and $\mathbf{y}$ in $\mathbb{R}^n$ is defined as

$$\cos \varphi = \frac{(\mathbf{x}, \mathbf{y})}{|\mathbf{x}||\mathbf{y}|}, \quad 0 \leq \varphi \leq \pi. \tag{2.3}$$

We will see below that this definition of $\cos \varphi$ is correct, that is, the right hand side of the above formula belongs to the interval $[-1, 1]$.

Let us show first that the angle between two vectors $\mathbf{x}$ and $\mathbf{y}$ in the Cartesian plane is the geometric angle (Fig. 2.9).

Take any two vectors $\mathbf{x} = (x_1, x_2)$ and $\mathbf{y} = (y_1, y_2)$ in $\mathbb{R}^2$. Then $\mathbf{y} - \mathbf{x} = (y_1 - x_1, y_2 - x_2)$. By the law of cosines we have

$$|\mathbf{y} - \mathbf{x}|^2 = |\mathbf{y}|^2 + |\mathbf{x}|^2 - 2|\mathbf{y}||\mathbf{x}| \cos \varphi,$$

or

$$(y_1 - x_1)^2 + (y_2 - x_2)^2 = y_1^2 + x_1^2 + y_2^2 + x_2^2 - 2\sqrt{y_1^2 + y_2^2}\sqrt{x_1^2 + x_2^2} \cos \varphi.$$

2.2 Dot Product of Two Vectors

Fig. 2.9 The angle between two vectors

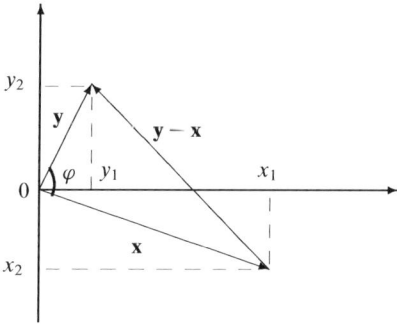

Then

$$\cos\varphi = \frac{y_1 x_1 + y_2 x_2}{\sqrt{y_1^2 + y_2^2}\sqrt{x_1^2 + x_2^2}} = \frac{(\mathbf{x},\mathbf{y})}{|\mathbf{x}||\mathbf{y}|}.$$

Definition 2.3. Two vectors $\mathbf{x}$ and $\mathbf{y}$ in $\mathbb{R}^n$ are called *orthogonal* (notation: $\mathbf{x}\perp\mathbf{y}$) if the angle between them is $\pi/2$, i.e. $(\mathbf{x},\mathbf{y})=0$.

Theorem 2.1 (Pythagoras). *Let $\mathbf{x}$ and $\mathbf{y}$ be two orthogonal vectors in $\mathbb{R}^n$. Then*

$$|\mathbf{x}+\mathbf{y}|^2 = |\mathbf{x}|^2 + |\mathbf{y}|^2. \tag{2.4}$$

Proof. $|\mathbf{x}+\mathbf{y}|^2 = (\mathbf{x}+\mathbf{y},\mathbf{x}+\mathbf{y}) = (\mathbf{x},\mathbf{x}) + (\mathbf{x},\mathbf{y}) + (\mathbf{y},\mathbf{x}) + (\mathbf{y},\mathbf{y}) = |\mathbf{x}|^2 + |\mathbf{y}|^2$ since $\mathbf{x}$ and $\mathbf{y}$ are orthogonal. □

The immediate generalization of the above theorem is the following one.

Theorem 2.2. *Let $\mathbf{z}_1,\ldots,\mathbf{z}_s$ be a set of mutually orthogonal vectors in $\mathbb{R}^n$, i.e., for all i,j and $i\neq j$, $(\mathbf{z}_i,\mathbf{z}_j)=0$. Then*

$$|\mathbf{z}_1+\mathbf{z}_2+\cdots+\mathbf{z}_s|^2 = |\mathbf{z}_1|^2 + |\mathbf{z}_2|^2 + \cdots + |\mathbf{z}_s|^2. \tag{2.5}$$

From the definition of the angle (2.3), it follows that

$$-1 \leq \frac{(\mathbf{x},\mathbf{y})}{|\mathbf{x}||\mathbf{y}|} \leq 1,$$

since $\varphi \in [0,\pi]$. The above inequalities can be rewritten as

$$\frac{(\mathbf{x},\mathbf{y})^2}{|\mathbf{x}|^2|\mathbf{y}|^2} \leq 1,$$

or

$$(\mathbf{x}, \mathbf{y})^2 \leq (\mathbf{x}, \mathbf{x}) \cdot (\mathbf{y}, \mathbf{y}). \tag{2.6}$$

The inequality (2.6) is called *Cauchy[5] inequality*.

Let us prove it so that we can better understand why the angle φ between two vectors can take any value in the interval of $[0, \pi]$.

Proof. Given any two vectors $\mathbf{x}$ and $\mathbf{y}$ in $\mathbb{R}^n$, consider the vector $\mathbf{x} - \lambda \mathbf{y}$, where λ is a real number. By axiom 4 of dot product we must have

$$(\mathbf{x} - \lambda \mathbf{y}, \mathbf{x} - \lambda \mathbf{y}) \geq 0,$$

that is,
$$\lambda^2 (\mathbf{y}, \mathbf{y}) - 2\lambda (\mathbf{x}, \mathbf{y}) + (\mathbf{x}, \mathbf{x}) \geq 0.$$

But then the discriminant of the quadratic equation

$$\lambda^2 (\mathbf{y}, \mathbf{y}) - 2\lambda (\mathbf{x}, \mathbf{y}) + (\mathbf{x}, \mathbf{x}) = 0$$

can not be positive. Therefore, it must be true that

$$(\mathbf{x}, \mathbf{y})^2 - (\mathbf{x}, \mathbf{x}) \cdot (\mathbf{y}, \mathbf{y}) \leq 0.$$

□

Corollary 2.2. *For all* $\mathbf{x}$ *and* $\mathbf{y}$ *in* $\mathbb{R}^n$,

$$|\mathbf{x} + \mathbf{y}| \leq |\mathbf{x}| + |\mathbf{y}|. \tag{2.7}$$

Proof. Note that

$$|\mathbf{x} + \mathbf{y}|^2 = (\mathbf{x} + \mathbf{y}, \mathbf{x} + \mathbf{y}) = (\mathbf{x}, \mathbf{x}) + 2(\mathbf{x}, \mathbf{y}) + (\mathbf{y}, \mathbf{y})$$

Now using $2(\mathbf{x}, \mathbf{y}) \leq 2 |(\mathbf{x}, \mathbf{y})| \leq 2 |\mathbf{x}| \, |\mathbf{y}|$ by Cauchy inequality, we obtain

$$|\mathbf{x} + \mathbf{y}|^2 \leq (\mathbf{x}, \mathbf{x}) + 2 |\mathbf{x}| \cdot |\mathbf{y}| + (\mathbf{y}, \mathbf{y})$$
$$= (|\mathbf{x}| + |\mathbf{y}|)^2$$

implying the desired result.

□

---

[5] Augustin Louis Cauchy (1789–1857) was a great French mathematician. In addition to his works in algebra and determinants, he had created a modern approach to calculus, so-called epsilon–delta formalism.

Exercise 2.3. Plot the vectors $\mathbf{u} = (1, 2)$, $\mathbf{v} = (-3, 1)$ and their sum $\mathbf{w} = \mathbf{u} + \mathbf{v}$ and check visually the above inequality.

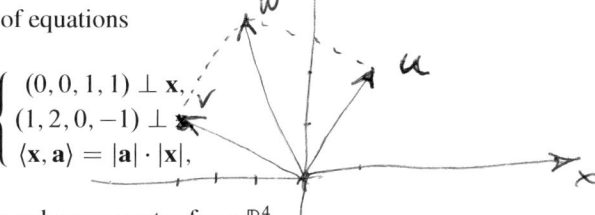

Exercise 2.4. Solve the system of equations

$$\begin{cases} (0, 0, 1, 1) \perp \mathbf{x}, \\ (1, 2, 0, -1) \perp \mathbf{x} \\ \langle \mathbf{x}, \mathbf{a} \rangle = |\mathbf{a}| \cdot |\mathbf{x}|, \end{cases}$$

where $\mathbf{a} = (2, 1, 0, 0)$ and $\mathbf{x}$ is an unknown vector from $\mathbb{R}^4$.

Exercise 2.5. Two vectors $\mathbf{a}$ and $\mathbf{b}$ are called *parallel* if they are linearly dependent (notation: $\mathbf{a} \| \mathbf{b}$). Solve the system of equations

$$\begin{cases} (0, 0, -3, 4) \| \mathbf{x}, \\ |\mathbf{x}| = 15. \end{cases}$$

Exercise 2.6. Find the maximal angle of the triangle ABC, where $A = (0, 1, 2, 0)$, $B = (0, 1, 0, -1)$ and $C = (1, 0, 0, 1)$ are three points in $\mathbb{R}^4$.

Exercise 2.7. Given three points $A(0, 1, 2, 3)$, $B(1, -1, 1, -1)$ and $C(1, 1, 0, 0)$ in $\mathbb{R}^4$, find the length of the median AM of the triangle ABC.

2.3 An Economic Example: Two Plants

Consider a firm operating two plants in two different locations. They both produce the same output (say, 10 units) using the same type of inputs. Although the amounts of inputs vary between the plants the output level is the same.

The firm management suspects that the production cost in Plant 2 is higher than in Plant 1. The following information was collected from the managers of these plants.

PLANT 1		
Input	Price	Amount used
Input 1	3	9
Input 2	5	10
Input 3	7	8

PLANT 2		
Input	Price	Amount used
Input 1	4	8
Input 2	7	12
Input 3	3	9

Question 1. Does this information confirm the suspicion of the firm management?

Answer. In order to answer this question one needs to calculate the cost function. Let w_{ij} denote the price of the i th input at the j th plant and x_{ij} denote the quantity of i th input used in production j th plant ($i = 1, 2, 3$ and $j = 1, 2$). Suppose both of these magnitudes are perfectly divisible, therefore can be represented by real numbers. The cost of production can be calculated by multiplying the amount of each input by its price and summing over all inputs.

This means price and quantity vectors (**p** and **q**) are defined on real space and inner product of these vectors are defined. In other words, both **p** and **q** are in the space $\mathbb{R}^3$. The cost function in this case can be written as an inner product of price and quantity vectors as

$$c = (\mathbf{w}, \mathbf{q}), \tag{2.8}$$

where c is the cost, a scalar. Using the data in the above tables cost of production can be calculated by using (2.8) as:

In Plant 1 the total cost is 133, which implies that unit cost is 13.3.

In Plant 2, on the other hand, cost of production is 143, which gives unit cost as 14.3 which is higher than the first plant.

That is, the suspicion is reasonable.

Question 2. The manager of the Plant 2 claims that the reason of the cost differences is the higher input prices in her region than in the other. Is the available information supports her claim?

Answer. Let the input price vectors for Plant 1 and 2 be denoted as $\mathbf{p}_1$ and $\mathbf{p}_2$. Suppose that the latter is a multiple λ of the former, i.e.,

$$\mathbf{p}_2 = \lambda \mathbf{p}_1.$$

Since both vectors are in the space $\mathbb{R}^3$, length is defined for both. From the definition of length one can obtain that

$$|\mathbf{p}_2| = \lambda |\mathbf{p}_1|.$$

In this case, however as can be seen from the tables this is not the case. Plant I enjoys lower prices for inputs 2 and 3, whereas Plant 2 enjoys lower price for input 3. For a rough guess, one can still compare the lengths of the input price vectors which are

$$|\mathbf{p}_1| = 9.11, |\mathbf{p}_2| = 8.60,$$

which indicates that price data does not support the claim of the manager of the Plant 2. When examined more closely, one can see that the Plant 2 uses the most expensive input (input 2) intensely. In contrast, Plant 2 managed to save from using the most expensive input (in this case input 3). Therefore, the manager needs to explain the reasons behind the choice mixture of inputs in her plant.

2.4 Another Economic Application: Index Numbers

One of the problems that applied economists deal with is how exactly the microeconomic information concerning many (in fact in millions) prices and quantities of goods can be aggregated into smaller number of price and quantity variables? Consider an economy which produces many different (in terms of quality, location and time) goods. This means there will thousands, if not millions, of prices to be considered.

Suppose, for example, one wants to estimate the rate of inflation for this economy. Inflation is the rate of change in the general price level, i.e., it has to be calculated by taking into account the changes in the prices of all goods. Assume that there are n different goods. Let p_i be the price and q_i is the quantity of the good i. Consider two points in time, 0 and t. Denote the aggregate value of all goods at time 0 and t, respectively, as

$$V^0 = \sum_i^n p_i^0 q_i^0 \qquad (2.9)$$

and

$$V^t = \sum_i^n p_i^t q_i^t. \qquad (2.10)$$

If $\mathbf{p}^0 = (p_1^0, \ldots, p_n^0)$ and $\mathbf{q}^0 = (q_1^0, \ldots, q_n^0)$ are the (row) vectors characterizing prices and quantities of goods, respectively, then $V^0 = (\mathbf{p}^0, \mathbf{q}^0)$ is just the dot product of vectors $\mathbf{p}^0$ and $\mathbf{q}^0$. Then V^t is the dot product of the vectors p^t and q^t, i.e. $V^t = (\mathbf{p}^t, \mathbf{q}^t)$.

Notice that, in general, between time 0 (initial period) and t (end period) both the prices and the quantities of goods vary. So simply dividing (2.10) by (2.9) will not give the rate of inflation. One needs to eliminate the effect of the change in the quantities. This is the index number problem which has a long history.[6]

In 1871, Laspeyres[7] proposed the following index number formula to deal with this problem

$$P_L = \frac{\sum_{i=1}^n p_i^t q_i^0}{\sum_{i=1}^n p_i^0 q_i^0} \qquad (2.11)$$

Notice that in this formula prices are weighted by initial period quantity weights, in other words, Laspeyres assumed that price changes did not lead to a change in the composition of quantities.

---

[6] Charles de Ferrare Dutot is credited with the introduction of first price index in his book *Refléxions politiques sur les finances et le commerce* in 1738. He used the averages of prices, without weights.

[7] Ernst Louis Etienne Laspeyres (1834–1913) was a German economist and statistician, a representative of German historical school in economics.

In 1874, Paasche[8], suggested using end-period weights, instead of the initial period's

$$P_p = \frac{\sum_{i=1}^n p_i^t q_i^t}{\sum_{i=1}^n p_i^0 q_i^t}$$

Laspeyeres index underestimates, whereas Paasche index overestimates the actual inflation.

Exercise 2.8. Formulate Laspeyres and Paasche indices in term of price and quantity vectors.

Outline of the answer:

$$P_L = \frac{(\mathbf{p}^t, \mathbf{q}^0)}{(\mathbf{p}^0, \mathbf{q}^0)},$$

$$P_P = \frac{(\mathbf{p}^t, \mathbf{q}^t)}{(\mathbf{p}^0, \mathbf{q}^t)}.$$

Exercise 2.9. Consider a three good economy. The initial ($t = 0$) and end period's ($t = 1$) prices and quantities of goods are as given in the following table:

	Price ($t = 0$)	Quantity ($t = 0$)	Price ($t = 1$)	Quantity ($t = 1$)
Good 1	2	50	1,8	90
Good 2	1,5	90	2,2	70
Good 3	0,8	130	1	100

i. Estimate the inflation (i.e. percentage change in overall price level) for this economy by calculating Laspeyres index
ii. Repeat the same exercise by calculating Paasche index.

For further information on index numbers, we refer the reader to [9, 23].

2.5 Matrices

A matrix is a rectangular array of real numbers

$$\begin{bmatrix} a_{11} & a_{12} & \cdots & a_{1n} \\ a_{21} & a_{22} & \cdots & a_{2n} \\ \vdots & \vdots & & \vdots \\ a_{m1} & a_{m2} & \cdots & a_{mn} \end{bmatrix}.$$

---

[8]Hermann Paasche (1851–1925), German economist and statistician, was a professor of political science at Aachen University.

2.5 Matrices

We will denote matrices with capital letters $A, B, \ldots$. The generic element of a matrix A is denoted by a_{ij}, $i = 1, \ldots, m$; $j = 1, \ldots, n$, and the matrix itself is denoted briefly as $A = \|a_{ij}\|_{m \times n}$. Such a matrix with m rows and n columns is said to be of order $m \times n$. If the matrix is square (that is, $m = n$), it is simply said to be of order n.

We denote by $\mathbf{0}$ the *null* matrix which contains zeros only. The *identity* matrix is a matrix $I = I_n$ of size $n \times n$ whose elements are $i_{k,k} = 1$ and $i_{k,m} = 0$ for $k \neq m$, $k = 1, \ldots, n$ and $m = 1, \ldots, n$, that is, it has units on the diagonal and zeroes on the other places. The notion of the identity matrix will be discussed in Sect. 3.2.

Example 2.6. Object – property: Consider m economic units each of which is described by n indices. Units may be firms, and indices may involve the output, the number of employees, the capital stock, etc., of each firm.

Example 2.7. Consider an economy consisting of $m = n$ sectors, where for all $i, j \in \{1, 2, \ldots, n\}$, a_{ij} denotes the share of the output produced in sector i and used by sector j, in the total output of sector i. (Note that in this case the row elements add up to one.)

Example 2.8. Consider $m = n$ cities. Here a_{ij} is the distance between city i and city j. Naturally, $a_{ii} = 0$, $a_{ij} > 0$, and $a_{ij} = a_{ji}$ for all $i \neq j$, and $i, j \in \{1, 2, \ldots, n\}$.

We say that a matrix $A = \|a_{ij}\|_{m \times n}$ is non-negative if $a_{ij} \geq 0$ for all $i = 1, \ldots, m$; $j = 1, \ldots, n$. This case is simply denoted by $A \geq \mathbf{0}$.

Analogously is defined a positive matrix $A > \mathbf{0}$.

2.5.1 Operations on Matrices

Let $A = \|a_{ij}\|_{m \times n}$ and $B = \|b_{ij}\|_{m \times n}$ be two matrices. The sum of these matrices is defined as
$$A + B = \|a_{ij} + b_{ij}\|_{m \times n}.$$

Example 2.9.
$$\begin{bmatrix} 1 & 0 \\ 4 & 2 \\ 7 & 1 \end{bmatrix} + \begin{bmatrix} 3 & 2 \\ 7 & 3 \\ 4 & 1 \end{bmatrix} = \begin{bmatrix} 4 & 2 \\ 11 & 5 \\ 11 & 2 \end{bmatrix}.$$

Let $A = \|a_{ij}\|_{m \times n}$ and $\lambda \in \mathbb{R}$. Then
$$\lambda A = \|\lambda a_{ij}\|_{m \times n}.$$

Example 2.10.

$$2 \begin{bmatrix} 3 & 0 \\ 2 & 4 \\ 1 & 9 \end{bmatrix} = \begin{bmatrix} 6 & 0 \\ 4 & 8 \\ 2 & 18 \end{bmatrix}$$

Properties of Matrix Summation and Multiplication by a Scalar

(1-a) $A + B = B + A$.
(1-b) $A + (B + C) = (A + B) + C$.
(1-c) $A + (-A) = \mathbf{0}$, where $-A = (-1)A$.
(1-d) $A + \mathbf{0} = A$.
(2-a) $1A = A$.
(2-b) $\lambda(\mu A) = (\lambda\mu)A$, $\lambda, \mu \in \mathbb{R}$.
(3-a) $0A = \mathbf{0}$.
(3-b) $(\lambda + \mu)A = \lambda A + \mu A$, $\lambda, \mu \in \mathbb{R}$.
(3-c) $\lambda(A + B) = \lambda A + \lambda B$, $\lambda, \mu \in \mathbb{R}$.

The properties of these two operations are the same as for vectors from $\mathbb{R}^n$. We will clarify this later in Chap. 6.

2.5.2 Matrix Multiplication

Let $A = \|a_{ij}\|_{m \times n}$ and $B = \|b_{jk}\|_{n \times p}$ be two matrices. Then the matrix AB of order $m \times p$ is defined as

$$AB = \left[\sum_{j=1}^{n} a_{ij} b_{jk} \right]_{m \times p}$$

In other words, a product $C = AB$ of the above matrices A and B is a matrix $C = \|c_{ij}\|_{m \times p}$, where c_{ij} is equal to the dot product (A_i, B^j) of the i-th row A_i of the matrix A and the j-th column B^j of the matrix B considered as vectors from $\mathbb{R}^n$.

Consider 2×2 case. Given

$$A = \begin{bmatrix} a_{11} & a_{12} \\ a_{21} & a_{22} \end{bmatrix} \text{ and } B = \begin{bmatrix} b_{11} & b_{12} \\ b_{21} & b_{22} \end{bmatrix},$$

2.5 Matrices

Fig. 2.10 A rotation

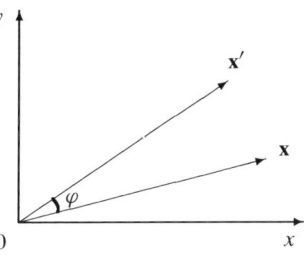

we have

$$AB = \begin{bmatrix} a_{11}b_{11} + a_{12}b_{21} & a_{11}b_{12} + a_{12}b_{22} \\ a_{21}b_{11} + a_{22}b_{21} & a_{21}b_{12} + a_{22}b_{22} \end{bmatrix}.$$

Example 2.11.

$$\begin{bmatrix} 0 & 1 & 2 \\ 2 & 1 & 5 \end{bmatrix} \begin{bmatrix} 3 & 6 & 4 \\ 2 & 5 & 8 \\ 7 & 1 & 9 \end{bmatrix} = \begin{bmatrix} 16 & 7 & 26 \\ 43 & 22 & 61 \end{bmatrix}.$$

$[(0 \cdot 3)+(1 \cdot 2)+(2 \cdot 7)]$
$[(0 \cdot 6)+(1 \cdot 5)+(2 \cdot 1)]$
$[(0 \cdot 4)+(1 \cdot 8)+(2 \cdot 9)]$

Example 2.12. Rotation of a vector $\mathbf{x} = (x, y)$ in $\mathbb{R}^2$ around the origin by a fixed angle φ (Fig. 2.10) can be expressed as a matrix multiplication. If $\mathbf{x}' = (x', y')$ is the rotated vector, then its coordinates can be expressed as

$$\begin{bmatrix} x' \\ y' \end{bmatrix} = R_\alpha \begin{bmatrix} x \\ y \end{bmatrix}, \tag{2.12}$$

where

$$R_\alpha = \begin{bmatrix} \cos\alpha & -\sin\alpha \\ \sin\alpha & \cos\alpha \end{bmatrix}$$

is called a *rotation matrix*.

Note that if we consider the vectors $\mathbf{x}$ and $\mathbf{x}'$ as 1×2 matrices, then (2.12) may be briefly re-written as $\mathbf{x}'^T = R_\alpha \mathbf{x}^T$.

Exercise 2.10. Using elementary geometry and trigonometry, prove the equality (2.12).

Properties of Matrix Multiplication

(1-a) $\alpha(AB) = ((\alpha A)B) = A(\alpha B)$.
(1-b) $A(BC) = (AB)C$.
(1-c) $A\mathbf{0} = \mathbf{0}$.
(2-a) $A(B + C) = AB + AC$.
(2-b) $(A + B)C = AC + BC$.

Remark 2.2. Warning. $AB \neq BA$, in general.

Indeed, let A and B be matrices of order $m \times n$ and $n \times p$, respectively. To define the multiplication BA, we must have $p = m$. But matrices A and B may not commute even if both of them are square matrices of order $m \times m$. For example, consider

$$A = \begin{bmatrix} 1 & 2 \\ 0 & 3 \end{bmatrix} \text{ and } B = \begin{bmatrix} -1 & 2 \\ 1 & 3 \end{bmatrix}.$$

We have

$$AB = \begin{bmatrix} 1 & 8 \\ 3 & 9 \end{bmatrix} \text{ while } BA = \begin{bmatrix} -1 & 4 \\ 1 & 11 \end{bmatrix}.$$

Exercise 2.11. Let A and B be square matrices such that $AB = BA$. Show that:
1. $(A + B)^2 = A^2 + 2AB + B^2$.
2. $A^2 - B^2 = (A - B)(A + B)$.

Exercise 2.12.* Prove the above properties of matrix multiplication.

Hint. To deduce the property 1-b), use the formula $\sum_{i=1}^{n} \left(\sum_{j=1}^{m} x_{ij} \right) = \sum_{j=1}^{m} \left(\sum_{i=1}^{n} x_{ij} \right)$.

Remark 2.3. The matrix multiplication defined above is one of the many concepts that are counted under the broader term "matrix product". It is certainly the most widely used one. However, there are two other matrix products that are of some interest to economists.

Kronecker Product of Matrices

Let $A = \|a_{ij}\|$ be an $m \times n$ matrix and $B = \|b_{ij}\|$ be a $p \times q$ matrix. Then the *Kronecker*[9] *product* of these two matrices is defined as

$$A \otimes B = \begin{bmatrix} a_{11}B & \ldots & a_{1n}B \\ \ldots & \ldots & \ldots \\ a_{m1}B & \ldots & a_{mn}B \end{bmatrix}$$

which is an $mp \times nq$ matrix. Kronecker product is also referred to as *direct product* or *tensor product* of matrices. For its use in econometrics, see [1, 8, 14].

---

[9] Leopold Kronecker (1823–1891) was a German mathematician who made a great contribution both to algebra and number theory. He was one of the founders of so-called constructive mathematics.

Hadamard Product of Matrices

The *Hadamard*[10] *product of matrices* (or *elementwise product*, or *Shur*[11] *product*) of two matrices $A = \|a_{ij}\|$ and $B = \|b_{ij}\|$ of the same dimensions $m \times n$ is a submatrix of the Kronecker product

$$A \circ B = \|a_{ij} b_{ij}\|_{m \times n}.$$

See [1, p. 340] and [24, Sect. 36] for the use of Hadamard product in matrix inequalities.

2.5.3 Trace of a Matrix

Given an $n \times n$ matrix $A = \|a_{ij}\|$, the sum of its diagonal elements $\mathrm{Tr}\, A = \sum_{i=1}^{n} a_{ii}$ is called the *trace* of the matrix A.

Example 2.13.

$$\mathrm{Tr} \begin{bmatrix} 1 & 2 & 3 \\ 10 & 20 & 30 \\ 100 & 200 & 300 \end{bmatrix} = 321$$

Exercise 2.13. Let A and B be two matrices of order n. Show that:

(a) $\mathrm{Tr}(A + B) = \mathrm{Tr}\, A + \mathrm{Tr}\, B$.
(b)* $\mathrm{Tr}(AB) = \mathrm{Tr}(BA)$.

2.6 Transpose of a Matrix

Let $A = \|a_{ij}\|_{m \times n}$. The matrix $B = \|b_{ij}\|_{n \times m}$ is called the *transpose* of A (and denoted by A^T) if $b_{ij} = a_{ji}$ for all $i \in \{1, 2, \ldots, m\}$ and $j \in \{1, 2, \ldots, n\}$.

Example 2.14.

$$\begin{bmatrix} 3 & 0 \\ 2 & 4 \\ 1 & 9 \end{bmatrix}^T = \begin{bmatrix} 3 & 2 & 1 \\ 0 & 4 & 9 \end{bmatrix}$$

---

[10] Jacques Salomon Hadamard (1865–1963), a famous French mathematician who contributed in many branches of mathematics such as number theory, geometry, algebra, calculus and dynamical systems, as well as in optics, mechanics and geodesy. His most popular book *The psychology of invention in the mathematical field* (1945) gives a nice description of mathematical thinking.

[11] Issai Schur (1875–1941), an Israeli mathematician who was born in Belarus and died in Israel, made fundamental contributions to algebra, integral and algebraic equations, theory of matrices and number theory.

The transpose operator satisfies the following properties:
1. $(A^T)^T = A$.
2. $(A + B)^T = A^T + B^T$.
3. $(\alpha A)^T = \alpha A^T$.
4. $(AB)^T = B^T A^T$.

Proof. To prove the above properties, note that one can formally write

$$A^T = \|a_{ij}\|_{m \times n}^T = \|a_{ji}\|_{n \times m}.$$

Then $(A^T)^T = \|a_{ji}\|_{n \times m}^T = \|a_{ij}\|_{m \times n} = A$. This proves the property 1.

Now, $(A + B)^T = \|a_{ij} + b_{ij}\|_{m \times n}^T = \|a_{ji} + b_{ji}\|_{n \times m} = \|a_{ji}\|_{n \times m} + \|b_{ji}\|_{n \times m} = A^T + B^T$. This gives the second property.

To check the third one, we deduce that $(\alpha A)^T = \|\alpha a_{ij}\|_{m \times n}^T = \|\alpha a_{ji}\|_{n \times m} = \alpha \|a_{ji}\|_{n \times m}^T = \alpha A^T$.

Now, it remains to check the fourth property. Let $M = AB$ and $N = B^T A^T$, where the matrices A and B are of orders $m \times n$ and $n \times p$, respectively. Then $M = \|\alpha m_{ji}\|_{m \times p}$ with $m_{ij} = (A_i, B^j)$ and $N = \|\alpha n_{ij}\|_{p \times n}$ with $n_{ij} = ((B^T)_i, (A^T)^j)$. Since the transposition changes rows and columns, we have the equalities of vectors

$$(B^T)_i = B^i, (A^T)^j = A_j.$$

Hence, $m_{ij} = n_{ji}$ for all $i = 1, \ldots, m$ and $j = 1, \ldots, p$. Thus $M^T = N$, as desired. □

A matrix A is called *symmetric* if $A = A^T$. A simple example of symmetric matrices is the distance matrix $A = [a_{ij}]$, where a_{ij} is the distance between the cities i and j. Obviously, $a_{ij} = a_{ji}$ or $A = A^T$.

Theorem 2.3. *For each matrix A of order $n \times n$, the matrix AA^T is symmetric.*

Proof. Consider $(AA^T)^T$. By the properties 3) and 4), we have $(AA^T)^T = (A^T)^T A^T = AA^T$. □

Exercise 2.14. Let A and B be two matrices of order n. Show that $\operatorname{Tr} A^T = \operatorname{Tr} A$.

2.7 Rank of a Matrix

Are the vectors $\mathbf{x}, \mathbf{y}$, and $\mathbf{z}$ in the Fig. 2.11 linearly dependent?
It is obvious that there exists γ such that

$$\mathbf{u} + \gamma \mathbf{z} = \mathbf{0},$$

or

$$\alpha \mathbf{x} + \beta \mathbf{y} + \gamma \mathbf{z} = \mathbf{0},$$

2.7 Rank of a Matrix

Fig. 2.11 Are the vectors **x**, **y**, and **z** linearly dependent?

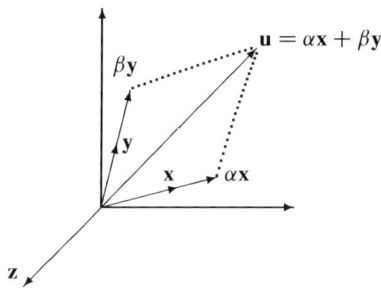

i.e., these three vectors are linearly dependent.

Let us recall the notion of linear dependence of vectors. Consider vectors

$$\alpha = (2, -5, 1, -1)$$
$$\beta = (1, 3, 6, 5)$$
$$\gamma = (-1, 4, 1, 2).$$

Are they linearly dependent? To answer, we construct a system of linear equations as follows: suppose the above vectors are linearly dependent. Then

$$a\alpha + b\beta + c\gamma = \mathbf{0}$$

for some parameters a, b, c, which are not all zero. In component-wise form, we obtain a homogeneous system of linear equation:

$$\begin{cases} 2a + 1b - c = 0 \\ -5a + 3b + 4c = 0 \\ a + 6b + c = 0 \\ -a + 5b + 2c = 0 \end{cases}$$

Here the system of linear equations is called *homogeneous* if every equation has the form "a linear combination of variables is equal to zero".

One can check directly that a solution of the above system is given by $a = 7, b = -3$ and $c = 11$. Hence

$$7\alpha - 3\beta + 11\gamma = \mathbf{0}.$$

Consider now a matrix A

$$A = \begin{bmatrix} a_{11} & a_{12} & \cdots & a_{1n} \\ a_{21} & a_{22} & \cdots & a_{2n} \\ \vdots & \vdots & & \vdots \\ a_{s1} & a_{s2} & \cdots & a_{sn} \end{bmatrix}.$$

Columns of this matrix can be considered as s-dimensional vectors, and maximal number of linearly independent columns is called the *rank* of A.

Example 2.15. Consider the matrix A with columns being the above vectors α, β and γ

$$A = \begin{bmatrix} 2 & 1 & -1 \\ -5 & 3 & 4 \\ 1 & 6 & 1 \\ -1 & 5 & 2 \end{bmatrix}.$$

Since A has 3 columns and the columns are linearly dependent, we have rank $A \leq 2$. On the other hand, it is easy to see that the first two columns of A are linearly independent, hence rank $A \geq 2$. Thus we conclude that rank $A = 2$.

Example 2.16. For the null matrix $\mathbf{0}$, we have the rank $A = 0$. On the other hand, the unit matrix I of the order $n \times n$ has the rank n.

Theorem 2.4. *The maximal number of linearly independent rows of a matrix equals to the maximal number of its linearly independent columns. Recalling the notion of the transpose, we have*

$$\text{rank } A = \text{rank } A^T$$

for every matrix A.

The proof of this theorem is given in Corollary 4.6.

Exercise 2.15. Check this statement for the above matrix A.

2.8 Elementary Operations and Elementary Matrices

In this section, we give a method to find linear dependence of columns of a matrix, and hence, to calculate its rank.

Let A be a matrix of order $m \times n$. Recall that its rows are n-vectors denoted by $A_1, A_2, \ldots, A_m$. The following simple transformations of A are called *elementary (row) operations*. All of them transform A to another matrix A' of the same order one or two rows (say, i-th and j-th) of which slightly differs from those of A:
1. Row switching: $A'_i = A_j$, $A'_j = A_i$.
2. Row multiplication: $A'_i = \lambda A_i$, where $\lambda \neq 0$ is a number.
3. Row replacement: $A'_i = A_i + \lambda A_j$, where $\lambda \neq 0$ is a number.

2.8 Elementary Operations and Elementary Matrices

Example 2.17. Let us apply these operations to the unit matrix

$$I_n = \begin{bmatrix} 1 & 0 & \cdots & 0 \\ 0 & 1 & & 0 \\ & & \ddots & \\ 0 & 0 & \cdots & 1 \end{bmatrix}.$$

The resulting matrices are called *elementary transformation matrices*; they are:

1. $T_{i,j} = \begin{bmatrix} 1 & & & & & \\ & \ddots & & & & \\ & & 0 & \cdots & 1 & \\ & & \vdots & \ddots & \vdots & \\ & & 1 & \cdots & 0 & \\ & & & & & \ddots \\ & & & & & & 1 \end{bmatrix}$;

2. $T_i(\lambda) = \begin{bmatrix} 1 & & & & \\ & \ddots & & & \\ & & \lambda & & \\ & & & \ddots & \\ & & & & 1 \end{bmatrix}$;

3. $T_{i,j}(\lambda) = \begin{bmatrix} 1 & & & & & \\ & \ddots & & & & \\ & & 1 & \cdots & \lambda & \\ & & & \ddots & \vdots & \\ & & & & 1 & \\ & & & & & \ddots \\ & & & & & & 1 \end{bmatrix}$.

Exercise 2.16.* Show that any elementary operation of the second type is a composition of several operations of the first and the third type.

Theorem 2.5. *If A' is a result of an elementary operation of a matrix A, then*

$$A' = TA,$$

where T is a matrix of elementary transformation corresponding to the operation.

Exercise 2.17. Prove this Theorem 2.5.

(*Hint:* Use the definition of product of two matrices as a matrix entries of which are the dot products of rows and columns of the multipliers.)

Let t_1 and t_2 be elementary operations with corresponding matrices T_1 and T_2. The composition $t = t_1 t_2$ of these two operation is another (non-elementary) operation. It follows from Theorem 2.5 that t transforms any matrix A to a matrix

$$A' = t(A) = t_1(t_2(A)) = TA, \text{ where } T = T_1 T_2.$$

So, a matrix T corresponding to a composition of elementary operations is a product of matrices corresponding to the composers.

Another property of elementary operations is that all of them are invertible. This means that one can define the inverse operations for elementary operations of all three kinds listed above. For an operation of the first kind (switching), this inverse operation is the same switching; for the row multiplication by λ, it is a multiplication of the same row by $1/\lambda$; finally, for the replacement of A_i by $A'_i = A_i + \lambda A_j$ the inverse is a replacement of A'_i by $A'_i - \lambda A'_j = A_i$. Obviously, all these inverses are again elementary operations.

We obtain the following

Lemma 2.6. *Suppose that some elementary operation transforms a matrix A to A'. Then there is another elementary operation, which transforms the matrix A' to A.*

Another property of elementary operations is given in the following theorem.

Theorem 2.7. *Suppose that some columns $A^{i_1}, \ldots, A^{i_k}$ of a matrix A are linearly dependent, that is, their linear combination is equal to zero*

$$\alpha_1 A^{i_1} + \cdots + \alpha_k A^{i_k} = \mathbf{0}.$$

Let B be a matrix obtained from A by a sequence of several elementary operations. Then the corresponding linear combination of columns of B is also equal to zero

$$\alpha_1 B^{i_1} + \cdots + \alpha_k B^{i_k} = \mathbf{0}.$$

Proof. Let $T_1, \ldots, T_q$ be the matrices of elementary operations whose compositions transforms A to B. Then $B = TA$, where T is a matrix product $T = T_q \ldots T_2 T_1$. This means that every column B^j of the matrix B is equal to TA^j. Thus,

$$\alpha_1 B^{i_1} + \cdots + \alpha_k B^{i_k} = \alpha_1 TA^{i_1} + \cdots + \alpha_k TA^{i_k} = T(\alpha_1 A^{i_1} + \cdots + \alpha_k A^{i_k}) = T\mathbf{0} = \mathbf{0}.$$

□

2.8 Elementary Operations and Elementary Matrices

Corollary 2.8. *Let a matrix B be obtained from a matrix A by a sequence of several elementary operations. Then a collection $A^{i_1}, \ldots, A^{i_k}$ of columns of the matrix A is linearly dependent if and only if corresponding collection $B^{i_1}, \ldots, B^{i_k}$ is linearly dependent.*

In particular, this means that rank A = rank B.

Proof. The 'only if' statement immediately follows from Theorem 2.7.

According to Lemma 2.6, the matrix A as well may be obtained from B via a sequence of elementary operations (inverses of the given ones). Thus, we can apply the 'only if' part to the collection $B^{i_1}, \ldots, B^{i_k}$ of columns of the matrix B. This imply the 'if' part.

By the definition of rank, the equality rank A = rank B follows. $\square$

Example 2.18. Let us find the rank of the matrix

$$A = \begin{bmatrix} 1 & 2 & 3 \\ 4 & 5 & 6 \\ 7 & 8 & 9 \end{bmatrix}.$$

Before calculations, we apply some elementary operations. First, let us substitute the third row: A_3 with $A_3 - 2A_2$. We get the matrix

$$A' = \begin{bmatrix} 1 & 2 & 3 \\ 4 & 5 & 6 \\ -1 & -2 & -3 \end{bmatrix}.$$

Now, substitute again: $A'_3 \mapsto A'_3 + A'_1$ and then $A'_2 \mapsto A'_2 - 4A'_1$. We obtain the matrix

$$A'' = \begin{bmatrix} 1 & 2 & 3 \\ 0 & -3 & -6 \\ 0 & 0 & 0 \end{bmatrix}.$$

Finally, let us substitute $A''_1 \to A''_1 + (2/3)A''_2$ and multiply $A''_2 \to (-1/3)A''_2$. We obtain the matrix

$$B = \begin{bmatrix} 1 & 0 & -1 \\ 0 & 1 & 2 \\ 0 & 0 & 0 \end{bmatrix}.$$

It is obvious that the first two columns of this matrix B are linearly independent while $B^3 = -B^1 + 2B^2$. Hence rank A = rank B = 2.

Definition 2.4. A matrix A is said to have a (row) *canonical form* (see Fig. 2.12), if the following four conditions are satisfied:
1. All nonzero rows are above any rows of all zeroes.
2. The first nonzero coefficient of any row (called also *leading coefficient*) is always placed to the right of the leading coefficient of the row above it.

Fig. 2.12 Row echelon form of a matrix

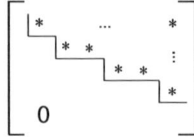

Fig. 2.13 Canonical form of a matrix

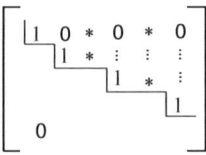

3. All leading coefficients are equal to 1.
4. All entries above a leading coefficient in the same column are equal to 0.

If only first two of the above conditions are satisfied, then the matrix is said to have a *row echelon form* (see Fig. 2.13).

Example 2.19. In Example 2.18 above, the marix A'' has a row echelon form while the matrix B has even a canonical form.

It is easy to calculate the rank of a matrix in an echelon form: it is simply equal to the number of nonzero rows in it.

Theorem 2.9. *Every matrix A can be transformed via a number of elementary operations to another matrix B in a row echelon form (and even in a canonical form). Then the rank of the matrix A is equal to the number of nonzero rows of the matrix B.*

Let us give an algorithm to construct an echelon form of the matrix. This algorithm is called the *Gaussian*[12] *elimination procedure*. It reduces all columns of the matrix one-by-one to the columns of some matrix in a row echelon form. In a recent step, we assume that a submatrix consisting of the first $(j-1)$ columns has an echelon form. Suppose that this submatrix has $(i-1)$ nonzero rows.

In the j-th step, we provide the following:
1. If all elements of the j-th column beginning with a_{ij} and below are equal to zero, the procedure is terminated. Then we go to the $(j+1)$-th step of the algorithm.
2. Otherwise, find the first nonzero element (say, a_{ij}) in the j-th column in the i-th row and below. If it is not a_{ij}, switch two rows A_i and A_j of the matrix. (see Fig. 2.14).

Now, we obtain a matrix A such that $a_{pk} \neq 0$ (Fig. 2.15).

---

[12]Carl Friedrich Gauss (1777–1855) was a great German mathematician and physicist. He made fundamental contribution to a lot of branches of pure and applied mathematics including geodesy, statistics, and astronomy.

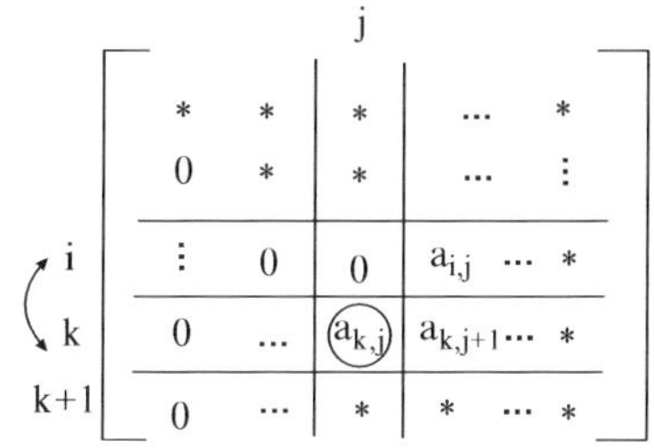

2.8 Elementary Operations and Elementary Matrices

Fig. 2.14 The Gaussian elimination, row switching

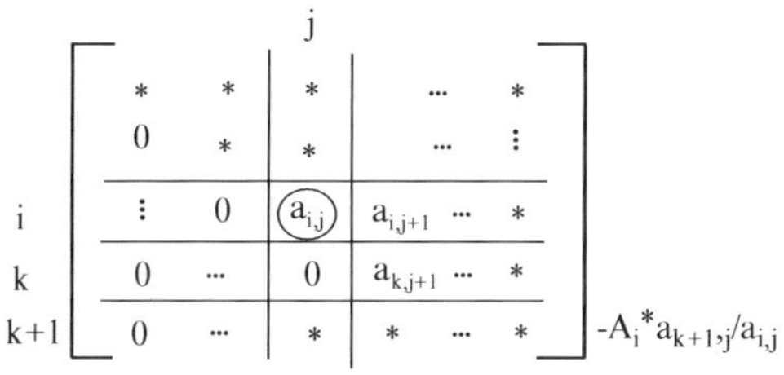

Fig. 2.15 The Gaussian elimination, row subtraction

Fig. 2.16 The Gaussian elimination, the result of row subtractions

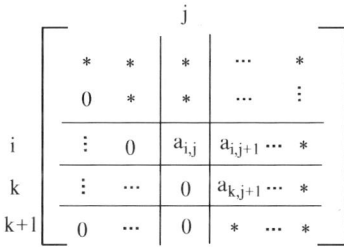

3. For every $p > i$, provide the following row replacement: the row $A_p \to A_p - (a_{pj}/a_{ij})A_j$ (Fig. 2.16).

These three types of operations are sufficient to construct a row echelon form. In the next step of the algorithm, we take $p + 1$ in place of p and $j + 1$ in place of j.

Note that in the above Example 2.18 we used this algorithm to transform A' to A''. Another example is given below.

Example 2.20. Using Gauss algorithm, let us construct a row echelon form of the matrix
$$A = \begin{bmatrix} 0 & 0 & 3 \\ 2 & 6 & -2 \\ 4 & 12 & -1 \end{bmatrix}.$$

In the beginning, $i = j = 1$, that is, we begin with the first column. In operation 1, we find the first nonzero element of this column, that is, a_{21}. In operation 2, we switch the first and the second rows and get the matrix

$$\begin{bmatrix} 2 & 6 & -2 \\ 0 & 0 & 3 \\ 4 & 12 & -1 \end{bmatrix}.$$

In operation 3, we subtract the doubled first row from the third one and get the matrix

$$\begin{bmatrix} 2 & 6 & -2 \\ 0 & 0 & 3 \\ 0 & 0 & 3 \end{bmatrix}.$$

Now, we provide the same three steps for the submatrix formed by the last two rows. The first nonzero column of the submatrix is the third one, so, in operation 1 we put $p = 3$. In operation 2, we find the first nonzero element of the column of the submatrix ($a_{23} = 3$). In operation 3, we replace the first row A_1 by $A_1 + (1/3)A_2$ and the third row A_3 by $A_3 - A_2$. We obtain a matrix in a row echelon form

$$\begin{bmatrix} 2 & 6 & -2 \\ 0 & 0 & 3 \\ 0 & 0 & 0 \end{bmatrix}.$$

The next theorem gives a stronger version of Gaussian elimination.

Theorem 2.10. *1. Every matrix A can also be transformed via elementary operations to a matrix C in a canonical form.*
2. The above canonical form C is unique for every matrix A, that is, it does not depend on the sequence of elementary operations which leads to this from.

Exercise 2.18. Prove the above Theorem 2.10.

Hint. To prove the first part, extend the above algorithm in the following way. To construct a canonical form, we need the same operations 1 and 2 as in the Gauss algorithm, a modified version of the above operation 3 and an additional operation 4.

(3') For every $p \neq i$, provide the following row replacement: the row $A_p \to A_p - (a_{pj}/a_{ij})A_j$.

(4) Replace the i-th row A_i by $(1/a_{ij})A_i$, that is, divide the i-th row by its first nonzero coefficient a_{ij}.

For the second part of the theorem, use Corollary 2.8.

Exercise 2.19. Find the canonical form of the matrix from Example 2.20.

2.9 Problems

1. Find a vector **x** such that:
   (a) $\mathbf{x} + \mathbf{y} = \mathbf{z}$, where $\mathbf{y} = (0, 3, 4, -2)$, and $\mathbf{z} = (3, 2, 1, -5)$.
   (b) $5\mathbf{x} = \mathbf{y} - \mathbf{z}$, where $\mathbf{y} = (-1, -1, 2)$ and $\mathbf{z} = (0, 1, 7)$.

2.9 Problems

2. Let **x** and **y** be two vectors in $\mathbb{R}^n$. Prove that:
   (a) $\mathbf{x} + \mathbf{y} = \mathbf{x}$ if and only if $\mathbf{y} = \mathbf{0}$.
   (b) $\lambda \mathbf{x} = \mathbf{0}$ and $\lambda \neq 0$ if and only if $\mathbf{x} = \mathbf{0}$.
3. Prove that vectors $\mathbf{z}_1, \ldots, \mathbf{z}_s$ in $\mathbb{R}^n$ are linearly dependent if one of them is the null vector.
4. Are the vectors below linearly dependent?
$$\begin{aligned} \mathbf{a}_1 &= (1, \quad 0, 0, \quad 2, \quad 5) \\ \mathbf{a}_2 &= (0, \quad 1, 0, \quad 3, \quad 4) \\ \mathbf{a}_3 &= (0, \quad 0, 1, \quad 4, \quad 7) \\ \mathbf{a}_4 &= (2, -3, 4, 11, 12) \end{aligned}$$
5. Let $\mathbf{z}_1, \ldots, \mathbf{z}_s$ be linearly independent vectors and $\mathbf{x}$ be a vector such that
$$\mathbf{x} = \lambda_1 \mathbf{z}_1 + \cdots + \lambda_s \mathbf{z}_s,$$
where $\lambda_i \in \mathbb{R}$ for all i. Show that this representation is unique.
6. Show that n vectors given by
$$\begin{aligned} \mathbf{x}_1 &= (1, 0, 0, \ldots 0, 0) \\ \mathbf{x}_2 &= (0, 1, 0, \ldots 0, 0) \\ &\cdot \quad \cdot \quad \cdot \quad \cdot \quad \cdot \quad \cdot \\ \mathbf{x}_n &= (0, 0, 0, \ldots 0, 1) \end{aligned}$$
are linearly independent in $\mathbb{R}^n$.
7. Find the rank of the following matrices:
(a) $\begin{bmatrix} 2 & -1 & 3 & -2 & 4 \\ 4 & -2 & 5 & 1 & 7 \\ 2 & -1 & 1 & 8 & 2 \end{bmatrix}$; (b) $\begin{bmatrix} 3 & -1 & 3 & 2 & 5 \\ 5 & -3 & 2 & 3 & 4 \\ 1 & -3 & -5 & 0 & -7 \\ 7 & -5 & 1 & 4 & 1 \end{bmatrix}$.
8. Show that n vectors given by
$$\begin{aligned} \mathbf{x}_1 &= (\eta_{11}, \eta_{12}, \ldots \eta_{1,n-1}, \eta_{1n}) \\ \mathbf{x}_2 &= (0, \quad \eta_{22}, \ldots \eta_{2,n-1}, \eta_{2n}) \\ &\cdot \quad \cdot \quad \cdot \quad \cdot \quad \cdot \quad \cdot \\ \mathbf{x}_n &= (0, \quad 0, \quad \ldots \quad 0, \quad \eta_{nn}) \end{aligned}$$
are linearly independent in $\mathbb{R}^n$ if $\eta_{ii} \neq 0$ for all i.
9. Check that in Definition 2.1 all axioms 1 – 4 are satisfied.
10. Show that the Cauchy inequality (2.6) holds with the equality sign if **x** and **y** are linearly dependent.
11. How many boolean (with components equal to 0 or 1) vectors exist in $\mathbb{R}^n$?
12. Find an example of matrices A, B and C such that $AB = AC$, $A \neq 0$, and $B \neq C$.
13. Find an example of matrices A and B such that $A \neq 0$, $B \neq 0$, but $AB = 0$.
14. Show that $A\mathbf{0} = \mathbf{0}A = \mathbf{0}$.

15. Prove that $(\alpha A)(\beta B) = (\alpha\beta) AB$ for all real numbers α and β, and for all matrices A and B such that the matrix products exist.
16. Prove that $(\alpha A) B = \alpha (AB) = A (\alpha B)$ for each real number α and for all matrices A and B such that the matrix products exist.
17. Let A, B and C be $n \times n$ matrices. Show that $ABC = CAB$ if $AC = CA$ and $BC = CB$.
18. Find a 2×3 matrix A and a 3×2 matrix B such that

$$AB = \begin{bmatrix} 1 & 0 \\ 0 & 1 \end{bmatrix}.$$

19. Let

$$A = \begin{bmatrix} 1 & -1 \\ -3 & 3 \end{bmatrix}.$$

   (a) Find $\mathbf{x} \neq \mathbf{0}$ such that $A\mathbf{x} = \mathbf{0}$.
   (b) Find $\mathbf{y} \neq \mathbf{0}$ such that $\mathbf{y}A = \mathbf{0}$.
20. Let α and β be two angles. Prove the following property of rotation matrices:

$$R_{\alpha+\beta} = R_\alpha R_\beta.$$

21. Prove the properties of matrix summation.
22. Calculate

$$\begin{bmatrix} 0 & 2 & -1 \\ -2 & -1 & 2 \\ 3 & -2 & -1 \end{bmatrix} \begin{bmatrix} 70 & 34 & -107 \\ 52 & 26 & -68 \\ 101 & 50 & -140 \end{bmatrix} \begin{bmatrix} 27 & -18 & 10 \\ -46 & 31 & -17 \\ 3 & 2 & 1 \end{bmatrix}.$$

23. How $A \cdot B$ will change if:
    (a) ith and jth rows of A are interchanged?
    (b) a constant c times jth row of A is added to its ith row?
    (c) ith and jth columns of B are interchanged?
    (d) a constant c times jth column of B is added to its ith column?
24.* Show that rank$(AB) \leq$ rank A and rank$(AB) \leq$ rank B.
25.* Show that the sum of the entries of the Hadamard product $A \circ B$ of two matrices A and B of order n (so-called a *Frobenius*[13] *product*) $(A, B)_F$ is equal to Tr AB^T.
26.* Prove that any matrix A can be represented as $A = B + C$, where B is symmetric matrix and C is an anti-symmetric matrix (i.e., $C^T = -C$).
27. Find all 2×2 matrices A satisfying $A^2 = \mathbf{0}$.
28. Find all 2×2 matrices A satisfying $A^2 = I_2$.

---

[13]Ferdinand Georg Frobenius (1849–1917) was a famous German algebraist. He made a great contribution to group theory and also proved a number of significant theorems in algebraic equations, geometry, number theory, and theory of matrices.

2.9 Problems

29. Find a row echelon form and the rank of the matrix

$$\begin{bmatrix} 0 & 0 & -1 & 3 \\ -2 & -1 & 2 & 1 \\ 2 & 1 & -4 & 5 \end{bmatrix}.$$

30. Find the canonical form of the matrix

$$\begin{bmatrix} 1 & 3 & 0 & 0 \\ 5 & 15 & 2 & 1 \\ -2 & -6 & 1 & 3 \end{bmatrix}.$$

Square Matrices and Determinants 3

The following section illustrates how a matrix can be used to represent a system of linear equations.

3.1 Transformation of Coordinates

Let our 'old' coordinate system be (x, y), and 'new' coordinate system be (x', y'), see Fig. 3.1.

It is obvious that this transformation from (x, y) to (x', y') consists of two moves: translation and rotation. Note that a translation after rotation also yields the same transformation. Let us study these two moves separately.

3.1.1 Translation

We move the origin $O = (0, 0)$ to $O' = (x_0, y_0)$, without changing the direction of axes (Fig. 3.2). Then each point X with coordinates (ξ, η) with respect to the first system get the coordinates (ξ, η) with respect to the second system, where

$$\begin{cases} \xi' = \xi - x_0, \\ \eta' = \eta - y_0. \end{cases}$$

In the vector form, the above system can be re-written as an equality

$$(\xi', \eta') = (\xi, \eta) - (x_0, y_0).$$

After a translation has been made, rotation is applied.

F. Aleskerov et al., *Linear Algebra for Economists*, Springer Texts in Business and Economics, DOI 10.1007/978-3-642-20570-5_3,
© Springer-Verlag Berlin Heidelberg 2011

Fig. 3.1 Two coordinate systems in a plane

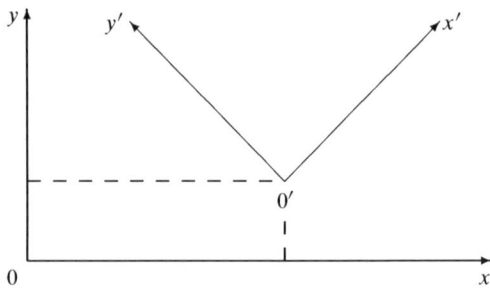

Fig. 3.2 The translation

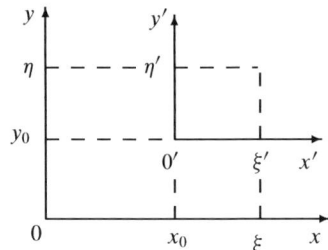

Fig. 3.3 The rotation

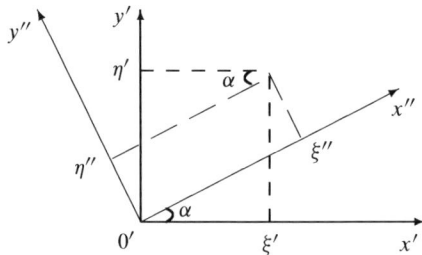

3.1.2 Rotation

We rotate the system by an angle α around O' (Fig. 3.3). Then the point X with the coordinates (ξ', η') (with respect to the old coordinate system) get the new coordinates (ξ'', η''), where

$$\begin{cases} \xi'' = \xi' \cos\alpha + \eta' \sin\alpha, \\ \eta'' = -\xi' \sin\alpha + \eta' \cos\alpha. \end{cases}$$

In matrix notation the above system can be rewritten as

$$\begin{bmatrix} \cos\alpha & \sin\alpha \\ -\sin\alpha & \cos\alpha \end{bmatrix} \begin{bmatrix} \xi' \\ \eta' \end{bmatrix} = \begin{bmatrix} \xi'' \\ \eta'' \end{bmatrix}.$$

Note that here the matrix

$$R_{-\alpha} = \begin{bmatrix} \cos\alpha & \sin\alpha \\ -\sin\alpha & \cos\alpha \end{bmatrix}$$

is the matrix of rotation by the opposite angle $-\alpha$, see Example 2.12.

Then the general transformation can be written as

$$\begin{cases} \xi'' = (\xi - x_0)\cos\alpha + (\eta - y_0)\sin\alpha, \\ \eta'' = -(\xi - x_0)\sin\alpha + (\eta - y_0)\cos\alpha, \end{cases}$$

or

$$\begin{bmatrix} \xi'' \\ \eta'' \end{bmatrix} = R_{-\alpha} \left(\begin{bmatrix} \xi \\ \eta \end{bmatrix} - \begin{bmatrix} x_0 \\ y_0 \end{bmatrix} \right).$$

3.2 Square Matrices

3.2.1 Identity Matrix

As a very special matrix form, we can define the identity matrix, which will be used in many matrix operations. In fact, we already defined the identity matrix but now we do it in general form. Before that, let us first introduce Kronecker delta function to be

$$\delta_{ij} = \begin{cases} 1, & \text{if } i = j, \\ 0, & \text{otherwise.} \end{cases}$$

Identity matrix can, then, be defined as

$$I_n = \|\delta_{ij}\|_{n \times n}$$

or

$$I_n = \begin{bmatrix} 1 & 0 & \ldots & \ldots & 0 \\ 0 & 1 & \ldots & \ldots & 0 \\ \ldots & \ldots & \ldots & \ldots & \ldots \\ 0 & 0 & \ldots & \ldots & 1 \end{bmatrix}.$$

Lemma 3.1. *Let* $A = \|a_{ij}\|_{n \times n}$, *then* $AI_n = I_n A = A$.

Proof. Consider $AI_n = \left[\sum_{j=1}^{n} a_{ij}\delta_{jk} \right]_{n \times n}$. If $j \neq k$ then $\delta_{jk} = 0$ and $a_{ij}\delta_{jk} = 0$; if $j = k$ then $\delta_{jk} = 1$ and $a_{ij}\delta_{jj} = a_{ij}$. Analogously can be proved $I_n A = A$. □

3.2.2 Power of a Matrix and Polynomial of a Matrix

Let A be a square matrix of order m. We can define the power of any square matrix A in the following way

$$A^0 = I_n$$
$$A^1 = A$$
$$A^2 = AA$$
$$\dots \dots \dots$$
$$A^n = A \cdot A^{n-1}$$

Usual polynomial is given by $f(x) = a_n x^n + a_{n-1} x^{n-1} + \cdots + a_1 x + a_0$. The polynomial of the matrix A can be similarly defined as

$$f(A) = a_n A^n + a_{n-1} A^{n-1} + \cdots + a_1 A + a_0 I_m.$$

Example 3.1. Let

$$A = \begin{bmatrix} 2 & 3 \\ 1 & 0 \end{bmatrix}$$

and $f(x) = x^2 - 2x - 1$. Then

$$f(A) = A^2 - 2A - I_2 = \begin{bmatrix} 7 & 6 \\ 2 & 3 \end{bmatrix} - \begin{bmatrix} 4 & 6 \\ 2 & 0 \end{bmatrix} - \begin{bmatrix} 1 & 0 \\ 0 & 1 \end{bmatrix} = \begin{bmatrix} 2 & 0 \\ 0 & 2 \end{bmatrix}.$$

3.3 Systems of Linear Equations: The Case of Two Variables

Consider an arbitrary system of two linear equations in two variables

$$\begin{cases} a_{11} x_1 + a_{12} x_2 = b_1 \\ a_{21} x_1 + a_{22} x_2 = b_2 \end{cases} \quad (3.1)$$

For the above system, let

$$A = \begin{bmatrix} a_{11} & a_{12} \\ a_{21} & a_{22} \end{bmatrix}_{2 \times 2}$$

be the matrix of the known coefficients,

$$\mathbf{x} = \begin{bmatrix} x_1 \\ x_2 \end{bmatrix}_{2 \times 1}$$

be the vector of unknown variables, and

$$\mathbf{b} = \begin{bmatrix} b_1 \\ b_2 \end{bmatrix}_{2\times 1}$$

be the vector of constants. Hence in the matrix form, we can rewrite our system of linear equations as

$$A\mathbf{x} = \mathbf{b}. \tag{3.2}$$

Now, let us try to solve the above system (3.1) in the 'generic' case provided that no division by zero will appear in our calculations. We will solve it by the exclusion of unknown variables.

Divide the first rows by a_{11} and the second row by a_{21} (we assume here that both these numbers are non-zero!).

$$\begin{cases} x_1 + (a_{12}/a_{11})\, x_2 = b_1/a_{11} \\ x_1 + (a_{22}/a_{21})\, x_2 = b_2/a_{21} \end{cases}$$

Subtract the second row from the first one.

$$\left(\frac{a_{12}}{a_{11}} - \frac{a_{22}}{a_{21}} \right) x_2 = \frac{b_1}{a_{11}} - \frac{b_2}{a_{21}}$$

Then x_2 is obtained as

$$x_2 = \frac{(b_1/a_{11}) - (b_2/a_{21})}{(a_{12}/a_{11}) - (a_{22}/a_{21})}$$

$$= \frac{b_2 a_{11} - b_1 a_{21}}{a_{11}a_{22} - a_{12}a_{21}}. \tag{3.3}$$

(We again assume that the denominator $a_{11}a_{22} - a_{12}a_{21}$ is non-zero!)
Inserting the value of x_2 into one of the original equations, and solving for x_1, we get

$$x_1 = \frac{b_1 a_{22} - b_2 a_{12}}{a_{11}a_{22} - a_{12}a_{21}}. \tag{3.4}$$

3.4 Determinant of a Matrix

For the coefficient matrix

$$A = \begin{bmatrix} a_{11} & a_{12} \\ a_{21} & a_{22} \end{bmatrix}$$

let us define the function

$$\det A = a_{11}a_{22} - a_{21}a_{12}$$

called *determinant* of the matrix A. It is equal to the denominators of the fractions in the right-hand sides of the formulae (3.3) and (3.4) for the solution of the system (3.1). Then the numerators of these fractions are equal to the determinants of other matrices, $b_1 a_{22} - b_2 a_{12} = \det A'$ and $b_2 a_{11} - b_1 a_{21} = \det A''$, where

$$A' = \begin{bmatrix} b_1 & a_{12} \\ b_2 & a_{22} \end{bmatrix}$$

and

$$A'' = \begin{bmatrix} a_{11} & b_1 \\ a_{21} & b_2 \end{bmatrix}.$$

So, we can rewrite x_1 and x_2 as

$$\begin{aligned} x_1 &= \det A'/\det A \\ x_2 &= \det A''/\det A \end{aligned} \qquad (3.5)$$

Now we will study the properties of the function

$$\det \begin{bmatrix} a_{11} & a_{12} \\ a_{21} & a_{22} \end{bmatrix}.$$

We can denote a matrix $A_{2\times 2}$ as a pair of its two columns $A_1 = \begin{bmatrix} a_{11} \\ a_{21} \end{bmatrix}$ and $A_2 = \begin{bmatrix} a_{12} \\ a_{22} \end{bmatrix}$, i.e., $A = [A_1, A_2]$. Then the following are true.

Linearity:
(a) $\det[A_1 + A_1', A_2] = \det[A_1, A_2] + \det[A_1', A_2]$.
(b) $\det[A_1, A_2 + A_2'] = \det[A_1, A_2] + \det[A_1, A_2']$.
(c) $\det[c A_1, A_2] = c \det[A_1, A_2]$.
(d) $\det[A_1, c A_2] = c \det[A_1, A_2]$.

Anti-symmetry:
(e) $\det[A_1, A_2] = -\det[A_2, A_1]$.
(f) $\det[A_1, A_1] = 0$.

Unitarity:
(g) $\det I_2 = 1$.

Geometrically, the absolute value of the determinant $\det[A_1, A_2]$ is the area of the parallelogram sides of which are the vectors A_1 and A_2 (Fig. 3.4).

Example 3.2. For example, let $A_1 = (-2, 0)^T$ and $A_2 = (1, 3)^T$ be two column vectors. The area of the parallelogram with sides A_1 and A_2 is the product of the length of the bottom side by the hight, that is, $2 \cdot 3 = 6$. On the other side, we have

3.4 Determinant of a Matrix

Fig. 3.4 The area of this parallelogram is $|\det[A_1, A_2]|$

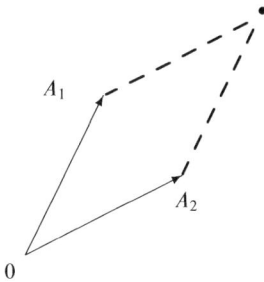

$$\det[A_1, A_2] = \det\begin{bmatrix} -2 & 0 \\ 1 & 3 \end{bmatrix} = -6,$$

a number with the same absolute value.

The determinant of an $n \times n$ matrix is defined as follows.

Definition 3.1. Let $A = [a_{ij}]_{n \times n}$. The *minor* M_{ij} of A is the determinant of the matrix formed from A by removing the i-th row and the j-th column; the product $A_{ij} = (-1)^{i+j} M_{ij}$ is called a *cofactor* of the element a_{ij}.

Then in terms of row i elements and the respective cofactors, the determinant of A is

$$\det A = \sum_{j=1}^{n} a_{ij} A_{ij} \tag{3.6}$$

or in terms of column j elements and their cofactors, it is

$$\det A = \sum_{i=1}^{n} a_{ij} A_{ij}. \tag{3.7}$$

One could prove[1] that all these formulae for different i give the same number $\det A$.

Example 3.3. Let

$$A = \begin{bmatrix} 1 & 2 & 3 \\ 5 & 0 & 6 \\ 7 & 8 & 9 \end{bmatrix}.$$

Using the decomposition by, say, the second row (the relation (3.6) with $i = 2$), we have

---

[1] See, e. g., [33, Theorem 2.4.5].

$$\det A = -5\det\begin{bmatrix} 2 & 3 \\ 8 & 9 \end{bmatrix} + 0\det\begin{bmatrix} 1 & 3 \\ 7 & 9 \end{bmatrix} - 6\det\begin{bmatrix} 1 & 2 \\ 7 & 8 \end{bmatrix} = 30 + 0 + 36 = 66.$$

Example 3.4. Given any matrix $A = [a_{ij}]_{3\times 3}$, we can compute its determinant (in terms of first row elements and their cofactors) as

$$\det\begin{bmatrix} a_{11} & a_{12} & a_{13} \\ a_{21} & a_{22} & a_{23} \\ a_{31} & a_{32} & a_{33} \end{bmatrix} = a_{11}\det\begin{bmatrix} a_{22} & a_{23} \\ a_{32} & a_{33} \end{bmatrix} - a_{12}\det\begin{bmatrix} a_{21} & a_{23} \\ a_{31} & a_{33} \end{bmatrix}$$

$$+ a_{13}\det\begin{bmatrix} a_{21} & a_{22} \\ a_{31} & a_{32} \end{bmatrix}$$

$$= a_{11}a_{22}a_{33} - a_{11}a_{23}a_{32} - a_{12}a_{21}a_{33} + a_{12}a_{23}a_{31} + a_{13}a_{21}a_{32} - a_{13}a_{22}a_{31}.$$

Example 3.5. Suppose that the matrix A is *lower triangular*, that is, all its entries above the main diagonal are zero,

$$A = \begin{bmatrix} a_{11} & 0 & \cdots & 0 \\ a_{21} & a_{22} & \cdots & 0 \\ \vdots & \vdots & \ddots & \vdots \\ a_{n1} & a_{n2} & \cdots & a_{nn} \end{bmatrix}.$$

To calculate its determinant, we apply the formula (3.6) with $j = 1$ recursively for the matrix A itself and its minors:

$$\det A = a_{11}\det\begin{bmatrix} a_{22} & \cdots & 0 \\ \vdots & \ddots & \vdots \\ a_{n2} & \cdots & a_{nn} \end{bmatrix} = a_{11}a_{22}\det\begin{bmatrix} a_{33} & \cdots & 0 \\ \vdots & \ddots & \vdots \\ a_{n3} & \cdots & a_{nn} \end{bmatrix}$$

$$= \cdots = a_{11}a_{22}\ldots a_{nn}.$$

$\square$

Exercise 3.1. Using (3.7), prove that the determinant of an *upper triangular matrix* (i.e., a matrix with zero entries *below* the main diagonal) is again equal to the product of the elements of the main diagonal.

Let us represent the matrix A as an n-tuple of its columns: $A = [A_1, \ldots, A_n]$. Then the properties of the determinant of an $n \times n$ matrix A are the following.

3.4.1 The Basic Properties of Determinants

Linearity:

(a) $\det[A_1, \ldots, A_{i-1}, A_i + A'_i, A_{i+1}, \ldots, A_n] =$
$\det[A_1, \ldots, A_{i-1}, A_i, A_{i+1}, \ldots, A_n] + \det[A_1, \ldots, A_{i-1}, A'_i, A_{i+1}, \ldots, A_n]$

(b) $\det[A_1, \ldots, A_{i-1}, cA_i, A_{i+1}, \ldots, A_n] = c \det[A_1, \ldots, A_{i-1}, A_i, A_{i+1}, \ldots, A_n]$

Anti-symmetry:

(c) $\det[A_1, \ldots, A_i, \ldots, A_j, \ldots, A_n] = -\det[A_1, \ldots, A_j, \ldots, A_i, \ldots, A_n]$

That is, switching the i-th and j-th columns of a matrix changes the sign of its determinant.

Unitarity:

(d) $\det I_n = 1$

Many other properties of determinants (including the formulae (3.6) and (3.7) from the definition) can be deduced from the above basic properties *(a)–(d)*. For example, one can obtain

Corollary 3.1. *The following properties of determinants are also hold:*

(e) $\det[A_1, \ldots, A_{i-1}, B, A_{i+1}, \ldots, A_{j-1}, B, A_{j+1}, \ldots, A_n] = 0$

(f) $\det[A_1, \ldots, A_{i-1}, A_i + cA_j, A_{i+1}, \ldots, A_n] = \det[A_1, \ldots, A_n]$

(g) $\det[A_1, \ldots, A_{i-1}, \mathbf{0}, A_{i+1}, \ldots, A_n] = 0$

Proof. (e) Let A be the matrix $[A_1, \ldots, A_{i-1}, B, A_{i+1}, \ldots, A_{j-1}, B, A_{j+1}, \ldots, A_n]$. According to the anti-symmetry property (e) above, we have

$$\det A = -\det A,$$

hence $\det A = 0$.

(f) Now, let $A = [A_1, \ldots, A_{i-1}, A_i + cA_j, A_{i+1}, \ldots, A_n]$. By the property (a), we have

$\det A = \det[A_1, \ldots, A_{i-1}, A_i, A_{i+1}, \ldots, A_n] + \det[A_1, \ldots, A_{i-1}, cA_j, A_{i+1}, \ldots, A_n].$

By the properties (b) and (e), the second component in the sum here is equal to

$$\det[A_1, \ldots, A_{i-1}, cA_j, A_{i+1}, \ldots, A_j, \ldots, A_n]$$

$$= c \det[A_1, \ldots, A_{i-1}, A_j, A_{i+1}, \ldots, A_j, \ldots, A_n] = c \cdot 0 = 0,$$

hence $\det A = \det[A_1, \ldots, A_n]$.

(g) Applying the property (b) with $c = 0$, we have

$$\det[A_1, \ldots, A_{i-1}, 0, A_{i+1}, \ldots, A_n] = 0 \cdot \det[A_1, \ldots, A_{i-1}, \mathbf{0}, A_{i+1}, \ldots, A_n] = 0.$$

□

Notation used below:

$$\begin{vmatrix} a_{11} & a_{12} & \cdots & a_{1n} \\ a_{21} & a_{22} & \cdots & a_{2n} \\ \cdot\cdot & \cdot & \cdot & \cdot \\ \cdot\cdot & \cdot & \cdot & \cdot \\ \cdot\cdot & \cdot & \cdot & \cdot \\ a_{n1} & a_{n2} & \cdots & a_{nn} \end{vmatrix} = \det \begin{bmatrix} a_{11} & a_{12} & \cdots & a_{1n} \\ a_{21} & a_{22} & \cdots & a_{2n} \\ \cdot\cdot & \cdot & \cdot & \cdot \\ \cdot\cdot & \cdot & \cdot & \cdot \\ \cdot\cdot & \cdot & \cdot & \cdot \\ a_{n1} & a_{n2} & \cdots & a_{nn} \end{bmatrix}.$$

Let us evaluate the following determinant using the properties given above.

Example 3.6. Let us evaluate the following determinant using the properties given above.

$$\begin{vmatrix} 3 & 2 & 7 \\ 0 & 1 & -3 \\ 3 & 4 & 1 \end{vmatrix} = \begin{vmatrix} 3+(-1)\cdot 2 & 2 & 7 \\ 0+(-1)\cdot 1 & 1 & -3 \\ 3+(-1)\cdot 4 & 4 & 1 \end{vmatrix} = \begin{vmatrix} 1 & 2 & 7 \\ -1 & 1 & -3 \\ -1 & 4 & 1 \end{vmatrix}$$

$$= \begin{vmatrix} 1 & 2+(-2)1 & 7+(-7)1 \\ -1 & 1+(-2)(-1) & -3+(-7)(-1) \\ -1 & 4+(-2)(-1) & 1+(-7)(-1) \end{vmatrix}$$

$$= \begin{vmatrix} 1 & 0 & 0 \\ -1 & 3 & 4 \\ -1 & 6 & 8 \end{vmatrix} = \begin{vmatrix} 3 & 4 \\ 8 & 6 \end{vmatrix} = 0$$

Here we have used the property (f) in order to obtain a matrix with many zeroes in the first row. Then we calculate the determinant by the definition using the first row, that is, we apply the formula (3.6) with $i = 1$.

The following two statements show what happens with the determinant after applying standard matrix operations.

(h) $\det(AB) = \det A \cdot \det B$

(i) $\det(A^T) = \det A$ \hfill (3.8)

Proof. We will illustrate that the both claims are true for 2×2 matrices. The complete proof can be obtained by induction on the number of elementary transformations which leads to a canonical form of the matrix A.

3.4 Determinant of a Matrix

(h) Let
$$A = \begin{bmatrix} a_{11} & a_{12} \\ a_{21} & a_{22} \end{bmatrix} \text{ and } B = \begin{bmatrix} b_{11} & b_{12} \\ b_{21} & b_{22} \end{bmatrix}.$$

Then
$$AB = \begin{bmatrix} a_{11}b_{11} + a_{12}b_{21} & a_{11}b_{12} + a_{12}b_{22} \\ a_{21}b_{11} + a_{22}b_{21} & a_{21}b_{12} + a_{22}b_{22} \end{bmatrix}$$

and
$$\det(AB) = \begin{vmatrix} a_{11}b_{11} & a_{11}b_{12} \\ a_{21}b_{11} & a_{21}b_{12} \end{vmatrix} + \begin{vmatrix} a_{12}b_{21} & a_{11}b_{12} \\ a_{22}b_{21} & a_{21}b_{12} \end{vmatrix}$$
$$+ \begin{vmatrix} a_{11}b_{11} & a_{12}b_{22} \\ a_{21}b_{11} & a_{22}b_{22} \end{vmatrix} + \begin{vmatrix} a_{12}b_{21} & a_{12}b_{22} \\ a_{22}b_{21} & a_{22}b_{22} \end{vmatrix}$$
$$= b_{11}b_{12} \begin{vmatrix} a_{11} & a_{11} \\ a_{21} & a_{21} \end{vmatrix} - b_{21}b_{12} \begin{vmatrix} a_{11} & a_{11} \\ a_{21} & a_{21} \end{vmatrix} + b_{11}b_{22} \begin{vmatrix} a_{11} & a_{11} \\ a_{21} & a_{21} \end{vmatrix}$$
$$+ b_{21}b_{22} \begin{vmatrix} a_{12} & a_{12} \\ a_{22} & a_{22} \end{vmatrix}$$
$$= (\det A)(b_{11}b_{22} - b_{21}b_{12})$$
$$= \det A \det B.$$

(i) Again, let
$$A = \begin{bmatrix} a_{11} & a_{12} \\ a_{21} & a_{22} \end{bmatrix}.$$

Then
$$A^T = \begin{bmatrix} a_{11} & a_{21} \\ a_{12} & a_{22} \end{bmatrix},$$

so that $\det A^T = a_{11}a_{22} - a_{21}a_{12} = \det A$.

For the complete proof see, e.g., [33, Sect. 2.4]. □

Warning. If A is an $n \times n$ matrix and c is a real number, then $\det(cA) = c^n \det A$, not $c \det A$!

The property *(i)* implies that in the above properties *(a)*–*(g)* one can replace the columns of the matrix A by the columns of its transpose A^T, that is, rows of A. We get

Corollary 3.2. *All the above properties (a)–(g) of determinants remains true after replacing the columns $A_1, \ldots, A_n$ of the matrix A by its rows.*

3.4.2 Determinant and Elementary Operations

Recall that there are three type of simplest operations under the rows of matrices called elementary operations, see Sect. 2.8, "Elementary operations and elementary matrices".

Proposition 3.1. *If a square matrix A is transformed to another matrix A' via an elementary operation e, then*

$$\det A' = q \det A,$$

where (according to the type of e) the number q is
1. -1, *if e is a row switching;*
2. λ, *if e is a row multiplication by a number λ;*
3. 1, *if e is a row replacement.*

Proof. By Corollary 3.2, we may apply the properties (a)–(g) of determinants to the rows of the matrix A.
1. By the anti-symmetry property (e), the row switching multiplies the determinant by -1.
2. By the linearity property (b), we have $\det A' = \lambda \det A$, hence $q = \lambda$.
3. By the linearity (f), we have in this case $\det A' = \det A$, i.e., $q = 1$.

$\square$

The above properties (a)–(g) and Proposition 3.1 allows to evaluate determinants of arbitrary high order.

Example 3.7. Let us evaluate the determinant of the matrix

$$A = \begin{bmatrix} 0 & 1 & 1 & \ldots & 1 & 1 \\ 1 & 1 & 1 & \ldots & 1 & 1 \\ 1 & 0 & 1 & \ldots & 1 & 1 \\ \vdots & \vdots & \vdots & & \vdots & \vdots \\ 1 & 0 & 0 & \ldots & 1 & 1 \\ 1 & 0 & 0 & \ldots & 0 & 1 \end{bmatrix}$$

of order n. Let us first apply to A elementary operations of the following kind: for each $i = 2, \ldots, n$, replace ith row A_i by $A_i - A_1$. Then

$$\det A = \begin{vmatrix} 0 & 1 & 1 & \ldots & 1 & 1 \\ 1 & 0 & 0 & \ldots & 0 & 0 \\ 1 & -1 & 0 & \ldots & 0 & 0 \\ \vdots & \vdots & \vdots & \vdots & \vdots & \vdots \\ 1 & -1 & -1 & \ldots & 0 & 0 \\ 1 & -1 & -1 & \ldots & -1 & 0 \end{vmatrix}.$$

Now, let us switch the first two rows; then switch the 2nd and the 3rd row, etc.

$$\begin{vmatrix} 0 & 1 & 1 & \ldots & 1 & 1 \\ 1 & 0 & 0 & \ldots & 0 & 0 \\ 1 & -1 & 0 & \ldots & 0 & 0 \\ \vdots & \vdots & \vdots & \vdots & \vdots & \vdots \\ 1 & -1 & -1 & \ldots & 0 & 0 \\ 1 & -1 & -1 & \ldots & -1 & 0 \end{vmatrix} = - \begin{vmatrix} 1 & 0 & 0 & \ldots & 0 & 0 \\ 0 & 1 & 1 & \ldots & 1 & 1 \\ 1 & -1 & 0 & \ldots & 0 & 0 \\ \vdots & \vdots & \vdots & \vdots & \vdots & \vdots \\ 1 & -1 & -1 & \ldots & 0 & 0 \\ 1 & -1 & -1 & \ldots & -1 & 0 \end{vmatrix}$$

$$= \begin{vmatrix} 1 & 0 & 0 & \ldots & 0 & 0 \\ 1 & -1 & 0 & \ldots & 0 & 0 \\ 0 & 1 & 1 & \ldots & 1 & 1 \\ \vdots & \vdots & \vdots & \vdots & \vdots & \vdots \\ 1 & -1 & -1 & \ldots & 0 & 0 \\ 1 & -1 & -1 & \ldots & -1 & 0 \end{vmatrix} = \cdots = (-1)^{n-2} \begin{vmatrix} 1 & 0 & 0 & \ldots & 0 & 0 \\ 1 & -1 & 0 & \ldots & 0 & 0 \\ \vdots & \vdots & \vdots & \vdots & \vdots & \vdots \\ 1 & -1 & -1 & \ldots & 0 & 0 \\ 0 & 1 & 1 & \ldots & 1 & 1 \\ 1 & -1 & -1 & \ldots & -1 & 0 \end{vmatrix}$$

$$= (-1)^{n-1} \begin{vmatrix} 1 & 0 & 0 & \ldots & 0 & 0 \\ 1 & -1 & 0 & \ldots & 0 & 0 \\ \vdots & \vdots & \vdots & \vdots & \vdots & \vdots \\ 1 & -1 & -1 & \ldots & 0 & 0 \\ 1 & -1 & -1 & \ldots & -1 & 0 \\ 0 & 1 & 1 & \ldots & 1 & 1 \end{vmatrix}.$$

This matrix is upper triangular, so, by Exercise 3.1 its determinant is equal to the product of its diagonal elements $1 \cdot (-1) \ldots (-1) \cdot 1 = (-1)^{n-2}$. Thus

$$\det A = (-1)^{n-1}(-1)^{n-2} = -1.$$

For the methods of evaluation of more complicated determinants, see Appendix B.

3.5 Problems

1. Evaluate the following determinants.

   (a) $\begin{vmatrix} 1 & 2 \\ 3 & 4 \end{vmatrix}$; (b) $\begin{vmatrix} a^2 & ab \\ ab & b^2 \end{vmatrix}$; (c) $\begin{vmatrix} n+1 & n \\ n & n-1 \end{vmatrix}$; (d) $\begin{vmatrix} a^2 + ab + b^2 & a^2 - ab + b^2 \\ a+b & a-b \end{vmatrix}$;

   (e) $\begin{vmatrix} \sin \alpha & \cos \alpha \\ \sin \beta & \cos \beta \end{vmatrix}$; (f) $\begin{vmatrix} \sin \alpha + \sin \beta & \cos \beta + \cos \alpha \\ \cos \beta - \cos \alpha & \sin \alpha - \sin \beta \end{vmatrix}$.

2. Evaluate the following determinants.

   (a) $\begin{vmatrix} 0 & 1 & 1 \\ 1 & 0 & 1 \\ 1 & 1 & 0 \end{vmatrix}$; (b) $\begin{vmatrix} 7 & 2 & -1 \\ -3 & 4 & 0 \\ 5 & -8 & 1 \end{vmatrix}$; (c) $\begin{vmatrix} a & x & x \\ x & b & x \\ x & x & c \end{vmatrix}$;

(d) $\begin{vmatrix} \alpha^2+1 & \alpha\beta & \alpha\zeta \\ \alpha\beta & \beta^2+1 & \beta\zeta \\ \alpha\zeta & \beta\zeta & \zeta^2+1 \end{vmatrix}$; (e) $\begin{vmatrix} \sin\alpha & \cos\alpha & 1 \\ \sin\beta & \cos\beta & 1 \\ \sin\zeta & \cos\zeta & 1 \end{vmatrix}$.

3. Consider a transformation of coordinates in the plane which translate the origin to the point $(2, -3)$ and then rotates the coordinate system by angle $\pi/4$. Give a matrix formula for the new coordinates (x', y') of a point X in terms of its old coordinates (x, y).

4. Evaluate the polynomial $f(x) = x^3 - 7x^2 + 13x - 5$ of the matrix

$$A = \begin{bmatrix} 5 & 2 & -3 \\ 1 & 3 & -1 \\ 2 & 2 & -1 \end{bmatrix}.$$

5. Let A be an arbitrary square matrix of order two, and let $f(x)$ be a polynomial $x^2 - ax + b$, where $a = \operatorname{Tr} A$ and $b = \det A$. Show that $f(A) = \mathbf{0}$.

6. Prove that

$$\begin{vmatrix} 0 & 1 & 1 & a \\ 1 & 0 & 1 & b \\ 1 & 1 & 0 & c \\ a & b & c & d \end{vmatrix} = a^2 + b^2 + c^2 - 2ab - 2bc - 2ac + 2d.$$

7. Evaluate the determinant

$$\begin{vmatrix} 1 & 2 & 3 & 4 & 5 \\ 2 & 3 & 4 & 5 & 1 \\ 3 & 4 & 5 & 1 & 2 \\ 4 & 5 & 1 & 2 & 3 \\ 5 & 1 & 2 & 3 & 4 \end{vmatrix}.$$

8. Given a number $d = \det(A_1, \ldots, A_n)$, find $\det(A_n, \ldots, A_1)$.

9. Evaluate the following determinants of matrices of order n.

(a) $\begin{vmatrix} 1 & 1 & 1 & \ldots & 1 & 1 \\ 1 & 1 & 1 & \ldots & 1 & 0 \\ 1 & 1 & 1 & \ldots & 0 & 0 \\ \vdots & & & & & \vdots \\ 1 & 1 & 0 & \ldots & 0 & 0 \\ 1 & 0 & 0 & \ldots & 0 & 0 \end{vmatrix}$; (b) $\begin{vmatrix} 2 & 1 & \ldots & 1 \\ 1 & 2 & \ldots & 1 \\ \vdots & & \ddots & \vdots \\ 1 & 1 & \ldots & 2 \end{vmatrix}$; (c) $\begin{vmatrix} 1 & 2 & 0 & \ldots & 0 & 0 \\ 0 & 1 & 2 & \ldots & 0 & 0 \\ \vdots & \vdots & \vdots & \ddots & \vdots & \vdots \\ 0 & 0 & 0 & \ldots & 1 & 2 \\ 2 & 0 & 0 & \ldots & 0 & 1 \end{vmatrix}$.

10. Suppose X is a matrix such that

$$X + X^2 = -I_n.$$

Find $\det X$.

3.5 Problems

11.* Suppose that a function $f(A)$ from the set of $n \times n$ matrices to the real numbers satisfies the basic properties of determinants (a), (b) and (c) above with 'det' replaced by 'f' (that is, f is a *multilinear anti-symmetric* function of the vectors $A_1, \ldots, A_n$). Show that there exist a number C such that

$$f(A) = C \det A$$

for all matrices A of order n.

This result leads to another definition of determinant: a determinant of an $n \times n$ matrix is a multilinear anti-symmetric function of its columns which is equal to one for the identity matrix I_n.

Hint. Follow the next plan.
(a) By the induction on $k = 0, \ldots, n-1$, for all square matrices B of order $n-k$ define the functions $f_{n-k}(B)$ such that

$$f_{s+1}\left(\begin{bmatrix} a_{11} & 0 \\ * & B \end{bmatrix}\right) = a_{11} f_s(B),$$

where B is an arbitrary matrix of order $s = n - k$ and the star denotes an arbitrary column.
(b) By the induction on $s = 1, \ldots, n-1$, prove the following formula analogous to (3.7):

$$f_{s+1}(A) = \sum_{j=1}^{n} a_{ij} (-1)^{i+j} f_s(M_{ij}),$$

where A is a matrix of order $s + 1$ and i is any row number from 1 to $s + 1$.
(c) By the induction on $s = 1, \ldots, n$, show that $f_s(A) = C \det A$, where $C = f_1(1)$.

Inverse Matrix

4

4.1 Inverse Matrix and Matrix Division

We have seen in Sect. 2.5 that matrices admit standard arithmetic operations, that is, addition and multiplication. Now we discuss a *division* of matrices.

The division $x = b/a$ for numbers gives the solution of the equation

$$ax = b.$$

Using reciprocal numbers, it can be explain via the multiplication: $x = a^{-1}b$. Similarly, for matrices one can consider an equation

$$AX = B, \qquad (4.1)$$

where the matrices A and B are given and X is an unknown matrix. Now we would like to define an 'inverse' matrix A^{-1} such that X is equal to $A^{-1}B$ for each B. If we substitute $B = I_n$, we have $X = A^{-1}I_n = A^{-1}$, so that $AA^{-1} = I_n$. This leads us to the following definition.

Definition 4.1. For any matrix A, the matrix C is called an inverse of A if

$$AC = I \text{ and } CA = I.$$

The inverse matrix is denoted as $A^{-1} = C$.

Exercise 4.1. Show that if A^{-1} exists, then the matrix A must be square matrix.

If the inverse A^{-1} exists, then the solution X of (4.1) exists and is unique. Indeed, multiplying the both sides of (4.1) by A^{-1} from the left, we obtain $A^{-1}AX = A^{-1}B$, that is,

$$X = I_n X = A^{-1}B.$$

Exercise 4.2. Solve the matrix equation $XA = B$ 'dual' to (4.1) under the assumption that A^{-1} does exist.

Theorem 4.1. *If the inverse matrix exists, then it is unique.*

Proof. Let A be a matrix with the inverse A^{-1}, i.e.

$$AA^{-1} = I.$$

Suppose there exists another matrix A' such that

$$A'A = I.$$

Then multiplying each side of the second equality by A^{-1} yields

$$A'AA^{-1} = IA^{-1}$$
$$A'I = A^{-1}$$
$$A' = A^{-1}.$$

□

Example 4.1. Let

$$A = \begin{bmatrix} 2 & 3 \\ 1 & 2 \end{bmatrix}.$$

Then it easy to see that the matrix

$$C = \begin{bmatrix} 2 & -3 \\ -1 & 2 \end{bmatrix}$$

satisfies the conditions of Definition 4.1, that is, $CA = AC = I_2$, so that $A^{-1} = C$.

Example 4.2. We have $I_n^{-1} = I_n$, since $I_n I_n = I_n$.

Let us now consider a matrix A, the inverse of which does not exist. Given

$$A = \begin{bmatrix} 0 & 1 \\ 0 & 1 \end{bmatrix}$$

and

$$B = \begin{bmatrix} b_{11} & b_{12} \\ b_{21} & b_{22} \end{bmatrix}$$

4.1 Inverse Matrix and Matrix Division

let us calculate BA.

$$BA = \begin{bmatrix} b_{11} & b_{12} \\ b_{21} & b_{22} \end{bmatrix} \begin{bmatrix} 0 & 1 \\ 0 & 1 \end{bmatrix}$$

$$= \begin{bmatrix} 0 & b_{11} + b_{12} \\ 0 & b_{21} + b_{22} \end{bmatrix}$$

We immediately note that $BA = I_2$ can hold for no B. Hence, the matrix A does not have the inverse. Such matrices which are not invertible are called *singular*.

Note that there exist a number of generalized versions of the concept of inverse matrix. One of the most important of them, so-called 'pseudoinverse', is discussed in Appendix D.

Theorem 4.2. *If* $\det A = 0$, *then the matrix A is singular.*

Proof. Let A be a matrix with $\det A = 0$ and suppose A^{-1} exists. Then

$$AA^{-1} = I, \text{ and}$$
$$\det A \det A^{-1} = 1$$
$$0 \det A^{-1} = 1,$$

which is a contradiction. □

Let us now give an algorithm to construct A^{-1} for a given non-singular matrix A. Recall that for a matrix

$$A = \begin{bmatrix} a_{11} & a_{12} & \ldots & a_{1n} \\ a_{21} & a_{22} & \ldots & a_{2n} \\ \vdots & & & \vdots \\ a_{n1} & a_{n2} & \ldots & a_{nn} \end{bmatrix},$$

the sign M_{ij} denotes the submatrix (minor) obtained from matrix A by deleting ith row and jth column and $A_{ij} = (-1)^{i+j} \det(M_{ij})$ denotes the cofactor of the element a_{ij}, see Definition 3.1. Let us construct the *adjoint* matrix A^* of the matrix A, using cofactors, as follows:

$$A^* = \begin{bmatrix} A_{11} & A_{21} & \ldots & A_{n1} \\ A_{12} & A_{22} & \ldots & A_{n2} \\ \vdots & & & \vdots \\ A_{1n} & A_{2n} & \ldots & A_{nn} \end{bmatrix} \quad (4.2)$$

Note that the number A_{ij} here replaces the element a_{ji} of A, not a_{ij}!
Let d denote $\det A$. Using the definition of the determinant one can obtain

$$AA^* = A^*A = \begin{bmatrix} d & 0 & \ldots & 0 \\ 0 & d & \ldots & 0 \\ \vdots & \vdots & & \vdots \\ 0 & 0 & \ldots & d \end{bmatrix} \qquad (4.3)$$

(Check the above equality for 2×2 case!)

One can also check that

$$\det(AA^*) = \det A \det A^* = d^n.$$

Hence,
$$\det A^* = d^{n-1}. \qquad (4.4)$$

In other words, if $\det A \neq 0$ then $\det A^* \neq 0$ as well, and (4.4) holds.

Now from (4.3) we immediately obtain

$$AA^* = dI,$$

and

$$A^{-1} = \frac{1}{d}A^* = \begin{bmatrix} A_{11}/d & A_{21}/d & \ldots & A_{n1}/d \\ A_{12}/d & A_{22}/d & \ldots & A_{n2}/d \\ \vdots & & & \vdots \\ A_{1n}/d & A_{2n}/d & \ldots & A_{nn}/d \end{bmatrix} \qquad (4.5)$$

Note that we have constructed the inverse for any matrix A such that $\det A \neq 0$. In view of Theorem 4.2, we get

Corollary 4.3. *A matrix A is singular if and only if $\det A = 0$.*

Example 4.3.

$$A = \begin{bmatrix} 3 & -1 & 0 \\ -2 & 1 & 1 \\ 2 & -1 & 4 \end{bmatrix}, \quad \det A = 5 \neq 0$$

$$A^* = \begin{bmatrix} 5 & 4 & -1 \\ 10 & 12 & -3 \\ 0 & 1 & 1 \end{bmatrix}$$

4.1 Inverse Matrix and Matrix Division

$$A^{-1} = \begin{bmatrix} 1 & 4/5 & -1/5 \\ 2 & 12/5 & -3/5 \\ 0 & 1/5 & 1/5 \end{bmatrix}$$

In Example 2.12, the rotation of a vector in $\mathbb{R}^2$ around the origin is determined by a matrix. Then one can say that the inverse matrix defines the inverse rotation around the origin, due to

Exercise 4.3. For each angle α, we have

$$R_{-\alpha} = R_{\alpha}^{-1}.$$

Example 4.4. The matrix

$$A = \begin{bmatrix} 1 & 0 \\ 1 & 1 \end{bmatrix}$$

is applied to the vector $\mathbf{x} = [1, 0]^T$ to yield the vector $\mathbf{y} = [1, 1]^T$ (Fig. 4.1), that is, $\mathbf{y} = A\mathbf{x}$. Then one obtains $\mathbf{x} = A^{-1}\mathbf{y}$. Here, A^{-1} exists and is equal to

$$A^{-1} = \begin{bmatrix} 1 & 0 \\ -1 & 1 \end{bmatrix}$$

Note that $\mathbf{y}$ can also be obtained from $\mathbf{x}$ by the transformation matrix

$$B = \begin{bmatrix} 1 & 1 \\ 1 & 1 \end{bmatrix}.$$

However, $\mathbf{x}$ cannot be obtained from $\mathbf{y}$ by the inverse rotation B^{-1}, since B^{-1} does not exist.

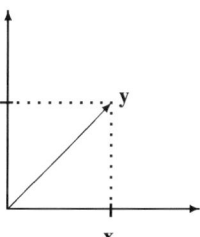

Fig. 4.1 $\mathbf{y} = \begin{bmatrix} 1 & 0 \\ 1 & 1 \end{bmatrix} \mathbf{x}$

4.2 Rank and Determinants

Consider a matrix

$$A = \begin{bmatrix} a_{11} & a_{12} & \cdots & a_{1n} \\ a_{21} & a_{22} & \cdots & a_{2n} \\ \vdots & \vdots & & \vdots \\ a_{s1} & a_{s2} & \cdots & a_{sn} \end{bmatrix}.$$

Recall that the columns of this matrix can be considered as s-dimensional vectors, and maximal number of linearly independent columns is called the rank of A.

Let us choose an arbitrary set containing k columns and k rows of A. Elements in these columns and rows altogether make a square matrix the determinant of which is called a kth-ordered minor of A.

Lemma 4.4. *If for any matrix A all minors of order k are equal to 0, then all minors of higher orders are also equal to 0.*

Proof. Assume that the hypothesis in the Lemma holds for some k, and consider a submatrix A^{k+1} of A which contains $k+1$ columns and $k+1$ rows. Then decomposing the determinant of A^{k+1} by any of its rows, we obtain that $\det A^{k+1} = 0$. □

Theorem 4.5. *For any matrix, the highest order of non-zero minors is equal to the rank of the matrix.*

Proof. Take any matrix $A = \|a_{ij}\|_{s \times n}$. Let the highest order of non-zero minors of A be equal to r. Clearly $r \leq \min\{s, n\}$. Without loss of generality assume that the first r rows and r columns of A yield a nonzero minor D, i.e., $D = \det \|a_{ij}\|_{r \times r}$. Then first r columns of A are linearly independent. (Why ?) If $r = \min\{s, n\}$, then the proof is complete. Consider the case in which $r < \min\{s, n\}$. Let us show that any lth column of A, with $r < l \leq n$, is linearly dependent on the first r columns. Consider the matrix

$$B^{i,l} = \begin{bmatrix} a_{11} & \cdots & a_{1r} & a_{1l} \\ \vdots & & \vdots & \vdots \\ a_{r1} & \cdots & a_{rr} & a_{rl} \\ a_{i1} & \cdots & a_{ir} & a_{il} \end{bmatrix}.$$

If $i > r$, then $\det B^{i,l}$ is a minor of order greater than r. So, by the initial assumption $\det B^{i,l} = 0$. On the other hand, if $i \leq r$, then $\det B^{i,l}$ is not a minor of A, but we still have that $\det B^{i,l} = 0$ for $B^{i,l}$ contains two identical rows. Using $\det B^{i,l} = 0$, we can consider its decomposition by the last row of $B^{i,l}$ to obtain,

$$a_{i1}A_1 + a_{i2}A_2 + \cdots + a_{ir}A_r + a_{il}D = 0,$$

where A_k is the minor associated with the element a_{ik} of the matrix $B^{i,l}$. Note that the minor A_k is formed by the elements of the first r rows of A, so, the tuple of coefficients $A_1, \ldots, A_r$ here is the same for all i. Since $D \neq 0$, we get

$$a_{il} = -\frac{A_1}{D}a_{i1} - \cdots - \frac{A_r}{D}a_{ir}.$$

This equality holds for all $i = 1, \ldots, s$. So, ith column of A is the sum of first r columns multiplied by coefficients $(-A_1/D, \ldots, -A_r/D)$, respectively. □

In particular, a square matrix of order n has the maximal rank n if and only if its determinant is non-zero. Such matrix is called *a matrix of full rank*.

Corollary 4.6. *The maximal number of linearly independent rows of a matrix equals the maximal number of its linearly independent columns, that is, to the rank of the matrix.*

Exercise 4.4. Consider the transpose matrix A^T of A to prove the above claim yourself.

4.3 Problems

1. For each of the following matrix, find the inverse.

   (a) $\begin{bmatrix} 1 & 2 \\ 3 & 4 \end{bmatrix}$; (b) $\begin{bmatrix} \cos\alpha & -\sin\alpha \\ \sin\alpha & \cos\alpha \end{bmatrix}$; (c) $\begin{bmatrix} 2 & 7 & 3 \\ 3 & 9 & 4 \\ 1 & 5 & 3 \end{bmatrix}$;

   (d) $\begin{bmatrix} 1 & 2 & 2 \\ 2 & 1 & -2 \\ 2 & -2 & 1 \end{bmatrix}$; (e) $\begin{bmatrix} 0 & -2 & 1 & 5 \\ 10 & 0 & -1 & 0 \\ 0 & 1 & -1 & 2 \\ 1 & 2 & -2 & 3 \end{bmatrix}$.

2. Find the inverse of

   (a) $\begin{bmatrix} 1 & 1 & \ldots & 1 & 1 \\ 0 & 1 & \ldots & 1 & 1 \\ \vdots & \vdots & \ddots & \vdots & \vdots \\ 0 & 0 & \ldots & 0 & 1 \end{bmatrix}$; (b) $\begin{bmatrix} 1 & 1 & 1 & \ldots & 1 & 1 \\ 1 & 0 & 1 & \ldots & 1 & 1 \\ \vdots & \vdots & \vdots & \ddots & \vdots & \vdots \\ 1 & 1 & 1 & \ldots & 1 & 0 \end{bmatrix}$; (c) $\begin{bmatrix} 0 & 1 & 1 & \ldots & 1 \\ 1 & 0 & 1 & \ldots & 1 \\ \vdots & \vdots & \vdots & \ddots & \vdots \\ 1 & 1 & 1 & \ldots & 0 \end{bmatrix}$.

3. Solve for the matrix X in the following equations

   (a) $\begin{bmatrix} 1 & 2 \\ 3 & 4 \end{bmatrix} X = \begin{bmatrix} 3 & 5 \\ 5 & a \end{bmatrix}$; (b) $X \begin{bmatrix} 3 & -2 \\ 5 & -4 \end{bmatrix} = \begin{bmatrix} -1 & 5 \\ -2 & 6 \end{bmatrix}$;

(c) $X \begin{bmatrix} 5 & 3 & 1 \\ 1 & -3 & -2 \\ -5 & 2 & 1 \end{bmatrix} = \begin{bmatrix} -8 & 30 \\ -5 & 90 \\ -2 & 150 \end{bmatrix}$;

(d) $\begin{bmatrix} 2 & -3 & 1 \\ 4 & -5 & 2 \\ 5 & -7 & 3 \end{bmatrix} X \begin{bmatrix} 9 & 7 & 6 \\ 1 & 1 & 2 \\ 1 & 1 & 1 \end{bmatrix} = \begin{bmatrix} 2 & 0 & -2 \\ 18 & 12 & 9 \\ 23 & 15 & 11 \end{bmatrix}$.

4. Let A and B be non-singular matrices of the order n. Show that the following statements are equivalent.
   (a) $AB = BA$,
   (b) $AB^{-1} = B^{-1}A$,
   (c) $A^{-1}B = BA^{-1}$,
   (d) $A^{-1}B^{-1} = B^{-1}A^{-1}$.

5. Let A be symmetric non-singular matrix. Show that A^{-1} is also symmetric.

6. Let A be a lower triangular (or an upper triangular) matrix. Show that A^{-1} is also lower (respectively, upper) triangular.

7. Find the rank of the matrix below for all possible values of λ

(a) $\begin{bmatrix} 1 & \lambda & -1 & 2 \\ 2 & -1 & \lambda & 5 \\ 1 & 10 & 6 & 1 \end{bmatrix}$;

(b) $\begin{bmatrix} 0 & -1 & 1 & -1 \\ \lambda & 0 & -1 & 0 \\ 0 & 0 & 2\lambda & 2 \\ 1 & 2 & -2 & 3 \end{bmatrix}$.

8. Prove that adding a row to a matrix A either does not change the rank of A or increases it by 1.

9. Find conditions which are necessary and sufficient for any three points (x_1, y_1), (x_2, y_2) and (x_3, y_3) to be on the same line.

10. What are the necessary and sufficient conditions for the three lines

$$\begin{cases} a_1x + b_1y + c_1 = 0 \\ a_2x + b_2y + c_2 = 0 \\ a_3x + b_3y + c_3 = 0 \end{cases}$$

to intersect at one point?

11. Let $A = \|a_{ij}\|_{m \times n}$ be a matrix, and write it as a set of columns

$$A = (A_1, A_2, \ldots, A_n).$$

4.3 Problems

Consider a matrix A' such that
(a) $A' = (A_1, A_2, \ldots, A_{i-1}, cA_i, A_{i+1}, \ldots, A_n)$ where $c \neq 0$,
(b) $A' = (A_1, \ldots, A_{i-1}, A_j, A_{i+1}, \ldots, A_{j-1}, A_i, A_{j+1}, \ldots, A_n)$.

Show that in both cases, matrix A' has the same rank as A.

12.* Prove that the rank of the sum of two matrices is not greater than the sum of their ranks.

13.* Matrix A is called *nilpotent* of degree k if $A^k = 0$.
Prove that if a square matrix $A = \|a_{ij}\|_{2\times 2}$ of order two is nilpotent of some positive degree k then $A^2 = 0$.

Systems of Linear Equations 5

Remember the economic model given in Introduction. It is one of the main problems in economics to find when (i.e, under what price) the demand and the supply of the economic system will be in equilibrium. We now begin to find these conditions for our model in the most general way.

A set of linear equations is called a *system*. It is of interest to find conditions under which (a) the solution of a system exists, (b) the solution of a system is unique.

Example 5.1.

$$\begin{cases} x_1 + 5x_2 = 1 \\ x_1 + 5x_2 = 7 \end{cases}$$ There is no solution of this system. Why?

$$\begin{cases} x_1 + 2x_2 = 7 \\ x_1 + x_2 = 4 \end{cases}$$ The solution is unique: $x_1 = 1$, $x_2 = 3$.

$$\begin{cases} 3x_1 - x_2 = 1 \\ 6x_1 - 2x_2 = 2 \end{cases}$$ This system has many solutions: $x_1 = k \in \mathbb{R}$, $x_2 = 3k - 1$.

We will discuss three approaches to solutions of such systems. First of them, so-called Cramer method, gives simple formulae for the solution, see Sect. 5.1. However, this method is used only for special systems. The second approach is based on a universal algorithm for solution of the system, see Sect. 5.2. However, the dependence of the solution on the data is not clear in the algorithm. The third approach gives an explicit formula for a particular or an approximate solution. It is based on the concept of a pseudoinverse of a matrix, see Appendix D.

Consider now a system of linear equations

$$\begin{cases} a_{11}x_1 + a_{12}x_2 + \ldots + a_{1n}x_n = b_1 \\ a_{21}x_1 + a_{22}x_2 + \ldots + a_{2n}x_n = b_2 \\ \ldots \quad \ldots \quad \ldots \quad \ldots \quad \ldots \\ a_{s1}x_1 + a_{s2}x_2 + \ldots + a_{sn}x_n = b_s \end{cases} \quad (5.1)$$

Let us discuss the conditions when the above system does have a solution.

For the above system, let A be the matrix containing coefficients of all unknown variables

$$A = \|a_{ij}\|_{s \times n},$$

and let $\widetilde{A}$ be the matrix of all coefficients in the system

$$\widetilde{A} = \begin{bmatrix} a_{11} & a_{12} & \dots & a_{1n} & b_1 \\ \dots & \dots & \dots & \dots & \dots \\ a_{s1} & a_{s2} & \dots & a_{sn} & b_s \end{bmatrix}.$$

Lemma 5.1. *Either* $\operatorname{rank}(A) = \operatorname{rank}(\widetilde{A})$ *or* $\operatorname{rank}(\widetilde{A}) = \operatorname{rank}(A) + 1$.

Proof. Left as an exercise. □

Now we can state the main theorem about the systems of linear equations.

Theorem 5.2 (Kronecker–Capelli[1]). *A system of linear equations (5.1) has a solution if and only if* $\operatorname{rank}(A) = \operatorname{rank}(\widetilde{A})$.

Proof. Assume that the system (5.1) has a solution $k_1, \dots, k_n$. Insert it in the system for the unknown variables $x_1, \dots, x_n$. Then, we have a system of s equalities such that the column $\mathbf{b} = [b_1, \dots, b_s]^T$ is represented as a linear combination of columns in A. Since all columns in $\widetilde{A}$, except for $\mathbf{b}$, are columns in A, we must have the same set of linearly independent columns in A and in $\widetilde{A}$, which means that $\operatorname{rank}(A) = \operatorname{rank}(\widetilde{A})$.

Assume now that $\operatorname{rank}(A) = \operatorname{rank}(\widetilde{A})$, i.e. the maximal number of linearly independent columns in A and in $\widetilde{A}$ are equal. Then, the last column of $\widetilde{A}$ can be represented in terms of the submatrix containing maximal number of linearly independent columns, i.e., there exist $k_1, \dots, k_n$ such that $\sum_{j=1}^{n} k_j a_{ij} = b_i$ for all $i = 1, \dots, s$. Hence, $k_1, \dots, k_n$ are the solution of (5.1). □

By the above theorem, we have a criterion of the existence of the solution. But we still have to answer the following questions: If a solution exists, how can we find it? If there are several solutions, how can we find all of them?

Consider the system (5.1). Suppose the matrix A has rank r, i.e. r equals the maximal number of linearly independent rows in A, and all other rows are linear combinations of those r rows. Without loss of generality, assume that those are first r rows of A. Then first r rows of $\widetilde{A}$ are linearly independent as well (why?).

Since $\operatorname{rank}(A) = \operatorname{rank}(\widetilde{A})$, the first r rows in $\widetilde{A}$ give a set of maximal number of linearly independent rows. Then any equation in (5.1) can be represented as a linear combination of first r equations, i.e., as a sum of those r equations multiplied by some coefficients.

Moreover, any solution of first r equations will satisfy all other equations in (5.1), as well. Hence it is sufficient to find the solutions of the system

---

[1] Alfredo Capelli (1855–1910), an Italian algebraist. He gave a modern definition of rank of matrix.

5 Systems of Linear Equations

$$\begin{cases} a_{11}x_1 + a_{12}x_2 + \ldots + a_{1n}x_n = b_1 \\ a_{21}x_1 + a_{22}x_2 + \ldots + a_{2n}x_n = b_2 \\ \ldots \\ a_{r1}x_1 + a_{r2}x_2 + \ldots + a_{rn}x_n = b_r \end{cases} \quad (5.2)$$

We can rewrite the above system as $A'x' = b'$, where $A' = \|a_{ij}\|_{r \times n}$, $x' = [x_1, \ldots, x_n]^T$ and $b' = [b_1, \ldots, b_r]^T$. The matrix A' has a rank r. (Why?)

If $r = n$ then the number of equations is equal to the number of variables, and the system (5.2) has a unique solution

$$x' = \left(A'\right)^{-1} b'.$$

If $r < n$, then $det \|a_{ij}\|_{r \times r}$ is not equal to 0. Then reconstruct (5.2) as

$$\begin{cases} a_{11}x_1 + \ldots + a_{1r}x_r = b_1 - a_{1,r+1}x_{r+1} - \cdots - a_{1n}x_n \\ \ldots \\ a_{r1}x_1 + \ldots + a_{rr}x_r = b_r - a_{r,r+1}x_{r+1} - \cdots - a_{rn}x_n \end{cases} \quad (5.3)$$

or $A''x'' = b''$ where $A'' = \|a_{ij}\|_{r \times r}$, $x'' = [x_1, \ldots, x_r]^T$ and $b'' = [\beta_1, \ldots, \beta_n]^T$ with $\beta_i = b_i - a_{i,r+1}x_{r+1} - \cdots - a_{in}x_n$.

Choose some arbitrary (!) values $c_{r+1}, \ldots, c_n$ for variables $x_{r+1}, \ldots, x_n$. Then we obtain the system of r equations in r unknown variables with $\det A'' \neq 0$.

Hence the system (5.3) has a solution

$$x'' = \left(A''\right)^{-1} b''. \quad (5.4)$$

Let $x'' = [c_1, \ldots, c_r]^T$. It is obvious that $[c_1, \ldots, c_r, c_{r+1}, \ldots, c_n]^T$ is then a solution of (5.2).

We can now formulate the general rule of obtaining the solutions of a system.

Let a system (5.1) be given such that $\text{rank}(A) = \text{rank}(\widetilde{A}) = r$. Choose r linearly independent rows in A and retain in (4.3) only the equations associated with r linearly independent rows that you chose. Keep at the left side of these equations only r variables such that the determinant of the associated coefficients for these variables is not equal to 0. For each equation, put all other variables, that are called free, in the right side of the equations. Assign to the free variables arbitrary values, and calculating values of all other variables by (5.4), we obtain all solutions of (4.3).

Example 5.2. Consider the following system of equations:

$$\begin{cases} x_1 + x_2 - 2x_3 - x_4 + x_5 = 1 \\ 3x_1 - x_2 + x_3 + 4x_4 + 3x_5 = 4 \\ x_1 + 5x_2 - 9x_3 - 8x_4 + x_5 = 0 \end{cases}$$

Check that $\text{rank}(A) = \text{rank}(\widetilde{A}) = 2$. First and third equations are linearly independent. So, we have to consider only:

$$\begin{cases} x_1 + x_2 = 1 + 2x_3 + x_4 - x_5 \\ x_1 + 5x_2 = 9x_3 + 8x_4 - x_5 \end{cases}$$

Then

$$x_1 = \frac{5}{4} + \frac{1}{4}x_3 - \frac{3}{4}x_4 - x_5$$

$$x_2 = -\frac{1}{4} + \frac{7}{4}x_3 + \frac{7}{4}x_4$$

where $x_1, x_2, x_3 \in \mathbb{R}$. This is the general solution.

Corollary 5.3. *The solution of system (4.3), whenever exists, is unique if and only if* $\text{rank}(A)$ *is equal to the number of variables.*

5.1 The Case of Unique Solution: Cramer's Rule

We have seen in (3.5) that the components of a solution for a system of two linear equations with two variables is expressed as ratios of determinants of certain matrices. Let us give analogous formulae for larger systems.

Theorem 5.4 (Cramer's[2] rule). *Let*

$$AX = B$$

be a system of linear equations such that the number of equations and the number of variables are the same, that is, A is a square matrix of order n. Suppose that A is not singular. Then the solution of the system is

$$x_i = \frac{\det A_i}{\det A}$$

for all $i = 1, \ldots, n$, *where* A_i *is the matrix formed from A by replacing its i-th column by the column B.*

Proof. From the equation

$$AX = B$$

---

[2]Gabriel Cramer (1704–1752) was a Swiss mathematician who introduced determinants and used them to solve algebraic equations.

5.1 The Case of Unique Solution: Cramer's Rule

we have

$$X = A^{-1}B = \frac{1}{\det A}A^*B,$$

so,

$$x_i = \frac{(i\text{-th row of }A^*, B)}{\det A}.$$

By the definition of the adjoint matrix (4.2), the denominator of the above fraction is $A_{1i}b_1 + \cdots + A_{ni}b_n = \det A_i$. □

Example 5.3. Solve the system

$$\begin{cases} x + 3y - z = 4, \\ 2x - y + z = 3, \\ 3x - 2y + 2z = 5. \end{cases}$$

Here

$$A = \begin{bmatrix} 1 & 3 & -1 \\ 2 & -1 & 1 \\ 3 & -2 & 2 \end{bmatrix}, \det A = -2 \neq 0.$$

The matrix A is not singular, so the system has a unique solution. We have

$$\det A_1 = \begin{vmatrix} 4 & 3 & -1 \\ 3 & -1 & 1 \\ 5 & -2 & 2 \end{vmatrix} = -2, \det A_2 = \begin{vmatrix} 1 & 4 & -1 \\ 2 & 3 & 1 \\ 3 & 5 & 2 \end{vmatrix} = -4,$$

$$\det A_3 = \begin{vmatrix} 1 & 3 & 4 \\ 2 & -1 & 3 \\ 3 & -2 & 5 \end{vmatrix} = -6,$$

hence

$$x_1 = -2/(-2) = 1,$$
$$x_2 = -4/(-2) = 2,$$
$$x_3 = -6/(-2) = 3.$$

5.2 Gauss Method: Sequential Elimination of Unknown Variables

Consider a system of linear equations

$$\begin{cases} a_{11}x_1 + a_{12}x_2 + \ldots + a_{1n}x_n = b_1 \\ a_{21}x_1 + a_{22}x_2 + \ldots + a_{2n}x_n = b_2 \\ a_{31}x_1 + a_{32}x_2 + \ldots + a_{3n}x_n = b_3 \\ \vdots \qquad \vdots \qquad \vdots \qquad \vdots \qquad \vdots \\ a_{m1}x_1 + a_{m2}x_2 + \ldots + a_{mn}x_n = b_m \end{cases} \quad (5.5)$$

Let us apply some operation to it – subtract from the second equation of (5.5), the first one multiplied by some constant $c \neq 0$. Then we get

$$\begin{cases} a_{11}x_1 + a_{12}x_2 + \ldots + a_{1n}x_n = b_1 \\ a'_{21}x_1 + a'_{22}x_2 + \ldots + a'_{2n}x_n = b'_2 \\ a_{31}x_1 + a_{32}x_2 + \ldots + a_{3n}x_n = b_3 \\ \vdots \qquad \vdots \qquad \vdots \qquad \vdots \qquad \vdots \\ a_{m1}x_1 + a_{m2}x_2 + \ldots + a_{mn}x_n = b_m \end{cases} \quad (5.6)$$

where

$$a'_{2k} = a_{2k} - ca_{1k},$$
$$b'_2 = b_2 - cb_1.$$

Example 5.4. Consider the system:

$$\begin{cases} x_1 + 2x_2 + 5x_3 = -9 \\ x_1 - x_2 + 3x_3 = 2 \\ 3x_1 - 6x_2 - x_3 = 25 \end{cases}$$

Let us subtract from the second equation the first one multiplied by one to obtain

$$\begin{cases} x_1 + 2x_2 + 5x_3 = -9 \\ \quad\quad -3x_2 - 2x_3 = 11 \\ 3x_1 - 6x_2 - x_3 = 25. \end{cases}$$

We have the following question: Is the system (5.5) equivalent to the system (5.6)? That is, is it true that if one of the systems does not have a solution, then the other system does not have a solution as well? Do the two systems always have the same solution(s)? The answer to both these questions is 'yes'!

5.2 Gauss Method: Sequential Elimination of Unknown Variables

Lemma 5.5. *Systems (5.5) and (5.6) are equivalent.*

Note that the matrix $\widetilde{A}$ for the system (5.6) is obtained from the matrix $\widetilde{A}$ of the system (5.5) via an elementary transformation of the third type (see Sect. 2.8). So, it follows from Lemma 5.5 that any elementary transformation, being applied to the matrix $\widetilde{A}$ of a system of linear equations, gives the matrix of an equivalent system.

Proof. Let $(k_1, \ldots, k_n)$ be a solution of (5.5). Let us show that it is also a solution of (5.6). Then it remains to prove that $(k_1, \ldots, k_n)$ satisfies the second equation in (5.6), for it clearly satisfies all other equations. Note that we have

$$a_{11}k_1 + a_{12}k_2 + \cdots + a_{1n}k_n = b_1$$
$$a_{21}k_1 + a_{22}k_2 + \cdots + a_{2n}k_n = b_2.$$

Multiplying the first equality by c and subtracting it from the second one gives

$$(a_{21} - ca_{11})k_1 + (a_{22} - ca_{12})k_2 + \cdots + (a_{2n} - ca_{1n})k_n = b_2 - cb_1$$

or

$$a'_{21}k_1 + a'_{22}k_2 + \cdots + a'_{2n}k_n = b'_2.$$

So $(k_1, \ldots, k_n)$ is also a solution of (5.6). It is left as an exercise to show that any solution of (5.6) is a solution of (5.5) as well. $\square$

It is obvious that such transformations of a system, if applied several times, still yield an equivalent system.

Remark 5.1. By this method, we can exclude step by step some variables from each equation in the system. If, at the end, we obtain an equation, say the ith one, such that all coefficients of unknown variables in it are equal to 0 as well as the right hand side of it (call it $\bar{b}_i$), then we can just omit this equation. But, if $\bar{b}_i$ is not zero while all unknown variables have zero coefficients in the ith equation, then that equation can not be satisfied by any values of variables. In that case, the system (5.5) is called *inconsistent*.

Now we are ready to give the Gauss method.

Let the system (5.5) be given. Relabel all coefficients a_{ij} by a_{ij}^0 and b_i by b_i^0, for all i and j. Let i and j be two counters (i for rows and j for variables) with the initial values being equal to 1, i.e., $i := 1$ and $j := 1$.

Procedure begins at

Step 1: Determine if $a_{ij}^{i-1} \neq 0$. If $a_{ij}^{i-1} = 0$, then look for a row $k \in \{i+1, \ldots, m\}$ such that $a_{kj} \neq 0$. If there is no such k, go to Step 4. Otherwise, choose such a k; interchange rows i and k, and switch the labels of the coefficients, i.e., $a_{il}^{i-1} \leftrightarrow a_{kl}^{i-1}$ and $b_i^{i-1} \leftrightarrow b_k^{i-1}$, for all $l = 1, \ldots, n$.

Step 2: To exclude x_j from all equations indexed by $k = i+1, \ldots, m$, subtract from the kth equation the ith one multiplied by $a_{kj}^{i-1}/a_{ij}^{i-1}$, and obtain the reduced coefficients

$$a_{kl}^i = a_{kl}^{i-1} - a_{il}^{i-1} a_{kj}^{i-1} / a_{ii}^{i-1},$$
$$b_k^i = b_k^{i-1} - b_i^{i-1} a_{kj}^{i-1} / a_{ij}^{i-1},$$

for all $l = 1, \ldots, n$.
Step 3: Increase i by 1 ($i := i + 1$).
Step 4: Increase j by 1 ($j := j + 1$).
Step 5: If either $i = m$ or $j = n$ then stop the procedure and solve for unknown variables starting from the last equation. If $i < m$ and $j < n$, go to Step 1.
Procedure ends.

Notice that after the procedure terminates, we get the following reduced system

$$\begin{cases} a_{11}^0 x_1 + & a_{12}^0 x_2 + \ldots + & a_{1n}^0 x_n = b_1^0 \\ a_{22}^1 x_2 + & a_{23}^1 x_3 + \ldots + & a_{2n}^1 x_n = b_2^1 \\ \vdots & \vdots & \vdots \\ a_{m-1,m-1}^{m-2} x_{m-1} + & a_{m-1,m}^{m-2} x_m + \ldots + & a_{m-1,n}^{m-2} x_n = b_{m-1}^{m-2} \\ a_{mm}^{m-1} x_m + a_{m,m+1}^{m-1} x_{m+1} + \ldots + & a_{mn}^{m-1} x_n = b_m^{m-1} \end{cases}$$

If in the above system there exists some row k ($1 \leq k \leq m$) such that $a_{kj}^{k-1} = 0$ for all $j = k+1, \ldots, n$ while $b_k^{k-1} \neq 0$, then we conclude that system (5.5) is inconsistent.

Now, assume that in the reduced system, $a_{kk}^{k-1} \neq 0$ for all $k = 1, \ldots, m$. Then system (5.5) is consistent, and its solution can be obtained from the reduced system as follows:

If $m = n$, then we can find x_n from the last equation, x_{n-1} from the $(n-1)$st equation which contains only x_n and x_{n-1}, and so on.

If $m < n$, then we have free variables $x_{m+1}, \ldots, x_n$. Assigning to free variables arbitrary values, we can then solve the system for the unknown variables $x_1, \ldots, x_m$.

Remark 5.2. Note that in the above algorithm we construct the row echelon form of the matrix extended matrix $[A|B]$, where $A\mathbf{x} = B$ is the matrix form of the system (5.5). Then the algorithm itself is analogous to the one used in the proof of Theorem 2.9.

Example 5.5. Problem. Solve the following system of linear equations:

$$\begin{cases} x_1 + 2x_2 + 5x_3 = -9 \\ x_1 - x_2 + 3x_3 = 2 \\ 3x_1 - 6x_2 - x_3 = 25 \end{cases}$$

5.2 Gauss Method: Sequential Elimination of Unknown Variables

Solution. Let us construct extended matrix of this system and apply Gauss elimination procedure to get

$$\begin{bmatrix} 1 & 2 & 5 & -9 \\ 1 & -1 & 3 & 2 \\ 3 & -6 & -1 & 25 \end{bmatrix} \to \begin{bmatrix} 1 & 2 & 5 & -9 \\ 0 & -3 & -2 & 11 \\ 0 & -12 & -16 & 52 \end{bmatrix} \to$$

$$\begin{bmatrix} 1 & 2 & 5 & -9 \\ 0 & -3 & -2 & 11 \\ 0 & 0 & -8 & 8 \end{bmatrix}.$$

Thus, we have:

$$x_1 + 2x_2 + 5x_3 = -9$$
$$-3x_2 - 2x_3 = 11$$
$$-8x_3 = 8$$

From the last equation, we have $x_3 = -1$. Substitute the value $x_3 = -1$ in the second equation and get $-3x_2 + 2 = 11$, hence $x_2 = -3$. Then substitute these values of x_2 and x_3 to the first equation and get $x_1 - 6 - 5 = -9$, or $x_1 = 2$.

To avoid non-necessary fractions, one can multiply each row by a nonzero number (say, by the denominator a_{ij} of the fractions in Step 2) in each stage of algorithm.

Example 5.6. *Problem.* Solve the system of linear equations

$$\begin{cases} x_1 - 5x_2 - 8x_3 + x_4 = 3 \\ 3x_1 + x_2 - 3x_3 - 5x_4 = 1 \\ x_1 - 7x_3 + 2x_4 = -5 \\ -11x_2 + 20x_3 - 9x_4 = 2 \end{cases}$$

Solution. Applying Gauss elimination procedure on the extended coefficient matrix we obtain

$$\begin{bmatrix} 1 & -5 & -8 & 1 & 3 \\ 3 & 1 & -3 & -5 & 1 \\ 1 & 0 & -7 & 2 & -5 \\ 0 & 11 & 20 & -9 & 2 \end{bmatrix} \to$$

$$\begin{bmatrix} 1 & -5 & -8 & 1 & 3 \\ 0 & 16 & 21 & -8 & -8 \\ 0 & 5 & 1 & 1 & -8 \\ 0 & 11 & 20 & -9 & 2 \end{bmatrix} \to$$

$$\begin{bmatrix} 1 & -5 & -8 & 1 & | & 3 \\ 0 & 16 & 21 & -8 & | & -8 \\ 0 & 0 & -89 & -56 & | & -88 \\ 0 & 0 & 89 & 56 & | & 120 \end{bmatrix} \rightarrow$$

$$\begin{bmatrix} 1 & -5 & -8 & 1 & | & 3 \\ 0 & 16 & 21 & -8 & | & -8 \\ 0 & 0 & -89 & -56 & | & -88 \\ 0 & 0 & 0 & 0 & | & 32 \end{bmatrix}.$$

From the last row, we conclude that the system is inconsistent.

Example 5.7. **Problem.** Solve the following system of linear equations:

$$\begin{cases} -x_2 + 2x_3 = 1 \\ x_1 - 2x_2 + x_3 = 0 \\ x_1 - x_2 - x_3 = -1 \\ 2x_1 - x_2 - 4x_3 = -3 \end{cases}$$

Solution. Applying Gauss elimination procedure on the extended coefficient matrix we obtain

$$\begin{bmatrix} 0 & -1 & 2 & | & 1 \\ 1 & -2 & 1 & | & 0 \\ 1 & -1 & -1 & | & -1 \\ 2 & -1 & -4 & | & -3 \end{bmatrix} \rightarrow$$

$$\begin{bmatrix} 1 & -2 & 1 & | & 0 \\ 0 & -1 & 2 & | & 1 \\ 1 & -1 & -1 & | & -1 \\ 2 & -1 & -4 & | & -3 \end{bmatrix} \rightarrow$$

$$\begin{bmatrix} 1 & -2 & 1 & | & 0 \\ 0 & -1 & 2 & | & 1 \\ 0 & 1 & -2 & | & -1 \\ 0 & 3 & -6 & | & -3 \end{bmatrix} \rightarrow$$

$$\begin{bmatrix} 1 & -2 & 1 & | & 0 \\ 0 & -1 & 2 & | & 1 \\ 0 & 0 & 0 & | & 0 \\ 0 & 0 & 0 & | & 0 \end{bmatrix}$$

Thus, we have

$$\begin{cases} x_1 - 2x_2 + x_3 = 0 \\ - x_2 + 2x_3 = 1 \end{cases}$$

Take one of the variables in the last equation, say, x_3, as a free variable. Then we have $x_2 = 2x_3 - 1$, where x_3 is an arbitrary real number. From the first equation, we get $x_1 = 2x_2 - x_3 = 3x_3 - 2$.

5.3 Homogeneous Equations

Consider the system:

$$\begin{cases} a_{11}x_1 + a_{12}x_2 + \ldots + a_{1n}x_n = 0 \\ a_{21}x_1 + a_{22}x_2 + \ldots + a_{2n}x_n = 0 \\ \vdots \quad\quad \vdots \quad\quad \vdots \quad\quad \vdots \quad\quad \vdots \\ a_{s1}x_1 + a_{s2}x_2 + \ldots + a_{sn}x_n = 0 \end{cases} \quad (5.7)$$

By Kronecker-Capelli Theorem, it follows that the above system has a solution. Indeed, it has a trivial solution given by $(x_1, x_2, \ldots, x_n) = (0, 0, \ldots, 0)$.

Any system which admits the zero vector as a solution is called a *homogeneous system*. Obviously, if the right hand-side of any equation in (5.7) were not zero, then the system could not admit the trivial solution, and will be called *nonhomogeneous*.

In the case of homogeneous system, the Kronecker-Capelli Theorem has a simpler form.

Theorem 5.6. *Let*

$$A\mathbf{x} = \mathbf{0}$$

be a homogeneous system of linear equations of n variables. Then either rank $A = n$ *and the system has the unique zero solution or* rank $A < n$ *and the system has infinitely many solutions.*

Proof. Left as an exercise. (*Hint.* Use Gauss elimination procedure. How the row echelon form of the matrix $\widetilde{A}$ depends on rank A?) □

Consider an economy with m agents exchanging n kinds of goods. Let a_{ij} be the quantity of good i received (sold) by an agent j. If a_{ij} is positive, then we consider that the good i is sold by agent j; if a_{ij} is negative, then the amount $|a_{ij}|$ is bought by agent j.

Then

$$(a_{1j}, a_{2j}, \ldots, a_{nj})$$

is the vector showing the 'exchanges' of agent j of any good $i = 1, \ldots, n$. Similarly, the components of the vector

$$(a_{i1}, a_{i2}, \ldots, a_{im})$$

shows the exchanges of the good i by each agent.

Let now $p_1, \ldots, p_m$ be the prices of some units of the goods $1, \ldots, m$. If an agent j sells first k goods and purchases the goods $k+1, \ldots, m$, then the expression

$$p_1 a_{1j} + p_2 a_{2j} + \cdots + p_k a_{kj} - p_{k+1} a_{k+1,j} - \cdots - p_m a_{mj}$$

exactly means the net revenue of the agent j from her trading activity.

If we put net revenue to be equal to 0, then we obtain

$$p_1 a_{1j} + p_2 a_{2j} + \cdots + p_k a_{kj} - p_{k+1} a_{k+1,j} - \cdots - p_m a_{mj} = 0.$$

Returning back to the example one can say that the set of homogeneous linear equations describes the problem how to find the prices of goods so that the net revenue of each agent being equal to 0.

Example 5.8. Consider the following system:

$$\begin{cases} 4x_1 + x_2 - 3x_3 - x_4 = 0 \\ 2x_1 + 3x_2 + x_3 - 5x_4 = 0 \\ x_1 - 2x_2 - 2x_3 + 3x_4 = 0 \end{cases}$$

The system is homogeneous since $(x_1, x_2, x_3, x_4) = (0, 0, 0, 0)$ is a solution. Since the rank of the system is less than the number of unknown variables, we can seek for other solutions, as well.

Let us construct the coefficient matrix (we can omit the last column of zeros) and perform some elementary operations on it.

$$\begin{bmatrix} 4 & 1 & -3 & -1 \\ 2 & 3 & 1 & -5 \\ 1 & -2 & -2 & 3 \end{bmatrix} \to \begin{bmatrix} 0 & 9 & 5 & -13 \\ 0 & 7 & 5 & -11 \\ 1 & -2 & -2 & 3 \end{bmatrix} \to$$

$$\begin{bmatrix} 0 & 2 & 0 & -2 \\ 0 & 7 & 5 & -11 \\ 1 & -2 & -2 & 3 \end{bmatrix}$$

Then, we have the reduced system of linear equations:

$$\begin{cases} -2x_2 - 2x_4 = 0 \\ -7x_2 + 5x_3 - 11x_4 = 0 \\ x_1 - 2x_2 - 2x_3 + 3x_4 = 0 \end{cases}$$

Take one of the variables, say x_4, as a free variable, i.e., $x_4 = \alpha \in \mathbb{R}$. Then $x_2 = \alpha$, $x_3 = (4/5)\alpha$ and $x_1 = (3/5)\alpha$. (Note that for $\alpha = 0$ we get the trivial solution.)

5.4 Problems

5.4.1 Mathematical Problems

1. Find the solution of the following systems of linear equations.
   (a)
   $$\begin{cases} 2x_1 + 2x_2 - x_3 + x_4 = 4 \\ 4x_1 + 3x_2 - x_3 + 2x_4 = 6 \\ 8x_1 + 5x_2 - 3x_3 + 4x_4 = 12 \\ 3x_1 + 3x_2 - 2x_3 + 2x_4 = 6 \end{cases}$$
   (b)
   $$\begin{cases} 2x_1 + 7x_2 + 3x_3 + x_4 = 6 \\ 3x_1 + 5x_2 + 2x_3 + 2x_4 = 4 \\ 9x_1 + 4x_2 + x_3 + 7x_4 = 2 \end{cases}$$

2. Find the solution of the following system in terms of λ.
   $$\begin{cases} \lambda x_1 + x_2 + x_3 = 1 \\ x_1 + \lambda x_2 + x_3 = 1 \\ x_1 + x_2 + \lambda x_3 = 1 \end{cases}$$

   Solve the following systems of linear equations using different methods:
3.
   $$\begin{cases} 3x_1 - 2x_2 - 5x_3 + x_4 = 3 \\ 2x_1 - 3x_2 + x_3 + 5x_4 = -3 \\ x_1 + 2x_2 + x_3 - 4x_4 = -3 \\ x_1 - x_2 - 4x_3 + 9x_4 = 22 \end{cases}$$

4.
   $$\begin{cases} 2x_1 + 3x_2 + x_3 + 2x_4 = 4 \\ 4x_1 + 3x_2 + x_3 + x_4 = 5 \\ 5x_1 + 11x_2 + 3x_3 + 2x_4 = 2 \\ 2x_1 + 5x_2 + x_3 + x_4 = 1 \\ x_1 - 7x_2 - x_3 + 2x_4 = 7 \end{cases}$$

5.
$$\begin{cases} 8x_1 + 6x_2 + 5x_3 + 2x_4 = 21 \\ 3x_1 + 3x_2 + 2x_3 + x_4 = 10 \\ 4x_1 + 2x_2 + 3x_3 + x_4 = 8 \\ 3x_1 + 5x_2 + x_3 + x_4 = 15 \\ 7x_1 + 4x_2 + 5x_3 + 2x_4 = 18 \end{cases}$$

6.
$$\begin{cases} x_1 + 2x_2 + 3x_3 + 4x_4 + 5x_5 = 2 \\ 2x_1 + 3x_2 + 7x_3 + 10x_4 + 13x_5 = 12 \\ 3x_1 + 5x_2 + 11x_3 + 16x_4 + 21x_5 = 17 \\ 2x_1 - 7x_2 + 7x_3 + 7x_4 + 2x_5 = 57 \\ x_1 + 4x_2 + 5x_3 + 3x_4 + 10x_5 = 7 \end{cases}$$

7.
$$\begin{cases} 6x_1 + 6x_2 + 5x_3 + 18x_4 + 20x_5 = 14 \\ 10x_1 + 9x_2 + 7x_3 + 24x_4 + 30x_5 = 18 \\ 12x_1 + 12x_2 + 13x_3 + 27x_4 + 35x_5 = 32 \\ 8x_1 + 6x_2 + 6x_3 + 15x_4 + 20x_5 = 16 \\ 4x_1 + 5x_2 + 4x_3 + 15x_4 + 15x_5 = 11 \end{cases}$$

8.
$$\begin{cases} 2x_1 + 3x_2 - x_3 + x_4 = 1 \\ 8x_1 + 12x - 9x_3 + 8x_4 = 3 \\ 4x_1 + 6x_2 + 3x_3 - 2x_4 = 3 \\ 2x_1 + 3x_2 + 9x_3 - 7x_4 = 3 \end{cases}$$

9. Find a polynomial $f(x) = ax^2 + bx + c$ such that

$$f(1) = -1$$
$$f(-1) = 9$$
$$f(2) = -3.$$

10. Find a polynomial $f(x)$ of degree 3 such that

$$f(-1) = 0$$
$$f(1) = 4$$
$$f(2) = 3$$
$$f(3) = 16$$

11. Show that for any two different n-tuples $x_0, x_1, \ldots, x_n$ and any $y_0, y_1, \ldots, y_n$ there exists unique polynomial $f(x)$ of degree at most n such that

5.4 Problems

$$f(x_i) = y_i, \quad i = 0, 1, \ldots, n.$$

Give a geometric explanation of this problem.

5.4.2 Economic Problems

12. Consider the simple supply and demand model introduced in Chap. 1 by (1.1)–(1.2). It can be written as a system of linear equations as follows

$$\begin{cases} q_i^d = \alpha_0 - \alpha_1 p_i \\ q_i^s = -\beta_0 + \beta_1 p_i \\ q_i^d = q_i^s \end{cases} \quad (5.8)$$

   i. Using the Kronecker–Capelli theorem, write the conditions for this system to have a solution;
   ii. Solve the system by using Cramer's rule;
   iii. Find the solution of (5.8) using Gauss method.

13. Consider an economy with three goods, (q_1, q_2, q_3). Let Y denote the income. Suppose the demand functions for these goods are as follows:

$$q_1^d = -0.05 p_1 + 0.02 p_2 - 0.01 p_3 + 0.02 Y,$$
$$q_2^d = 0.01 p_1 - 0.04 p_2 + 0.01 p_3 + 0.04 Y,$$
$$q_3^d = -0.03 p_1 + 0.02 p_2 - 0.06 p_2 + 0.01 Y,$$

and the supply functions for these goods are given as

$$q_1^s = -20 + 0.2 p_1$$
$$q_2^s = -14 + 0.3 p_2 \quad (5.9)$$
$$q_3^s = -25 + 0.1 p_3$$

   i. Interpret the signs of the coefficients of price and income variables in the demand functions?
   Hint. Why should the demand for a good is affected by a change in the price of another good? Notice the symmetric nature of the signs of coefficients.
   ii. Suppose $Y = 1{,}000$. Find the equilibrium prices for this three good economy.
   iii. What happens to the prices when income (Y) increases to 1,200.

14. Following the model described in Chapter 1 by (1.13)–(1.18), consider the following macroeconomic model:

$$\begin{cases} Y = C + I + G + X - M, \\ C = 0.75 Y_d, \\ Y_d = Y - T, \\ T = G, \\ G = G_C + G_I, \\ I = 0.25(Y - Y_{-1}) + 0.1 G_I, \\ M = 0.02C + 0.08I + 0.06 G_I + 0.03 X, \\ N = 0.8 Y, \\ B = 1.05 X - 1.07 M, \end{cases}$$

where Y – GDP, Y_{-1} – GDP of the previous year, C – Private Consumption, I – Private Investment, M – Imports, X – Exports, Y_d – Disposable Income, T – Taxes, G – Government Expenditure, G_C – Public Consumption Expenditures, G_I – Public Investment Expenditures, N – Employment, B – Current Account of the Balance of Payments. All variables, except N, are measured in \$ billion. Employment (N) unit of measurement is 1,000 persons.

Suppose the following data is given: $Y_{-1} = 1{,}200$, $X = 200$.
i. Identify the endogenous and exogenous variables of this model.
ii. Suppose that the government wants to target balance of payments and employment. Is the structure of the model suitable for this purpose? Why?
iii. Can government use private investment (I) as a policy instrument? Why?
iv. Can government use only public investment to reach its balance of payments and employment targets? Why?
v. Eliminate the irrelevant endogenous variables and express the remaining target variables in terms of the exogenous ones.
vi. Suppose that the government aims at achieving \$160 billion surplus in the balance of payments and an employment level of 1,100 by using public consumption and public investment. How much should government spend?

Linear Spaces

6

We now construct a generalization of the notions studied in the previous chapters. Consider some set $\mathcal{L}$ of elements $\mathbf{x}, \mathbf{y}, \mathbf{z}, \ldots$. We call $\mathcal{L}$ a *linear space*[1] if
i. for any $\mathbf{x}, \mathbf{y} \in \mathcal{L}$, there exists some $\mathbf{z} \in \mathcal{L}$ such that

$$\mathbf{z} = \mathbf{x} + \mathbf{y};$$

ii. for any $\mathbf{x} \in \mathcal{L}$ and any real number λ, we have $\lambda \mathbf{x} \in \mathcal{L}$.

Elements of linear spaces are called vectors. (Here, the word 'vector' is used in the abstract sense; it can represent a matrix, a function, a real vector, etc.) The linear space is also called a *vector space*.

The vector operations $+$ (vector addition) and $\cdot$ (dot multiplication) must satisfy the following axioms on $\mathcal{L}$

I-(i) $\mathbf{x} + \mathbf{y} = \mathbf{y} + \mathbf{x}$ (commutativity),
I-(ii) $(\mathbf{x} + \mathbf{y}) + \mathbf{z} = \mathbf{x} + (\mathbf{y} + \mathbf{z})$ (associativity),
I-(iii) there exists a (null) element $\mathbf{0}$ such that $\mathbf{x} + \mathbf{0} = \mathbf{0} + \mathbf{x} = \mathbf{x}$ for all $\mathbf{x}$,
I-(iv) for all $\mathbf{x}$, there exists the additive inverse, $(-\mathbf{x})$, such that $\mathbf{x} + (-\mathbf{x}) = \mathbf{0}$.
II-(i) $1(\mathbf{x}) = \mathbf{x}$,
II-(ii) $\alpha(\beta \mathbf{x}) = (\alpha \beta)\mathbf{x}$.
III-(i) $(\alpha + \beta)\mathbf{x} = \alpha \mathbf{x} + \beta \mathbf{x}$,
III-(ii) $\alpha(\mathbf{x} + \mathbf{y}) = \alpha \mathbf{x} + \alpha \mathbf{y}$.

Note that $+$ and $\cdot$ are abstract operations, they are not necessarily addition and multiplication defined on real numbers.

Example 6.1. Consider the set of all vectors in $\mathbb{R}^n$. Show that under the standard operations of vector addition and multiplication by a number (Sect. 2.1), this set is a linear space; check that all 8 axioms above are satisfied.

---

[1] In mathematics, a *space* is an abstract set of points with some additional structure. Mathematical spaces are often considered as models for real physical spaces. In addition to linear spaces, there are so-called Euclidean spaces, metric spaces, topological spaces and many others.

Example 6.2. Consider a set M_n, elements of which are square matrices of order n. Summation of matrices, multiplication of a matrix by a scalar, and the null matrix are defined as usual. Check that M_n is a linear space.

Example 6.3. Consider $\mathcal{L} = C[a,b]$ to be the set of all continuous real-valued functions on an interval $[a,b]$. The sum of any two functions $f(x)$ and $g(x)$ is defined in the usual way as

$$f(x) + g(x).$$

Analogously is defined $\lambda f(x)$. Show that $\mathcal{L}$ is a linear space.

Example 6.4. Let P_n be the set of all polynomials on a variable x of degree at most n. Show that P_n is a linear space under the usual summation and dot multiplication operations.

Remark 6.1. The set of all polynomials of degree n is not a linear space since

$$(t^n + t^{n-1}) + (-t^n + t^{n-1}) = 2t^{n-1},$$

is not a polynomial of degree n, that is, the addition of vectors is not well-defined.

6.1 Linear Independence of Vectors

Definition 6.1. Let $\mathcal{L}$ be a linear space. Vectors $\mathbf{x}_1, \mathbf{x}_2, \ldots, \mathbf{x}_k$ are called *linearly dependent* if there are numbers $\alpha_1, \alpha_2, \ldots, \alpha_k$ such that at least one of them is different from 0 and

$$\alpha_1 \mathbf{x}_1 + \alpha_2 \mathbf{x}_2 + \cdots + \alpha_k \mathbf{x}_k = \mathbf{0}.$$

Vectors which are not linearly dependent are called *linearly independent*.

A *linear combination* of vectors $\mathbf{x}_1, \mathbf{x}_2, \ldots, \mathbf{x}_k$ is a vector of the from $\alpha_1 \mathbf{x}_1 + \alpha_2 \mathbf{x}_2 + \cdots + \alpha_k \mathbf{x}_k$, where $\alpha_1, \ldots, \alpha_k$ are some numbers (referred as coefficients of the linear combination). Thus the vectors are linearly independent if and only if none of their linear combinations with at least one nonzero coefficient is equal to null vector.

Definition 6.2 (Dimension of a Linear Space). The linear space $\mathcal{L}$ is called n-dimensional if the maximal number of linearly independent vectors in it is equal to n. We denote this as $\dim(\mathcal{L}) = n$.

Here are some examples.

Example 6.5. The system of vectors

$$\begin{aligned} \mathbf{e}_1 &= (1, 0, \ldots, 0, 0) \\ \mathbf{e}_2 &= (0, 1, 0, \ldots, 0) \\ &\cdot \cdot \cdot \cdot \cdot \cdot \\ \mathbf{e}_n &= (0, 0, \ldots, 0, 1) \end{aligned}$$

6.1 Linear Independence of Vectors

in $\mathbb{R}^n$ is linearly independent, since any linear combination

$$\alpha_1 \mathbf{e}_1 + \alpha_2 \mathbf{e}_2 + \cdots + \alpha_n \mathbf{e}_n = (\alpha_1, \alpha_2, \ldots, \alpha_n)$$

is not equal to the zero vector provided that at least one coordinate α_i is nonzero. It follows that $\dim \mathbb{R}^n \geq n$.

On the other side, any m vectors $\mathbf{v}_1, \ldots, \mathbf{v}_m$ in $\mathbb{R}^n$ with $m > n$ form an $n \times m$ matrix (as columns). Since the rank of the matrix is not greater than the number n of its rows, it follows that the columns $\mathbf{v}_1, \ldots, \mathbf{v}_m$ are linearly dependent. Thus $\dim R^n = n$.

Example 6.6. Consider the following matrices of order n:

$$A^{k,l} = [a^{k,l}_{i,j}]_{n \times n}, \quad k = 1, \ldots, n; \ l = 1, \ldots, n$$

where

$$a^{k,l}_{i,j} = \begin{cases} 1 \text{ if } i = k \text{ and } j = l, \\ 0 \text{ otherwise.} \end{cases}$$

Check that these vectors (matrices) are linearly independent.

Example 6.7. In the linear space P_n of all polynomials of degree $\leq n$ of variable t (Example 6.4), see that the vectors

$$1, t, t^2, \ldots, t^n$$

are linearly independent.

Now we give an insight to 'the construction' of linear spaces.

Lemma 6.1. *Let*

$$\mathbf{f}_1, \ldots, \mathbf{f}_k.$$

be a set of linearly independent vectors, and let each of the vectors $\mathbf{g}_1, \ldots, \mathbf{g}_l$ *be a linear combination of the vectors* $\mathbf{f}_1, \ldots, \mathbf{f}_k$. *If the vectors* $\mathbf{g}_1, \ldots, \mathbf{g}_l$ *are linearly independent, then* $l \leq k$.

Above lemma can be equivalently stated as follows: there cannot be more than k linearly independent vectors which are linear combinations of k linearly independent vectors $f_1, \ldots, f_k$.

Proof by induction. For $k = 1$ the claim is obviously true. Suppose that the claim is true for $k - 1$ vectors $\mathbf{f}_1, \ldots, \mathbf{f}_{k-1}$, and prove the statement for the case of k vectors.

Take any k linearly independent vectors $\mathbf{f}_1, \ldots, \mathbf{f}_k$. Consider l linear combinations of $\mathbf{f}_1, \ldots, \mathbf{f}_k$ such that

$$\mathbf{g}_1 = \alpha_{11}\mathbf{f}_1 + \cdots + \alpha_{1k}\mathbf{f}_k$$
$$\mathbf{g}_2 = \alpha_{21}\mathbf{f}_1 + \cdots + \alpha_{2k}\mathbf{f}_k$$
$$\vdots$$
$$\mathbf{g}_l = \alpha_{l1}\mathbf{f}_1 + \cdots + \alpha_{lk}\mathbf{f}_k.$$

Suppose that $\mathbf{g}_1, \ldots, \mathbf{g}_l$ are linearly independent. We have to show that $l \le k$. There are two cases to consider.

Case 1. If $\alpha_{jk} = 0$, for all $j = 1, \ldots, l$, then Lemma 6.1 is proved, since in that case $l \le k - 1 < k$.

Case 2. Suppose there exists $j \in \{1, \ldots, l\}$ such that $\alpha_{jk} \ne 0$. Without loss of generality assume that $\alpha_{lk} \ne 0$.

We will construct new $l - 1$ linearly independent vectors which in turn will be linear combinations of $\mathbf{f}_1, \ldots, \mathbf{f}_{k-1}$. To do this, we solve for f_k using the last equality to get

$$\mathbf{f}_k = \frac{1}{\alpha_{lk}}\mathbf{g}_l - \frac{\alpha_{l1}}{\alpha_{lk}}\mathbf{f}_1 - \cdots - \frac{\alpha_{l,k-1}}{\alpha_{lk}}\mathbf{f}_{k-1}.$$

Now inserting $\mathbf{f}_k$ into the above equalities and rearranging, we obtain

$$\mathbf{g}_1 - (\alpha_{1k}/\alpha_{lk})\mathbf{g}_l = \beta_{11}\mathbf{f}_1 + \cdots + \beta_{1,k-1}\mathbf{f}_{k-1},$$
$$\mathbf{g}_2 - (\alpha_{2k}/\alpha_{lk})\mathbf{g}_l = \beta_{21}\mathbf{f}_1 + \cdots + \beta_{2,k-1}\mathbf{f}_{k-1},$$
$$\vdots$$
$$\mathbf{g}_{l-1} - (\alpha_{l-1,k}/\alpha_{lk})\mathbf{g}_l = \beta_{l-1,1}\mathbf{f}_1 + \cdots + \beta_{l-1,k-1}\mathbf{f}_{k-1},$$

where $\beta_{i,j} = \alpha_{i,j} - \alpha_{i,k}\alpha_{l,j}/\alpha_{l,k}$, for all $i = 1, \ldots, l-1$ and $j = 1, \ldots, k-1$. Then the vectors

$$\mathbf{g}'_i = \mathbf{g}_i - \frac{\alpha_{ik}}{\alpha_{lk}}\mathbf{g}_l, \quad i = 1, \ldots, l-1,$$

are linear combinations of $\mathbf{f}_1, \ldots, \mathbf{f}_{k-1}$.

If we can prove that $\mathbf{g}'_1, \ldots, \mathbf{g}'_{l-1}$ are linearly independent, then by induction assumption we will prove that $l - 1 \le k - 1$, i.e. $l \le k$.

Suppose towards a contradiction that $\mathbf{g}'_1, \ldots, \mathbf{g}'_{l-1}$ are linearly dependent, i.e., there are $\lambda_1, \ldots, \lambda_{l-1}$ such that

$$\lambda_1 \mathbf{g}'_1 + \cdots + \lambda_{l-1}\mathbf{g}'_{l-1} = \mathbf{0}$$

or

$$\lambda_1\left(\mathbf{g}_1 - \frac{\alpha_{1k}}{\alpha_{lk}}\mathbf{g}_l\right) + \cdots + \lambda_{l-1}\left(\mathbf{g}_{l-1} - \frac{\alpha_{l-1,k}}{\alpha_{lk}}\mathbf{g}_l\right) = \mathbf{0}$$

6.1 Linear Independence of Vectors

which can be rearranged as

$$\lambda_1 \mathbf{g}_1 + \cdots + \lambda_{l-1}\mathbf{g}_{l-1} - \left(\lambda_1 \frac{\alpha_{1k}}{\alpha_{lk}} + \cdots + \lambda_{l-1}\frac{\alpha_{l-1,k}}{\alpha_{lk}}\right) \mathbf{g}_l = \mathbf{0}.$$

Since vectors $\mathbf{g}_1, \mathbf{g}_2, \ldots, \mathbf{g}_l$ are linearly independent (by assumption), all coefficients in the last equality must be equal to 0. That is, $\lambda_1 = \lambda_2 = \cdots = \lambda_{l-1} = 0$, and hence $\mathbf{g}'_1, \ldots, \mathbf{g}'_{l-1}$ are linearly independent.

Definition 6.3. A set of n linearly independent vectors $\mathbf{e}_1, \ldots, \mathbf{e}_n$ in an n-dimensional space $\mathcal{L}$ is called a *basis* for $\mathcal{L}$.

Definition 6.4. Two vectors $\mathbf{e}_1, \mathbf{e}_2$ are called *collinear* if they are parallel to the same line, and *non-collinear* otherwise.

Example 6.8. In $\mathbb{R}^2$ any two non-collinear vectors are linearly independent.

Example 6.9. In two-dimensional space $\mathbb{R}^2$ any two non-collinear vectors $\mathbf{x}$ and $\mathbf{y}$ form a basis.

Example 6.10. The n linearly independent vectors in Example 6.5 form a basis for $\mathbb{R}^n$.

This special basis is called the *canonical basis* for $\mathbb{R}^n$.

Lemma 6.2. *Any set containing $k < n$ linearly independent vectors $\mathbf{x}_1, \ldots, \mathbf{x}_k$ can be completed up to form a basis for an n-dimensional space $\mathcal{L}$.*

Proof. Let $\mathbf{e}_1, \ldots, \mathbf{e}_n$ be a basis for $\mathcal{L}$. If any of the vectors $\mathbf{e}_1, \ldots, \mathbf{e}_n$ is a linear combination of $\mathbf{x}_1, \ldots, \mathbf{x}_k$, then by Lemma 6.1, $n \leq k$, but $k < n$ by assumption. Hence among the basis vectors $\mathbf{e}_1, \ldots, \mathbf{e}_n$ there is at least one vector, say $\mathbf{e}_p$, which is not a linear combination of $\mathbf{x}_1, \ldots, \mathbf{x}_k$. Add $\mathbf{e}_p$ to the system $\mathbf{x}_1, \ldots, \mathbf{x}_k$. Consider now the system of $k+1$ vectors

$$\mathbf{x}_1, \ldots, \mathbf{x}_k, \mathbf{e}_p.$$

These vectors are linearly independent. (Why?)

If $k+1 = n$, then Lemma 6.2 is proved. If $k+1 < n$, then one can proceed until constructing the system of n linearly independent vectors which by construction contains $\mathbf{x}_1, \ldots, \mathbf{x}_k$. □

Theorem 6.3. *Given a basis in an n-dimensional linear space $\mathcal{L}$, any vector in $\mathcal{L}$ can be uniquely represented as a linear combination of basis vectors.*

Proof. Let $\mathbf{e}_1, \ldots, \mathbf{e}_n$ be a basis in $\mathcal{L}$. Add $\mathbf{x}$ to these vectors. Since we have now $n+1$ vectors, they are linearly dependent, i.e.,

$$\alpha_1 \mathbf{e}_1 + \cdots + \alpha_n \mathbf{e}_n + \beta \mathbf{x} = \mathbf{0}.$$

We can state that $\beta \neq 0$ (why ?). Then

$$\mathbf{x} = -\frac{\alpha_1}{\beta}\mathbf{e}_1 - \cdots - \frac{\alpha_n}{\beta}\mathbf{e}_n,$$

i.e., $\mathbf{x}$ is represented us linear combination of the vector $\mathbf{e}_1, \ldots, \mathbf{e}_n$.

Let us show that the representation is unique. Assume on the contrary that there exists two representations

$$\mathbf{x} = \xi_1\mathbf{e}_1 + \cdots + \xi_n\mathbf{e}_n$$

and

$$\mathbf{x} = \eta_1\mathbf{e}_1 + \cdots + \eta_n\mathbf{e}_n$$

Subtract the second equality from the first. Then

$$(\xi_1 - \eta_1)\mathbf{e}_1 + (\xi_2 - \eta_2)\mathbf{e}_2 + \cdots + (\xi_n - \eta_n)\mathbf{e}_n = \mathbf{0}.$$

Since $\mathbf{e}_1, \ldots, \mathbf{e}_n$ are linearly independent, we obtain $\xi_i = \eta_i$ for all i. □

Definition 6.5. Given a basis $E = \{\mathbf{e}_1, \ldots, \mathbf{e}_n\}$ in A and the vector

$$\mathbf{x} = \xi_1\mathbf{e}_1 + \cdots + \xi_n\mathbf{e}_n,$$

the numbers $\xi_1, \ldots, \xi_n$ are called *coordinates* of $\mathbf{x}$ in the basis E.

The statement of Theorem 6.3 means that in a given basis $\mathbf{e}_1, \ldots, \mathbf{e}_n$ any vector has uniquely defined coordinates.

6.1.1 Addition of Vectors and Multiplication of a Vector by a Real Number

Let $\mathbf{x} = (\xi_1, \ldots, \xi_n)$ and $\mathbf{y} = (\eta_1, \ldots, \eta_n)$ in the basis $\mathbf{e}_1, \ldots, \mathbf{e}_n$. Then

$$\mathbf{x} = \xi_1\mathbf{e}_1 + \cdots + \xi_n\mathbf{e}_n$$
$$\mathbf{y} = \eta_1\mathbf{e}_1 + \cdots + \eta_n\mathbf{e}_n$$

and

$$\mathbf{x} + \mathbf{y} = (\xi_1 + \eta_1)\mathbf{e}_1 + \cdots + (\xi_n + \eta_n)\mathbf{e}_n,$$

i.e., vector $\mathbf{x} + \mathbf{y}$ has coordinates $(\xi_1 + \eta_1, \ldots, \xi_n + \eta_n)$. Analogously, vector $\lambda\mathbf{x}$ has coordinates $(\lambda\xi_1, \ldots, \lambda\xi_n)$.

Example 6.11. i. For $\mathbb{R}^2$ our definition of coordinates of a vector coincides with the usual definition.
ii. Consider the space $\mathcal{L}$ with polynomials of degree $\leq n - 1$. As it was shown before, a simplest basis in it is a system

$$\mathbf{e}_1 = 1, \mathbf{e}_2 = t, \ldots, \mathbf{e}_n = t^{n-1}.$$

Then any polynomial is represented as

$$P(t) = a_0 t^{n-1} + a_1 t^{n-2} + \cdots + a_{n-1}.$$

What are coordinates of the vector in this basis?
   Consider now another basis

$$\mathbf{e}'_1 = 1, \; \mathbf{e}'_2 = t - a, \; \mathbf{e}'_3 = (t-a)^2, \ldots, \mathbf{e}'_n = (t-a)^{n-1}.$$

Any polynomial $P(t)$, by Taylor expansion, can be represented as

$$P(t) = P(a) + P'(a)(t-a) + \cdots + \frac{P^{(n-1)}(a)}{(n-1)!}(t-a)^{n-1}$$

What are the coordinates of $P(t)$ in this basis?

6.2 Isomorphism of Linear Spaces

Definition 6.6. Linear spaces $\mathcal{L}$ and $\mathcal{L}'$ are called isomorphic if between vectors $x \in \mathcal{L}$ and $x' \in \mathcal{L}'$ there exists a one-to-one and onto correspondence $x \leftrightarrow x'$ such that
(a) if $x \leftrightarrow x'$, $y \leftrightarrow y'$ then $x + y \leftrightarrow x' + y'$;
(b) $\lambda x \leftrightarrow \lambda x'$.

Lemma 6.4. *If $\mathcal{L}$ and $\mathcal{L}'$ are isomorphic then linearly independent vectors in $\mathcal{L}$ are linearly independent in $\mathcal{L}'$, and vice versa.*

Proof. Exercise. □

Lemma 6.5. *Two linear spaces $\mathcal{L}$ and $\mathcal{L}'$ of different dimensions are not isomorphic.*

Proof. Exercise (use Lemma 6.4). □

Theorem 6.2. *All linear spaces of dimension n are isomorphic.*

Proof. Let $\mathcal{L}, \mathcal{L}'$ be two n-dimensional spaces. Choose in $\mathcal{L}$ a basis $\mathbf{e}_1, \ldots, \mathbf{e}_n$, and in $\mathcal{L}'$ - a basis $\mathbf{e}'_1, \ldots, \mathbf{e}'_n$.
   For any vector

$$\mathbf{x} = \xi_1 \mathbf{e}_1 + \cdots + \xi_n \mathbf{e}_n \qquad (6.1)$$

construct the corresponding vector

$$\mathbf{x}' = \xi_1 \mathbf{e}'_1 + \cdots + \xi_n \mathbf{e}'_n. \qquad (6.2)$$

Note that coefficients are the same in both cases. This correspondence is a bijection (one-to-one and onto). Indeed, any $x \in \mathcal{L}$ can be represented as in (6.1). Then, all $\xi_1, \ldots, \xi_n$ and hence vector $\mathbf{x}'$ are defined by $\mathbf{x}$ uniquely.

Change $\mathcal{L}$ and $\mathcal{L}'$ in this reasoning and obtain a one-to-one and onto correspondence between $\mathcal{L}$ and $\mathcal{L}'$.

Now, if $\mathbf{x} \leftrightarrow \mathbf{x}'$, $\mathbf{y} \leftrightarrow \mathbf{y}'$ then by construction $\mathbf{x} + \mathbf{y} \leftrightarrow \mathbf{x}' + \mathbf{y}'$, and $\lambda \mathbf{x} \leftrightarrow \lambda \mathbf{x}'$. Hence, $\mathcal{L}$ and $\mathcal{L}'$ are isomorphic. □

6.3 Subspaces

Subspace $\mathcal{L}'$ of a linear space $\mathcal{L}$ is a subset of elements of $\mathcal{L}$ which itself is a linear space with respect to operations $+$ and $\cdot$ defined in $\mathcal{L}$.

In other words, if $\mathcal{L}' \subseteq \mathcal{L}$, and $+$, and $\cdot$ are operations of addition and multiplication by a real number defined in $\mathcal{L}$, then all axioms $I.(i) - III.(ii)$ are satisfied for vectors in $\mathcal{L}'$.

6.3.1 Examples of Subspaces

1. Null space and $\mathcal{L}$ itself.
2. Consider $\mathbb{R}^2$. Then the set of all vectors which lie on same line going through the origin is a linear subspace (Fig. 6.1).
3. Consider $\mathcal{L}$ to be a set of n−tuples of real numbers:

$$\mathbf{x} = (\xi_1, \ldots, \xi_n).$$

The subset $\mathcal{L}'$ of $\mathcal{L}$ such that

$$A = \{\mathbf{x} \mid (\mathbf{x} = (\xi_1, \ldots, \xi_n), \xi_1 = 0)\}$$

is a subspace of $\mathcal{L}$.

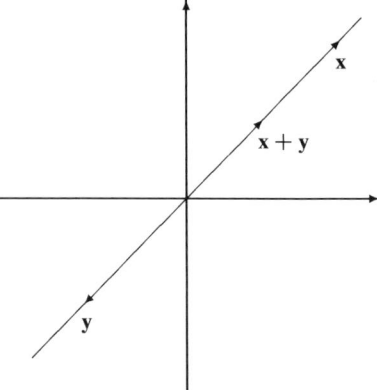

Fig. 6.1 A one-dimensional subspace in $\mathbb{R}^2$

6.3 Subspaces

4. Let $\mathcal{L}$ be a space of all continuous functions on $[a,b]$. Prove that the set of all polynomials of degree less or equal to $n-1$ is a subspace in $\mathcal{L}$.

Since subspace of a linear space is a linear space itself, all notions introduced above about linear independence, dimension, etc., are applied to the linear subspaces.

Hence, the dimension of a subspace $\mathcal{L}'$ of $\mathcal{L}$ can not be greater than the dimension of $\mathcal{L}$ itself.

6.3.2 A Method of Constructing Subspaces

Choose in $\mathcal{L}$ vectors $\mathbf{g}_1, \mathbf{g}_2, \ldots, \mathbf{g}_n$. Then the set S of all linear combinations

$$\alpha_1 \mathbf{g}_1 + \alpha_2 \mathbf{g}_2 + \cdots + \alpha_n \mathbf{g}_n$$

of these vectors is a subspace of A.

Exercise 6.1. Prove the last statement above.

This subspace S is called a subspace spanned by the vectors $\mathbf{g}_1, \mathbf{g}_2, \ldots, \mathbf{g}_n$ (or simply *span* of these vectors) and is the smallest linear space which contains $\mathbf{g}_1, \mathbf{g}_2, \ldots, \mathbf{g}_n$. Notation: $S = \langle \mathbf{g}_1, \mathbf{g}_2, \ldots, \mathbf{g}_n \rangle$.

Theorem 6.7. *Subspace $\mathcal{L}'$ spanned by linearly independent vectors $\mathbf{e}_1, \ldots, \mathbf{e}_k$ is k-dimensional, and these vectors make a basis in $\mathcal{L}'$.*

Proof. Indeed, in $\mathcal{L}'$ there is a set of linearly independent vectors, which are $\mathbf{e}_1, \ldots, \mathbf{e}_n$ themselves.

On the other hand, if $\mathbf{x}_1, \ldots, \mathbf{x}_l$ are arbitrary linearly independent vectors in $\mathcal{L}'$, they are linear combinations of $\mathbf{e}_1, \ldots, \mathbf{e}_k$, and by Lemma 6.1, $l \leq k$. Hence, $\mathcal{L}'$ is k-dimensional subspace. □

6.3.3 One-Dimensional Subspaces

A basis of such a subspace consists of one vector, say, e_1, and all subspace consists of vectors αe_1, where α is an arbitrary real number.

Consider a set of vectors

$$\mathbf{x} = \mathbf{x}_0 + \alpha \mathbf{e}_1,$$

where $\mathbf{x}_0$, $\mathbf{e}_1$ are fixed, α is an arbitrary real number. The set of such vectors can be called *a line* in $\mathcal{L}$.

6.3.4 Hyperplane

Analogously to a line in $\mathcal{L}$, we can define a set of vectors

$$\mathbf{x} = \mathbf{x}_0 + \alpha \mathbf{e}_1 + \beta \mathbf{e}_2$$

where $\mathbf{x}_0$, $\mathbf{e}_1$, $\mathbf{e}_2$, are fixed, and α and β are arbitrary real numbers. This set by analogy to the space $\mathbb{R}^3$ is called a plane in $\mathcal{L}$.

Extending this construction we can obtain a set of vectors

$$\mathbf{x} = \mathbf{x}_0 + \alpha_1 \mathbf{e}_1 + \alpha_2 \mathbf{e}_2 + \cdots + \alpha_k \mathbf{e}_k$$

which is called a *hyperplane* in $\mathcal{L}$.

6.4 Coordinate Change

Suppose a vector $\mathbf{x}$ has some coordinates with respect to a given basis. It is a common situation if we consider another basis in the same space. When the basis changes, the coordinates change too. For example, let $\mathbf{x}$ be a vector with coordinates (x, y) in some basis $\mathbf{e}_1, \mathbf{e}_2$ in the plane $\mathbb{R}^2$. After the rotation of coordinates by an angle α, we get another basis $\mathbf{e}'_1, \mathbf{e}'_2$. In this basis, same vector $\mathbf{x}$ has another coordinates (x', y'), see Fig. 6.2. By (2.12), these coordinates are given by

$$\begin{bmatrix} x' \\ y' \end{bmatrix} = R_{-\alpha} \begin{bmatrix} x \\ y \end{bmatrix}.$$

In general, how do the coordinates of a vector change when the basis changes?

Let $\mathbf{e}_1, \ldots, \mathbf{e}_n$ and $\mathbf{e}'_1, \ldots, \mathbf{e}'_n$ be two bases on n-dimensional space $\mathcal{L}$. Let each $\mathbf{e}'_i$ is represented through $\mathbf{e}_1, \ldots, \mathbf{e}_n$ as follows

$$\begin{aligned}
\mathbf{e}'_1 &= \alpha_{11}\mathbf{e}_1 + \alpha_{21}\mathbf{e}_2 + \cdots + \alpha_{n1}\mathbf{e}_n, \\
\mathbf{e}'_2 &= \alpha_{12}\mathbf{e}_1 + \alpha_{22}\mathbf{e}_2 + \cdots + \alpha_{n2}\mathbf{e}_n, \\
&\ldots\ldots\ldots \\
\mathbf{e}'_n &= \alpha_{1n}\mathbf{e}_1 + \alpha_{2n}\mathbf{e}_2 + \cdots + \alpha_{nn}\mathbf{e}_n.
\end{aligned} \quad (6.3)$$

In other words, the transformation from the basis $\{\mathbf{e}_i\}_1^n$ to the basis $\{\mathbf{e}'_i\}_1^n$ is defined by matrix $A = \|\alpha_{ik}\|$, with $det\ A \neq 0$ (Why?). The matrix A is called a *basis transformation matrix*. By definition, its j-th column consists of the coordinates of the j-th element of the new basis in the old one.

Let $\mathbf{x}$ has coordinates $(\xi_1, \ldots, \xi_n)$ in the first basis, and coordinates $(\xi'_1, \ldots, \xi'_n)$ – in the second one. Let us find how ξ'_i are expressed through $\xi_i, i = 1, \ldots, n$.

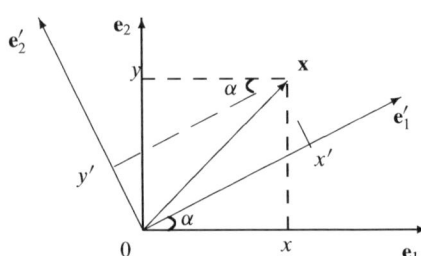

Fig. 6.2 Coordinates after and before rotation

So, we have

$$x = \xi_1 e_1 + \xi_2 e_2 + \cdots + \xi_n e_n = \xi'_1 e'_1 + \xi'_2 e'_2 + \cdots + \xi'_n e'_n$$
$$= \xi'_1(\alpha_{11} e_1 + \alpha_{21} e_2 + \cdots + \alpha_{n1} e_n) + \xi'_2(\alpha_{12} e_1 + \alpha_{22} e_2 + \cdots + \alpha_{n2} e_n) +$$
$$\cdots$$
$$+ \xi'_n(\alpha_{1n} e_1 + \alpha_{2n} e_2 + \cdots + \alpha_{nn} e_n).$$

Since $e_1, \ldots, e_n$ are linearly independent, coefficients for e_i, $i = 1, \ldots, n$ in the left side of the above equation are equal to those in the right side.

Then we have

$$\begin{aligned}\xi_1 &= \alpha_{11}\xi'_1 + \alpha_{12}\xi'_2 + \cdots + \alpha_{1n}\xi'_n, \\ \xi_2 &= \alpha_{21}\xi'_1 + \alpha_{22}\xi'_2 + \cdots + \alpha_{2n}\xi'_n, \\ &\cdots\cdots\cdots \\ \xi_n &= \alpha_{n1}\xi'_1 + \alpha_{n2}\xi'_2 + \cdots + \alpha_{nn}\xi'_n.\end{aligned} \quad (6.4)$$

Hence, this transformation is defined by A^T, the transpose matrix of A. We can now express ξ'_i, $i = 1, \ldots, n$, from (6.4) as

$$\xi'_1 = b_{11}\xi_1 + b_{12}\xi_2 + \cdots + b_{1n}\xi_n,$$
$$\cdots\cdots\cdots$$
$$\xi'_n = b_{n1}\xi_1 + b_{n2}\xi_2 + \cdots + b_{nn}\xi_n,$$

where $B = \|b_{ik}\|$ is a matrix inverse to A^T. In other words, all coordinates are transformed with respect to $(A^T)^{-1}$, where A is a matrix which defines a transformation between the bases $\{e_i\}_1^n$ and $\{e'_i\}_1^n$.

6.5 Economic Example: Production Technology Set

Consider a firm that operates three plants to produce the same good, say y_1. Each plant is located in a different region. Both the production capacities of these plants and their production techniques are different. The latter is reflected by differences in the combination of input requirements of production techniques. Each technique uses a fixed combination of inputs and these are invariant with respect to the scale of the operation. Both output and inputs (denoted by y_2, y_3 and y_4) are perfectly divisible (i.e. fractions are allowed)

Suppose the technological information concerning these plants is given as follows:

	Plant 1	Plant 2	Plant 3
Capacity Output Level y_1	100	80	120
Input y_2	26	17	38
Input y_3	25	19	30
Input y_4	24	21	25

In this table output and inputs are measured in their natural measurement units. Input figures given in the table correspond to the levels that enable the plant to produce its capacity output level. By assumption, plants can not produce output over the corresponding capacity output levels given in the first row of table. However if they need to produce less, they can do it by using less inputs, without changing their combination.

Question 1. Suppose all plants are used at their capacity level. Find the total amount of output and inputs used for its production.

Answer. This is simply calculated by summing capacity output and input levels over plants. In other words, summation of the columns of the table that gives numerical information is permitted.

Total capacity output y_1: $100 + 80 + 120 = 300$ units, total amount of input y_2 used: $26 + 17 + 38 = 81$ units, total amount of input y_3 used: $25 + 19 + 30 = 74$ units, total amount of input y_4 used: $24 + 21 + 25 = 70$ units.

Question 2. Suppose due to a decline in demand in the region where Plant 1 is located, its output declined to 65. What will be the new input levels?

Answer. By assumption, each technique of production can be employed at the desired level of intensity by keeping the input combinations intact. Therefore multiplying the column that gives capacity level of output in Plant 1 by $65/100 = 0.65$ we can find the new input levels:

Input y_2: $0.65 \cdot 26 = 16.9$ units;
Input y_3: $0.65 \cdot 25 = 16.3$ units;
Input y_4: $0.65 \cdot 24 = 15.6$ units.
Now let's rephrase what has been done in mathematical terms.
Let

$$\mathbf{y}^j = (y_1^j, -y_2^j, -y_3^j, -y_4^j), j = 1, 2, 3,$$

be the jth production technique for y_1. For convenience, inputs are distinguished by a negative sign. Let Y denote the production set for y_1, that is, the set of all possible value of the vector $\mathbf{y} = (y_1, -y_2, -y_3, -y_4)$. Then

$$\mathbf{y}_1^j \in Y, j = 1, 2, 3.$$

Note that the vectors $\mathbf{y}$ are 4-dimensional, so that $Y \subset \mathbb{R}^4$.

Question 3. Show that Y is a subset of some linear space $L \subset \mathbb{R}^4$ such that its dimension $\dim L$ is equal to 3.

Answer. Notice that the assumption made concerning the production techniques enabled us to sum the columns of the table above as well as multiplying them by a scalar. By doing these two operations we create new feasible input/output combinations, i.e. "production techniques". In formal terms, Y consists of the vectors

6.5 Economic Example: Production Technology Set

$$\sum_{j=1}^{3} \alpha_j \mathbf{y}^j \subset Y,$$

for $\alpha \subset [0, 1]$. Therefore, Y is a subset of the linear space L spanned by the vectors $\mathbf{y}^1, \mathbf{y}^2$ and $\mathbf{y}^3$.

It remains to show that dim $L = 3$. By Theorem 6.7, it is sufficient to show that three vectors $\mathbf{y}^1, \mathbf{y}^2$ and $\mathbf{y}^3$ are linearly independent. This means that the only solution of the equation

$$x_1 \mathbf{y}^1 + x_2 \mathbf{y}^2 + x_3 \mathbf{y}^3 = \mathbf{0}$$

is $x_1 = x_2 = x_3 = 0$. Otherwise (i.e., if a solution with at least some nonzero value exists), these vectors are dependent.

The vector equation above is equivalent to the matrix equation $A\mathbf{x} = \mathbf{0}$, where

$$\mathbf{x} = \begin{bmatrix} x_1 \\ x_2 \\ x_3 \end{bmatrix} \text{ and } A = \begin{bmatrix} 100 & 80 & 120 \\ -26 & -17 & -38 \\ -25 & -19 & -30 \\ -24 & -21 & -25 \end{bmatrix}$$

is a matrix formed by the vectors $\mathbf{y}^1, \mathbf{y}^2$ and $\mathbf{y}^3$ from the above table as the columns. By Theorem 5.6, we should prove that A is a matrix of full rank, that is, rank $A = 3$. Indeed, the minor determinant formed by the first three rows of A is

$$\begin{vmatrix} 100 & 80 & 120 \\ -26 & -17 & -38 \\ -25 & -19 & -30 \end{vmatrix} = 680 \neq 0,$$

thus, rank $A = 3$. Therefore, dim $L = 3$.

Interpretation of the result. In the above example each production technique was represented by four figures. However, since the dimension of the technology set is only three, which indicates that one of these figures is superfluous; it doesn't add any extra information to those supplied by the other three. This can be done by dividing all the entities in one column by the figure in the first row. The new figures will, then, give the input requirements per unit of output. These are called 'input/output coefficients'. In this example the set of input/output coefficients is given as

$$\begin{bmatrix} 0.260 & 0.213 & 0.317 \\ 0.250 & 0.276 & 0.250 \\ 0.240 & 0.263 & 0.208 \end{bmatrix}$$

Notice that this information is sufficient to calculate total input requirements at 'that particular output level'. The assumptions made so far, are not sufficient to ensure that by using these coefficients one can get the amounts of inputs required to produce the specified level of output. In order to reach that generalization level one needs to introduce another assumption, namely 'constant returns to scale technology'. This assumption means that these coefficient are independent from the scale of production. Therefore, once they are computed, they can be safely used to establish the input/output relation for any feasible production level.[2]

6.6 Problems

1. Does the set of all continuous functions on $[a, b]$, such that $|f(x)| \leq 1$, form a linear space?
2. Give geometric interpretation of axioms $I.(i) - III.(ii)$.
3. Find coordinates $\eta_1, \ldots, \eta_n$ of a vector $\mathbf{x} = (\xi_1, \ldots, \xi_n)$ in the basis

$$\mathbf{e}_1 = (1, 0, \ldots, 0)$$
$$\mathbf{e}_2 = (0, 1, 0 \ldots, 0)$$
$$\cdots \cdots \cdots$$
$$\mathbf{e}_n = (0, \ldots, 0, 1)$$

   if the linear space $\mathcal{L}$ consists of n-tuples of real numbers.
4. Repeat Problem 3 with the basis changed to:

$$\mathbf{e}_1 = (1, 1, \ldots, 1),$$
$$\mathbf{e}_2 = (0, 1, \ldots, 1),$$
$$\cdots \cdots \cdots$$
$$\mathbf{e}_n = (0, \ldots, 0, 1),$$

5. Let M_n be a set of all square matrices of order n with real elements. Define addition and multiplication on real number in a usual way. Prove that M_n is a linear space. Find a basis and dimension of M_n.
6. Let $\mathcal{L}$ be the linear space with n-tuples of real numbers $\mathbf{x} = (\xi_1, \ldots, \xi_n)$. Prove that

$$\mathcal{L}' = \{\mathbf{x} | \alpha_1 \xi_1 + \cdots + \alpha_n \xi_n = 0 \text{ with some fixed } \alpha_1, \ldots, \alpha_n\}$$

   is a subspace of $\mathcal{L}$.

---

[2] Although widely used in economics, constant returns to scale assumption does not correspond to reality in many instances. It has been observed, particularly in manufacturing industry, that as the scale of production increases per unit input requirements may decline. This is the case of 'increasing returns to scale'.

On the other hands, in some instances, production may also be subject to 'decreasing returns to scale'. Such phenomenon is observed in agriculture and occasionally in 'oversized enterprises'.

6.6 Problems

7. Let f be a map from $\mathbb{R}^2$ to $\mathbb{R}^2$ given by $f(x_1, x_2) = (2x_1, x_2 - x_1)$. Check if f is an isomorphism or not.
8. Let f be a map from $\mathbb{R}^3$ to $\mathbb{R}^3$ given by $f(x_1, x_2, x_3) = (x'_1, x'_2, x'_3)$, where

$$\begin{bmatrix} x'_1 \\ x'_2 \\ x'_3 \end{bmatrix} = \begin{bmatrix} 1 & 0 & -1 \\ 2 & 5 & 10 \\ 0 & -1 & 0 \end{bmatrix} \begin{bmatrix} x_1 \\ x_2 \\ x_3 \end{bmatrix}.$$

Check is f an isomorphism or not.
9. Prove that the vector space M_2 of 2×2 matrices and the vectors space P_3 of polynomials of degree at most 3 are isomorphic and give an explicit formula for an isomorphism from M_2 to P_3.
10. Show that the set S of symmetric matrices of order n is a subspace in M_n. Find the dimension of S.
11. Let $\mathcal{L}'$ be a subspace of $\mathcal{L}$, and dimension of $\mathcal{L}'$ be equal to the dimension of $\mathcal{L}$. Then show that $\mathcal{L} = \mathcal{L}'$.
12. Let $\mathcal{L}$ be a space with n-tuples $(\xi_1, \ldots, \xi_n)$ of real numbers. Show that the set of vectors which meet the condition

$$\alpha_1 \xi_1 + \alpha_2 \xi_2 + \cdots + \alpha_n \xi_n = 0,$$

where $\alpha_1, \ldots, \alpha_n$ are real numbers not all equal to 0, forms a linear subspace of $\mathcal{L}$.
13. Let $\mathcal{L}_1$ and $\mathcal{L}_2$ be two subspaces of $\mathcal{L}$ such that

$$\mathcal{L}_1 \cap \mathcal{L}_2 = \mathbf{0}.$$

Then show that $\dim \mathcal{L}_1 + \dim \mathcal{L}_2 \leq \dim \mathcal{L}$.
14. Let $\mathcal{L}'$ be a subspace spanned by vectors $\mathbf{a}, \mathbf{b}, \mathbf{c}, \ldots$. Show that $\dim \mathcal{L}'$ is equal to the maximal number of linearly independent vectors in $\mathbf{a}, \mathbf{b}, \mathbf{c}, \ldots$.
15. Find the coordinates of vector

$$\mathbf{x} = (7, 14, -1, 2)$$

in the basis

$$\mathbf{e}_1 = (1, 2, -1, 1),$$
$$\mathbf{e}_2 = (2, 3, 0, -1),$$
$$\mathbf{e}_3 = (1, 2, 1, 3),$$
$$\mathbf{e}_4 = (1, 3, -1, 0).$$

Show that $\mathbf{e}_1, \ldots, \mathbf{e}_4$ make a basis.
16. Let $\mathbf{e}_1, \mathbf{e}_2, \mathbf{e}_3$ and $\mathbf{e}'_1, \mathbf{e}'_2, \mathbf{e}'_3$ be two bases for a linear space, and vectors $\mathbf{e}'_1, \mathbf{e}'_2, \mathbf{e}'_3$ be defined trough $\mathbf{e}_1, \mathbf{e}_2, \mathbf{e}_3$ as:

$$e'_1 = 5e_1 - e_2 - 2e_3,$$
$$e'_2 = 2e_1 + 3e_2,$$
$$e'_3 = -2e_1 + e_2 + 3e_3.$$

Find the coordinates of the vector

$$x = -7e_1 + 11e_2 + 3e_3$$

with respect to the basis e'_1, e'_2, e'_3.

17. Prove that each of the following two systems of vectors make a basis. Find the coordinates in the basis e' of a vector y if its coordinates in the basis e are $y_e = (1, 2, 3, 4)$.

$$e_1 = (1, 1, 1, 1) \qquad e'_1 = (1, 0, 3, 3)$$
$$e_2 = (1, 2, 1, 1) \qquad e'_2 = (-2, -3, -5, -4)$$
$$e_3 = (1, 1, 2, 1) \qquad e'_3 = (2, 2, 5, 4)$$
$$e_4 = (1, 3, 2, 3) \qquad e'_4 = (-2, -3, -4, -4)$$

18. Is any of these a linear subspace?
    (a) All vectors of n-dimensional space with integer coordinates?
    (b) All vectors on plane which lie either on axis x or on axis y?
    (c) All vectors of n-dimensional space with first coordinate equal to a given number c?
19. List all geometric types of linear subspaces of a 3-dimensional subspace.
20. Prove that $\mathcal{L}'$ which consists of vectors with equal first and last coordinates is a linear subspace of n-dimensional space $\mathcal{L} = R^n$. Find a basis and dimension of this subspace. Give an illustration for $n = 2$ and $n = 3$.
21. Find a basis and dimension of linear space spanned by the following vectors

$$a_1 = (1, 1, 1, 1, 0),$$
$$a_2 = (1, 1, -1, -1, -1),$$
$$a_3 = (2, 2, 0, 0, -1),$$
$$a_4 = (1, 1, 5, 5, 2),$$
$$a_5 = (1, -1, -1, 0, 0).$$

Euclidean Spaces

7

In this chapter, we will analyse how various metric concepts, such as the length of a vector, the angle between two vectors etc., are related to linear spaces. It is to be noted that one cannot directly define these concepts in linear spaces, since first a dot product for a given linear space should be defined.

7.1 General Definitions

Definition 7.1. Let $\mathcal{L}$ be a linear space over the real numbers. Consider the vectors $\mathbf{x}, \mathbf{y}, \mathbf{z}$ and the scalars α and β. Then, a function $(\mathbf{x}, \mathbf{y})$ which transforms a pair of vectors $\mathbf{x}$ and $\mathbf{y}$ to real numbers is called a *dot product*[1] if it satisfies the following conditions:
1. (symmetry) $(\mathbf{x}, \mathbf{y}) = (\mathbf{y}, \mathbf{x})$.
2. (bilinearity) $(\alpha \mathbf{x} + \beta \mathbf{y}, \mathbf{z}) = \alpha(\mathbf{x}, \mathbf{z}) + \beta(\mathbf{y}, \mathbf{z})$.
3. (positivity) $(\mathbf{x}, \mathbf{x}) > 0$ for all but one $\mathbf{x}$.

Proposition 7.1. *The only* $\mathbf{x} \in \mathcal{L}$ *such that* $(\mathbf{x}, \mathbf{x}) = 0$ *is the null vector* $\mathbf{0}$.

Proof. By condition 2, for null vector we have

$$(\mathbf{0}, \mathbf{0}) = (0 \cdot \mathbf{0}, \mathbf{0}) = 0(\mathbf{0}, \mathbf{0}) = 0.$$

It follows from condition 3 that for each $x \neq \mathbf{0}$ we have $(\mathbf{x}, \mathbf{x}) > 0$. □

Definition 7.2. An n-dimensional linear space $\mathcal{L}$ over the real numbers with the inner-product function $(\mathbf{x}, \mathbf{y})$ is called a *Euclidean space*.

---
[1] The dot product sometimes referred also as *inner product* and *scalar product*.

Example 7.1. Consider the following spaces:
(a) Let $\mathbf{x} = (\xi_1, \ldots, \xi_n)$ be an n-tuple of real numbers, and $\mathbf{y} = (\eta_1, \ldots, \eta_n)$. Consider the dot product
$$(\mathbf{x}, \mathbf{y}) = \sum_{i=1}^{n} \xi_i \eta_i.$$
Check that the corresponding space is Euclidean.
(b) Let $\mathbf{x} = (\xi_1, \ldots, \xi_n)$ and $\mathbf{y} = (\eta_1, \ldots, \eta_n)$ be n-tuples in $\mathbb{R}^n$. Consider some matrix $\|a_{ik}\|_{n \times n}$ and the function
$$(\mathbf{x}, \mathbf{y}) = \sum_{i=1}^{n} \sum_{k=1}^{n} a_{ik} \xi_i \eta_i. \tag{7.1}$$

Let us find some restrictions over the matrix $\|a_{ik}\|$ so as to guarantee that the above function is an inner-product. First note that condition 1 is met only if $\|a_{ik}\|$ is symmetric. Condition 2 is always satisfied. Let us check condition 3. We note that
$$(\mathbf{x}, \mathbf{x}) = \sum_{i=1}^{n} \sum_{k=1}^{n} a_{ik} \xi_i \xi_i \geq 0 \tag{7.2}$$
and $(\mathbf{x}, \mathbf{x}) = 0$ if $\xi_1 = \xi_2 = \cdots = \xi_n = 0$. This property is called the *positive definiteness* of the polynomial
$$\sum_{i=1}^{n} \sum_{k=1}^{n} a_{ik} \xi_i \eta_i.$$

Thus, if $\|a_{ik}\|$ is symmetric and the corresponding polynomial (it is also called quadratic form) is positively defined, then the function defined by (7.1) is an inner-product.
(c) Consider the linear space defined by all continuous functions on $[a, b] \subset \mathbb{R}$. Note that the function
$$(f, g) = \int_a^b f(x) g(x) dx$$
is an inner-product.

Definition 7.3. Given a Euclidean space $\mathcal{E}$, the *length* of a vector $\mathbf{x} \in \mathcal{E}$ is denoted by $|\mathbf{x}|$ and defined as
$$|\mathbf{x}| = \sqrt{(\mathbf{x}, \mathbf{x})}.$$

Definition 7.4. Given a Euclidean space $\mathcal{E}$, the *distance* between two vectors $\mathbf{x}, \mathbf{y} \in \mathcal{E}$ is denoted by $d(\mathbf{x}, \mathbf{y})$ and defined as
$$d(\mathbf{x}, \mathbf{y}) = \sqrt{(\mathbf{x} - \mathbf{y}, \mathbf{x} - \mathbf{y})}.$$

7.2 Orthogonal Bases

Definition 7.5. Given a Euclidean space $\mathcal{E}$, the *angle* between two vectors $\mathbf{x}, \mathbf{y} \in \mathcal{E}$ is denoted by $\gamma_{\mathbf{x},\mathbf{y}}$ and defined as

$$\cos \gamma_{\mathbf{x},\mathbf{y}} = \frac{(\mathbf{x}, \mathbf{y})}{|\mathbf{x}| |\mathbf{y}|}.$$

The expression $\cos \gamma_{\mathbf{x},\mathbf{y}}$ is also called the *correlation coefficient* of the vectors $\mathbf{x}$ and $\mathbf{y}$.

Definition 7.6. The vectors $\mathbf{x}, \mathbf{y}$ of a Euclidean space $\mathcal{E}$ are said to be *orthogonal* if $(\mathbf{x}, \mathbf{y}) = 0$.

Lemma 7.2 (Cauchy inequality). *Let $\mathcal{E}$ be a Euclidean space. Then,*

$$(\mathbf{x}, \mathbf{y})^2 \leq (\mathbf{x}, \mathbf{x})(\mathbf{y}, \mathbf{y}).$$

for all $\mathbf{x}, \mathbf{y} \in \mathcal{E}$.

Proof. See the corresponding proof of equality 2.6 in Chap. 2. ☐

7.2 Orthogonal Bases

Consider a Euclidean space $\mathcal{E}$. We say that a set of n non-zero vectors $\mathbf{e}_1, \ldots, \mathbf{e}_n \in \mathcal{E}$ make an *orthogonal* basis for $\mathcal{E}$ if:
1. The vectors $\mathbf{e}_1, \ldots, \mathbf{e}_n$ span $\mathcal{E}$.
2. All pairs of this set are mutually orthogonal, i.e. $(\mathbf{e}_i, \mathbf{e}_j) = 0$ for all $i, j = 1, \ldots, n$ such that $i \neq j$.

See Fig. 7.1 for an orthogonal basis in the plane $\mathbb{R}^2$.

The orthogonal basis $\mathbf{e}_1, \ldots, \mathbf{e}_n$ is called *orthonormal* if pairs in $\mathbf{e}_1, \ldots, \mathbf{e}_n$ are mutually orthogonal and each vector of this set has the unit length, i.e.,

$$(\mathbf{e}_i, \mathbf{e}_j) = \begin{cases} 1, & \text{if } i = j, \\ 0, & \text{otherwise,} \end{cases}$$

for all $i, j = 1, \ldots, n$.

Two different orthonormal bases in the plane $\mathbb{R}^2$ are given in Fig. 7.2.

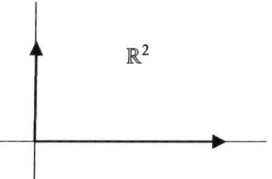

Fig. 7.1 An orthogonal basis in the plane

Fig. 7.2 Two orthonormal bases on the plane

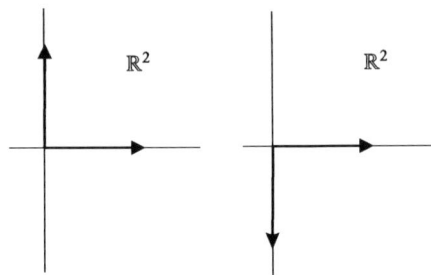

Now let us show that the elements of $e_1, \ldots, e_n$ are linearly independent if this set is an orthogonal basis. In the view of condition 1 above, this implies that this set of vectors forms also a basis of $\mathcal{E}$ as a linear space.

Suppose towards a contradiction that there exist $\lambda_1, \ldots, \lambda_n$, not all being equal to zero, such that

$$\lambda_1 e_1 + \cdots + \lambda_n e_n = 0,$$

i.e., $e_1, \ldots, e_n$ are linearly dependent. Then get the inner-product of the above equality with e_i for any $i = 1, \ldots, n$ to obtain

$$\lambda_1 (e_1, e_i) + \cdots + \lambda_n (e_n, e_i) = (0, e_i) = 0.$$

Hence, we get $\lambda_i (e_i, e_i) = 0$ by the assumption that $e_1, \ldots, e_n$ is an orthogonal set. Since $(e_i, e_i) \neq 0$, it must be true that $\lambda_i = 0$. Since this is true for all $i = 1, \ldots, n$, we have obtained a contradiction. Hence the vectors $e_1, \ldots, e_n$ are linearly independent.

Theorem 7.3. *For any n-dimensional Euclidean space, there exists an orthogonal basis.*

Proof. Let $\mathcal{E}$ be a Euclidean space. Consider any basis $\mathbf{f}_1, \ldots, \mathbf{f}_n$ for $\mathcal{E}$. From this basis, we will obtain an orthogonal basis $e_1, \ldots, e_n$.

Let $e_1 = \mathbf{f}_1$. Next construct $e_2 = \mathbf{f}_2 + \alpha e_1$ such that e_1 and e_2 are orthogonal, see Fig. 7.3. This means that $(e_1, e_2) = 0$, i.e., $(e_1, \mathbf{f}_2 + \alpha e_1) = 0$, that is,

$$\alpha = -\frac{(e_1, \mathbf{f}_2)}{(e_1, e_1)}.$$

Assume now that the orthogonal set $e_1, \ldots, e_{k-1}$ has already been constructed. Define e_k to be

$$e_k = \mathbf{f}_k + \lambda_1 e_1 + \cdots + \lambda_{k-1} e_{k-1},$$

where $\lambda_1, \ldots, \lambda_{k-1}$ will be determined by the fact that e_k must be orthogonal to each of the vectors $e_1, \ldots, e_{k-1}$, i.e.,

$$(e_i, \mathbf{f}_k + \lambda_1 e_1 + \cdots + \lambda_{k-1} e_{k-1}) = 0,$$

7.2 Orthogonal Bases

Fig. 7.3 A construction of the second orthogonal basis vector

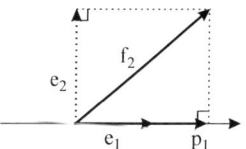

Fig. 7.4 A construction of the third orthogonal basis vector

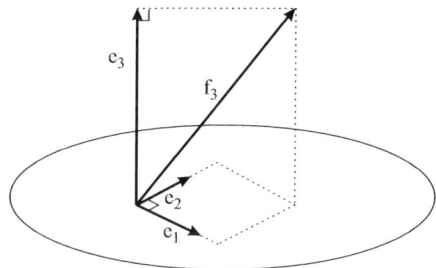

for all $i = 1, \ldots, k - 1$ (see Fig. 7.4 for the case $k = 3$). Since vectors $\mathbf{e}_1, \ldots, \mathbf{e}_{k-1}$ are pairwise orthogonal, we get

$$(\mathbf{e}_i, \mathbf{f_k}) + \lambda_i (\mathbf{e}_i, \mathbf{e}_i) = 0$$

for all $i = 1, \ldots, k - 1$. It then follows that

$$\lambda_i = -\frac{(\mathbf{e}_i, \mathbf{f_k})}{(\mathbf{e}_i, \mathbf{e}_i)}$$

for all $i = 1, \ldots, k - 1$.

To show that $\mathbf{e}_k \neq 0$, note that for all $i < k$

$$\mathbf{e}_i = \mathbf{f_i} + \lambda_1 \mathbf{e}_1 + \cdots + \lambda_{i-1} \mathbf{e}_{i-1},$$

with λ_i's being appropriately defined. Then using the above equalities, we have

$$\mathbf{e}_k = \mathbf{f_k} + \lambda_1 \mathbf{e}_1 + \cdots + \lambda_{k-1} \mathbf{e}_{k-1} = \alpha_1 \mathbf{f_1} + \alpha_2 \mathbf{f_2} + \cdots + \alpha_{k-1} \mathbf{f_{k-1}} + \mathbf{f_k}.$$

If $\mathbf{e}_k = 0$, then since the coefficient of $f_k = 1$, vectors $f_1, \ldots, f_k$ must be linearly dependent, yielding a contradiction. Therefore $\mathbf{e}_k \neq 0$.

By this process, we can construct n pairwise orthogonal vectors $\mathbf{e}_1, \ldots, \mathbf{e}_n$ which are also linearly independent. So, $\mathbf{e}_1, \ldots, \mathbf{e}_n$ is an orthogonal basis for the Euclidean space $\mathcal{E}$. □

Here are some remarks in order.

Remark 7.1. The process of constructing an orthogonal basis from any given basis is called the *Gram²–Schmidt³ orthogonalization process*.

Remark 7.2. Given any orthogonal basis $\mathbf{e}_1, \ldots, \mathbf{e}_n$, the set of vectors $\mathbf{e}_1/|\mathbf{e}_1|, \ldots, \mathbf{e}_n/|\mathbf{e}_n|$ form an orthonormal basis.

Example 7.2. Let L be a subspace in $\mathbb{R}^4$ spanned by three vectors $\mathbf{f}_1 = (1, 1, 0, 1)$, $\mathbf{f}_2 = (3, 1, 1, -1)$ and $\mathbf{f}_3 = (-1, 2, -3, 2)$. Let us construct an orthogonal and orthonormal bases of V.

Note that the vector f_1, f_2 and f_3 form a basis of L (since these vectors are linearly independent), so that we can apply the Gram–Schmidt procedure. First, let us construct an orthogonal basis. Let $\mathbf{e}_1 = \mathbf{f}_1 = (1, 1, 0, 1)$ be its first element. For the second element, we put $\mathbf{e}_2 = \mathbf{f}_2 + \alpha_{21}\mathbf{e}_1$, where

$$\alpha_{21} = -\frac{(\mathbf{e}_1, \mathbf{f}_2)}{(\mathbf{e}_1, \mathbf{e}_1)} = -\frac{3}{3} = -1,$$

so that $\mathbf{e}_2 = \mathbf{f}_2 - \mathbf{e}_1 = (2, 0, 1, -2)$. To construct the third element of the basis, we put

$$\mathbf{e}_3 = \mathbf{f}_3 + \alpha_{31}\mathbf{e}_1 + \alpha_{32}\mathbf{e}_2,$$

where

$$\alpha_{31} = -\frac{(\mathbf{e}_1, \mathbf{f}_3)}{(\mathbf{e}_1, \mathbf{e}_1)} = -\frac{3}{3} = -1 \text{ and } \alpha_{32} = -\frac{(\mathbf{e}_2, \mathbf{f}_3)}{(\mathbf{e}_2, \mathbf{e}_2)} = -\frac{-9}{9} = 1.$$

Thus $\mathbf{e}_3 = \mathbf{f}_3 - \mathbf{e}_1 + \mathbf{e}_2 = (0, 1, -2, -1)$. Now, the vectors $\mathbf{e}_1, \mathbf{e}_2$ and $\mathbf{e}_3$ form an orthogonal basis for L.

By Remark 7.2, the vectors

$$\mathbf{e}'_1 = \frac{1}{|\mathbf{e}_1|}\mathbf{e}_1 = \frac{1}{\sqrt{3}}\mathbf{e}_1 = \left(\frac{1}{\sqrt{3}}, \frac{1}{\sqrt{3}}, 0, \frac{1}{\sqrt{3}}\right), \mathbf{e}'_2 = \frac{1}{|\mathbf{e}_2|}\mathbf{e}_2 = \frac{1}{3}\mathbf{e}_2 = \left(\frac{2}{3}, 0, \frac{1}{3}, -\frac{2}{3}\right)$$

and

$$\mathbf{e}'_3 = \frac{1}{|\mathbf{e}_3|}\mathbf{e}_3 = \frac{1}{\sqrt{6}}\mathbf{e}_3 = \left(0, \frac{1}{\sqrt{6}}, -\frac{1}{\sqrt{3}}, -\frac{1}{\sqrt{6}}\right)$$

form an orthonormal basis of L.

---

[2] Jorgen Pedersen Gram (1850–1916) was a Danish mathematician who was famous because of his works both in pure mathematics (algebra and number theory) and applications (such as a mathematical model of forest management).

[3] Erhard Schmidt (1876–1959) was an Estonia born German mathematician, one of originators of functional analysis.

7.2 Orthogonal Bases

Remark 7.3. For any order of the basis vectors $\mathbf{f}_1, \ldots, \mathbf{f}_n$, we may get a different orthogonal basis by the Gram-Schmidt orthogonalization process. So, it is obvious that orthogonal bases for any Euclidean space are not necessarily unique.

Let $\mathcal{L}$ be a Euclidean space, and $\mathbf{e}_1, \ldots, \mathbf{e}_n$ be an orthonormal basis for it. Let $\mathbf{x}, \mathbf{y} \in \mathcal{L}$ be two vectors such that

$$\mathbf{x} = \xi_1 \mathbf{e}_1 + \cdots + \xi_n \mathbf{e}_n,$$
$$\mathbf{y} = \eta_1 \mathbf{e}_1 + \cdots + \eta_n \mathbf{e}_n.$$

Let us find the inner-product

$$(\mathbf{x}, \mathbf{y}) = (\xi_1 \mathbf{e}_1 + \cdots + \xi_n \mathbf{e}_n, \eta_1 \mathbf{e}_1 + \cdots + \eta_n \mathbf{e}_n)$$

in terms of the coordinates of these two vectors. From the fact

$$(\mathbf{e}_i, \mathbf{e}_j) = \begin{cases} 1 & \text{if } i = j \\ 0 & \text{otherwise} \end{cases}$$

it follows that

$$(\mathbf{x}, \mathbf{y}) = \sum_{i=1}^{n} \xi_i \eta_i,$$

i.e., the inner-product of two vectors that are defined with respect to an orthonormal basis is equal to the sum of the products of their coordinates.

Now, let us express any vector $\mathbf{x} \in \mathcal{L}$ with respect to the orthonormal basis $\mathbf{e}_1, \ldots, \mathbf{e}_n$. That is, we want to find ξ_i for all $i = 1, \ldots, n$ in the expression

$$\mathbf{x} = \xi_1 \mathbf{e}_1 + \cdots + \xi_n \mathbf{e}_n.$$

Multiplying $\mathbf{x}$ by $\mathbf{e}_i$, we get

$$(\mathbf{x}, \mathbf{e}_i) = \xi_1 (\mathbf{e}_1, \mathbf{e}_i) + \cdots + \xi_n (\mathbf{e}_n, \mathbf{e}_i) = \xi_i (\mathbf{e}_i, \mathbf{e}_i) = \xi_i.$$

Hence we have

$$\xi_i = (\mathbf{x}, \mathbf{e}_i)$$

for all $i = 1, \ldots, n$.

The vector $(\mathbf{x}, \mathbf{e}_i)\mathbf{e}_i$, where $\mathbf{e}_i$ is a vector of unit length, is called the *projection* of $\mathbf{x}$ on $\mathbf{e}_i$. Then, the coordinates of any vector with respect to an orthonormal basis can be regarded as the lengths of projections of the vector on each basis vector.

Definition 7.7. Let $\mathcal{L}$ be a Euclidean space. Then a vector $\mathbf{x} \in \mathcal{L}$ is called *orthogonal* to a subspace B of $\mathcal{L}$ if it is orthogonal to each vector $\mathbf{y} \in B$.

Lemma 7.4. *Let $\mathcal{L}$ be a Euclidean space. If a vector $\mathbf{x} \in \mathcal{L}$ is orthogonal to some vectors $\mathbf{e}_1, \ldots, \mathbf{e}_n \in \mathcal{L}$, then it is orthogonal to any linear combinations of these vectors as well.*

Proof. Since $(\mathbf{x}, \mathbf{e}_i) = 0$ for all $i = 1, \ldots, n$, we have

$$(\mathbf{x}, \lambda_1 \mathbf{e}_1 + \cdots + \lambda_n \mathbf{e}_n) = \lambda_1 (\mathbf{x}, \mathbf{e}_1) + \cdots + \lambda_n (\mathbf{x}, \mathbf{e}_n) = 0,$$

for any $\lambda_1, \ldots, \lambda_m \in \mathbb{R}$. □

An immediate result of the above Lemma is the following

Corollary 7.5. *Let $\mathcal{L}$ be a Euclidean space and B be a subspace of $\mathcal{L}$. If a vector $\mathbf{x} \in \mathcal{L}$ is orthogonal to a basis for B, then it is orthogonal to B.*

Now consider the following problem: given any (nontrivial) subspace B of a Euclidean space $\mathcal{L}$, and any vector $\mathbf{x}$ which is an element of $\mathcal{L} \setminus B$, can we find a vector $\mathbf{y}$ in B such that the difference $\mathbf{x} - \mathbf{y}$ is orthogonal to $\mathbf{x}$? Such vector $\mathbf{y}$ is called an (orthogonal) *projection* of $\mathbf{x}$ to B.

Lemma 7.6. *Let $\mathcal{L}$ be an n-dimensional Euclidean space and B be an m-dimensional subspace of $\mathcal{L}$, where $0 < m < n$. Pick any $\mathbf{x} \in \mathcal{L}$. Let $\mathbf{y}$ be the projection of $\mathbf{x}$ to B, that is, $\mathbf{y} \in B$ and $\mathbf{x} - \mathbf{y}$ is orthogonal to B. Then*

$$|\mathbf{x} - \mathbf{y}_1| > |\mathbf{x} - \mathbf{y}_0|$$

for all $\mathbf{y}_1 \in B \setminus \{\mathbf{y}\}$, see Fig. 7.5.

Proof. Assume $\mathbf{x} - \mathbf{y}$ is orthogonal to $\mathbf{y} - \mathbf{y}_1$. By Pythagoras' Theorem

$$|\mathbf{x} - \mathbf{y}|^2 + |\mathbf{y} - \mathbf{y}_1|^2 = |\mathbf{x} - \mathbf{y} + \mathbf{y} - \mathbf{y}_1|^2 = |\mathbf{x} - \mathbf{y}_1|^2$$

which implies $|\mathbf{x} - \mathbf{y}_1| > |\mathbf{x} - \mathbf{y}|$. □

The next Lemma shows that the projection of a vector to a subspace always exists.

Lemma 7.7. *Let $\mathcal{L}$ be an n-dimensional Euclidean space and B be an m-dimensional subspace of $\mathcal{L}$, with an orthonormal basis $\mathbf{e}_1, \ldots, \mathbf{e}_m$, where*

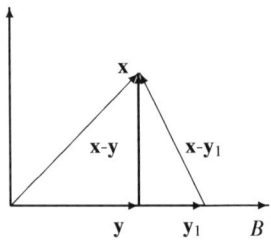

Fig. 7.5 The shortest way from $\mathbf{x}$ to a subspace B

7.2 Orthogonal Bases

Fig. 7.6 The projection on a line

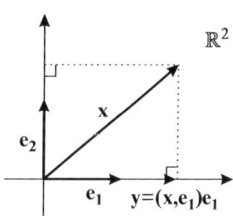

Fig. 7.7 The projection on a plane

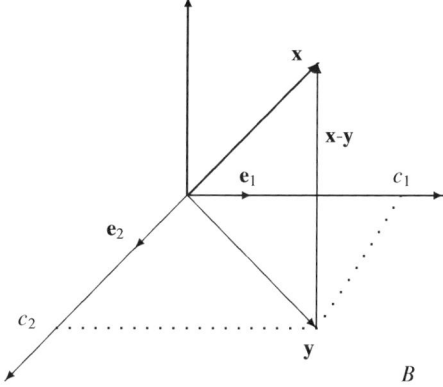

$0 < m < n$. Then the orthogonal projection of any vector $\mathbf{x} \in \mathcal{L} \setminus B$ on B is

$$\mathbf{y}_0 = c_1 \mathbf{e}_1 + \cdots + c_m \mathbf{e}_m$$

where $c_i = (\mathbf{x}, \mathbf{e}_i)$ for all $i = 1, \ldots, m$.

The statement of Lemma 7.7 is illustrated in Fig. 7.6 for the case $n = 2, m = 1$ and in Fig. 7.7 for the case $n = 3, m = 2$.

Proof. Let $\mathbf{y} = c_1 \mathbf{e}_1 + \cdots + c_m \mathbf{e}_m$ be the orthogonal projection of $\mathbf{x} \in \mathcal{L} \setminus B$ on B. Then $\mathbf{x} - \mathbf{y}$ is orthogonal to B, i.e., $(\mathbf{x} - \mathbf{y}, \mathbf{e}_i) = 0$, implying $(\mathbf{x}, \mathbf{e}_i) = (\mathbf{y}, \mathbf{e}_i)$ for all $i = 1, \ldots, m$. It follows that

$$(\mathbf{x}, \mathbf{e}_i) = c_1 (\mathbf{e}_1, \mathbf{e}_i) + \cdots + c_m (\mathbf{e}_m, \mathbf{e}_i) = c_i (\mathbf{e}_i, \mathbf{e}_i) = c_i, \quad i = 1, \ldots, m.$$

$\square$

The *distance* between a vector $\mathbf{x}$ and a subspace B in a Euclidean space $\mathcal{L}$ is the minimal length of a vector of the form $\mathbf{x} - \mathbf{y}$, where $\mathbf{y} \in B$, that is, the distance

$$d(\mathbf{x}, B) = \min_{\mathbf{y} \in B} \{|\mathbf{x} - \mathbf{y}|\}.$$

For example, we have $d(\mathbf{x}, B) = 0$ if and only if $\mathbf{x} \in B$. Then we can re-formulate Lemma 7.6 as follows

Corollary 7.8. *The distance between a vector $\mathbf{x}$ and a subspace B in a Euclidean space $\mathcal{L}$ is equal to $|\mathbf{x} - \mathbf{y}|$, where $\mathbf{y}$ is the projection of $\mathbf{x}$ to B.*

Remark 7.4. Let us now find the orthogonal projection of any vector $\mathbf{x} \in \mathcal{L}$ on B with respect to an arbitrary basis $\mathbf{e}_1, \ldots, \mathbf{e}_m$ for B. Then following the argument in the above proof, we obtain a system

$$\begin{bmatrix} (\mathbf{e}_1, \mathbf{e}_1) & (\mathbf{e}_2, \mathbf{e}_1) & \ldots & (\mathbf{e}_m, \mathbf{e}_1) \\ (\mathbf{e}_1, \mathbf{e}_2) & (\mathbf{e}_2, \mathbf{e}_2) & \ldots & (\mathbf{e}_m, \mathbf{e}_2) \\ \ldots & \ldots & \ldots & \ldots \\ (\mathbf{e}_1, \mathbf{e}_m) & (\mathbf{e}_2, \mathbf{e}_m) & \ldots & (\mathbf{e}_m, \mathbf{e}_m) \end{bmatrix} \begin{bmatrix} c_1 \\ c_2 \\ \ldots \\ c_m \end{bmatrix} = \begin{bmatrix} (\mathbf{x}, \mathbf{e}_1) \\ (\mathbf{x}, \mathbf{e}_2) \\ \ldots \\ (\mathbf{x}, \mathbf{e}_m) \end{bmatrix}.$$

The $m \times m$ matrix above, containing the inner-product of the basis vectors, is called the *Gram matrix* of the vectors $\mathbf{e}_1, \ldots, \mathbf{e}_m$. A system of m equations with m variables has a unique solution if and only if the determinant of the coefficient matrix is nonzero. On the other hand, the above equation system, when describes the orthogonal projection of a vector $\mathbf{x}$ on B, must have a unique solution, since the vector $\mathbf{y}$, according to Lemmas 7.6 and 7.7, always exists and is unique. From here we conclude that given any basis (linearly independent) vectors, the determinant of the associated Gram matrix is nonzero. Note that this is obvious when the basis vectors are orthonormal, since in that case the Gram matrix is the identity matrix whose determinant is one.

Example 7.3. Problem. Given a vector $\mathbf{x} = (25, 0, 25) \in \mathbb{R}^3$ and a subspace $B \subset \mathbb{R}^3$ spanned by the vectors $\mathbf{a} = (3, 4, 5)$ and $\mathbf{b} = (-4, 3, 5)$, find the orthogonal projection of $\mathbf{x}$ on B and the distance between $\mathbf{x}$ and B.

Solution. Let $\mathbf{y} = c_1 \mathbf{a} + c_2 \mathbf{b}$ be the projection. By the above, we have

$$G \begin{bmatrix} c_1 \\ c_2 \end{bmatrix} = \begin{bmatrix} (\mathbf{x}, \mathbf{a}) \\ (\mathbf{x}, \mathbf{b}) \end{bmatrix},$$

where G is the Gram matrix,

$$G = \begin{bmatrix} (\mathbf{a}, \mathbf{a}) & (\mathbf{a}, \mathbf{b}) \\ (\mathbf{b}, \mathbf{a}) & (\mathbf{b}, \mathbf{b}) \end{bmatrix} = \begin{bmatrix} 50 & 25 \\ 25 & 50 \end{bmatrix}.$$

We have

$$\begin{bmatrix} 50 & 25 \\ 25 & 50 \end{bmatrix} \begin{bmatrix} c_1 \\ c_2 \end{bmatrix} = \begin{bmatrix} 200 \\ 25 \end{bmatrix},$$

hence $c_1 = 5$ and $c_2 = -2$. Thus $\mathbf{y} = 5\mathbf{a} - 2\mathbf{b} = (23, 14, 15)$.

7.3 Least Squares Method

By Corollary 7.8, the distance between y and B is

$$d(\mathbf{y}, B) = |\mathbf{x} - \mathbf{y}| = |(2, -14, 10)| = 10\sqrt{3} \approx 17.32.$$

7.3 Least Squares Method

We give below the linear algebraic interpretation of the least square method that is widely used in econometrics. Another approach to the least square method is discussed in Appendix D.

Let y be a linear function of m variables $x_1, \ldots, x_m \in \mathbb{R}$ given by

$$y(x_1, \ldots, x_m) = c_1 x_1 + \cdots + c_m x_m,$$

where $c_1, \ldots, c_m$ are unknown real constants. Assume that $c_1, \ldots, c_m$ are obtained via n experiments, in which $x_1, \ldots, x_m$ and y are measured. We accept that the solution can be with error but we would like to minimize that error (see Fig. 7.8).

Let the result of k-th experiment give

$$x_{k1}, \ldots, x_{km} \text{ and } y_k.$$

Then we can construct a system of linear equations for $c_1, \ldots, c_m$:

$$\begin{cases} x_{11}c_1 + x_{12}c_2 + \cdots + x_{1m}c_m = y_1, \\ \cdots \cdots \cdots \\ x_{n1}c_1 + x_{n2}c_2 + \cdots + x_{nm}c_m = y_n, \end{cases} \quad (7.3)$$

or

$$A\mathbf{c} = \mathbf{y}, \quad (7.4)$$

where $A = (x_{ij})_{n \times m}$ is the matrix of the system, $\mathbf{c} = (c_1, \ldots, c_m)$ is the unknown vector and $\mathbf{y} = (y_1, \ldots, y_n)$ is the vector of right-hand sides.

Assume that $n > m$. Since x_i-s and y are measured with errors, hoping to obtain an exact solution for (7.3) or (7.4) is senseless. The aim should be to determine the

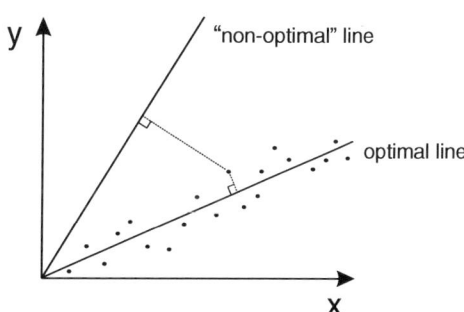

Fig. 7.8 The least square approximation

set of constants $c_1,\ldots,c_m$ which make the left and right sides of the equalities in (7.3) as close as possible. As a distance of closeness, consider the functional

$$d(c_1,\ldots,c_m) = \sum_{k=1}^{n}(x_{k1}c_1 + \cdots + x_{km}c_m - y_k)^2 \tag{7.5}$$

for all $k = 1,\ldots,n$. Then the problem is to find $c_1,\ldots,c_m$ such that $d(c_1,\ldots,c_m)$ is minimized.

To reformulate the problem, note that the sum $x_{k1}c_1 + \cdots + x_{km}c_m$ is the k-th component of the vector $A\mathbf{c}$ in $\mathbb{R}^n$. This means that $d(c_1,\ldots,c_m) = |A\mathbf{c} - \mathbf{y}|^2$, that is, the square of the distance between $\mathbf{y}$ and $A\mathbf{c}$.

Consider the columns $\mathbf{A}^1,\ldots,\mathbf{A}^m$ of the matrix A, where $\mathbf{A}^k = (x_{1k}, x_{2k},\ldots, x_{nk})$. It follows that minimizing (7.5) corresponds to finding $c_1,\ldots,c_m$ which make the distance between $\mathbf{y}$ and

$$\mathbf{y}_0 = A\mathbf{c} = c_1\mathbf{A}^1 + \ldots c_m\mathbf{A}^m$$

minimal. Let B be the linear span of the vectors $\mathbf{A}^1,\ldots,\mathbf{A}^m$ in $\mathbb{R}^n$. Then the problem is to find the projection of $\mathbf{y}_0$ on B.

Let us assume now that the vectors $\mathbf{A}^1,\ldots,\mathbf{A}^m$ are linearly independent.[4] An appeal to Lemma 7.7 implies

$$(\mathbf{A}^1,\mathbf{A}^k)c_1 + (\mathbf{A}^2,\mathbf{A}^k)c_2 + \cdots + (\mathbf{A}^m,\mathbf{A}^k)c_m = (\mathbf{y},\mathbf{A}^k) \tag{7.6}$$

for all $k = 1,\ldots,m$, where

$$(\mathbf{y},\mathbf{A}^k) = \sum_{j=1}^{n} y_j x_{jk}$$

and

$$(\mathbf{A}^j,\mathbf{A}^k) = \sum_{i=1}^{n} x_{ij} x_{ik}.$$

Thus the problem has been reduced from finding an approximated solution of system (7.3) to finding an exact solution of (7.6).

Example 7.4. Consider least-squares estimation in the plane (Fig. 7.8). Given the measurements $\mathbf{x} = (x_1, x_2,\ldots, x_n)$ and $\mathbf{y} = (y_1, y_2,\ldots, y_n)$ for the variables x and y, respectively, the aim is to find the slope c of the line $y = cx$ using the system

---

[4]This means that the matrix A is of full rank. The general case of arbitrary matrix A will be discussed in Appendix D.

$$\begin{cases} y_1 = cx_1, \\ y_2 = cx_2, \\ \cdots\cdots\cdots \\ y_n = cx_n. \end{cases}$$

Using (7.6) we obtain

$$c = \frac{(\mathbf{x}, \mathbf{y})}{(\mathbf{x}, \mathbf{x})} = \frac{\sum_{k=1}^{n} x_k y_k}{\sum_{k=1}^{n} x_k^2}.$$

For $n = 4$, $(x_1, x_2, x_3, x_4) = (1, 2, 3, 4)$ and $(y_1, y_2, y_3, y_4) = (2, 3, 4, 10)$, check that $c = 2$.

7.4 Isomorphism of Euclidean Spaces

Definition 7.8. Any two Euclidean spaces $\mathcal{L}$ and $\mathcal{L}'$ are said to be *isomorphic* if they are isomorphic as linear spaces and if $(\mathbf{x}, \mathbf{y}) = (\mathbf{x}', \mathbf{y}')$ whenever

$$\mathcal{L} \ni \mathbf{x} \longleftrightarrow \mathbf{x}' \in \mathcal{L}',$$
$$\mathcal{L} \ni \mathbf{y} \longleftrightarrow \mathbf{y}' \in \mathcal{L}',$$

Theorem 7.9. *Any two n-dimensional Euclidean spaces are isomorphic.*

Proof. It is sufficient to prove that any n-dimensional Euclidean space $\mathcal{L}$ is isomorphic to $\mathbb{R}^n$.

Isomorphism of $\mathcal{L}$ and $\mathcal{L}'$ as linear spaces is obvious, so we will prove the rest.

Given any orthonormal basis $e_1, \ldots e_n$ in $\mathcal{L}$, let the dot product of two vectors $\mathbf{x} = \xi_1 e_1 + \ldots + \xi_n e_n$ and $\mathbf{y} = \eta_1 e_1 + \cdots + \eta_n e_n$ in $\mathcal{L}$ be defined as

$$(\mathbf{x}, \mathbf{y}) = \xi_1 \eta_1 + \cdots + \xi_n \eta_n.$$

Now consider any n-dimensional Euclidean space $\mathcal{L}$. Choose for it an orthonormal basis $e_1, \ldots e_n$. (It must exist, why?) For any two vectors $\mathbf{x}' = \xi_1 e_1 + \cdots + \xi_n e_n$ and $\mathbf{y}' = \eta_1 e_1 + \cdots + \eta_n e_n$ in $\mathcal{L}$, pick the vectors $\mathbf{x} = (\xi_1, \ldots, \xi_n)$ and $\mathbf{y} = (\eta_1, \ldots, \eta_n)$ in $\mathbb{R}^n$, and vice versa. (Note that we use the canonical basis for $\mathbb{R}^n$.)

Then we have

$$(\mathbf{x}', \mathbf{y}') = \xi_1 \eta_1 + \cdots + \xi_n \eta_n.$$

On the other hand,

$$(\mathbf{x}, \mathbf{y}) = \xi_1 \eta_1 + \cdots + \xi_n \eta_n.$$

Thus,

$$(\mathbf{x}', \mathbf{y}') = (\mathbf{x}, \mathbf{y}),$$

completing the proof. □

7.5 Problems

1. Check whether the function defined by equation (7.1) is an inner product for the matrices:

   (a) $\begin{bmatrix} 1 & 1 \\ 1 & 2 \end{bmatrix}$; (b) $\begin{bmatrix} 0 & 1 \\ 1 & 0 \end{bmatrix}$.

2. Consider any two vectors
$$\mathbf{x} = (\xi_1, \ldots, \xi_n),$$
$$\mathbf{y} = (\eta_1, \ldots, \eta_n)$$
in $\mathbb{R}^n$. Check whether any of the following functions is an inner-product:

   (a) $(\mathbf{x}, \mathbf{y}) = \sum_{j=1}^{n} |\xi_i| |\eta_i|$; (b) $(\mathbf{x}, \mathbf{y}) = \sum_{j=1}^{n} \xi_i^2 \eta_i^2$.

3. Consider the basis given by the vectors
$$\mathbf{f}_1 = (1, 1, \ldots, 1, 1),$$
$$\mathbf{f}_2 = (0, 1, \ldots, 1, 1),$$
$$\vdots$$
$$\mathbf{f}_n = (0, 0, \ldots, 0, 1).$$

   Is this basis orthogonal? If not, construct an orthogonal basis using Gram-Schmidt process. For this basis, find the associated orthonormal basis.

4. Prove that for any Euclidean space $\mathcal{L}$, any vectors $\mathbf{x}, \mathbf{y} \in \mathcal{L}$ and $\lambda \in \mathbb{R}$, if $\mathbf{x} = \lambda \mathbf{y}$ then
$$|\mathbf{x}| = |\lambda| \, |\mathbf{y}|.$$

5. Let $\mathcal{L}_1$ and $\mathcal{L}_2$ be linear subspaces of the Euclidean space $\mathbb{R}^4$ such that
$$\dim \mathcal{L}_1 = \dim \mathcal{L}_2.$$

   Prove that there exists a non-zero vector $\mathbf{x} \in \mathcal{L}_2$ which is orthogonal to $\mathcal{L}_1$.

6. Check whether the following vectors are orthogonal and complete them up to a basis for $\mathbb{R}^4$.

   (a) $\mathbf{f}_1 = (1, -2, 2, -3)$ and $\mathbf{f}_2 = (2, -3, 2, 4)$.

   (b) $\mathbf{f}_1 = (1, 1, 1, 2)$ and $\mathbf{f}_1 = (1, 2, 3, -3)$.

7. Consider the Euclidean space $\mathbb{R}^4$. Find the angles between the line $\xi_1 = \xi_2 = \xi_3 = \xi_4$ and the vectors $e_1 = (1, 0, 0, 0)$, $e_2 = (0, 1, 0, 0)$, $e_3 = (0, 0, 1, 0)$ and $e_4 = (0, 0, 0, 1)$.

8. Let $\mathcal{L}_1$ and $\mathcal{L}_2$ be two orthogonal subspaces of a Euclidean space $\mathcal{L}$. Prove that $\mathcal{L}_1 \cap \mathcal{L}_2 = \mathbf{0}$.

7.5 Problems

9. Given the space of polynomials of degree at most n, and any two polynomials P and Q in this space, consider the mapping

$$(P(t), Q(t)) = \int_a^b P(t)Q(t)dt.$$

   (a) Check that the above mapping is an inner-product.

   (b) Find the distance between the polynomials $P(t) = 3t^2 + 6$ and $Q(t) = 2t^2 + t + 1$ according to the above inner-product.

10. Using Gramm-Shmidt orthogonalization process construct an orthonormal basis from the vectors

$$\mathbf{f}_1 = (2, 1, 3, -1),$$
$$\mathbf{f}_2 = (7, 4, 3, -3),$$
$$\mathbf{f}_3 = (1, 1, -6, 0),$$
$$\mathbf{f}_4 = (5, 7, 7, 8).$$

11. Find the distance between the vector $\mathbf{x} = (4, -1, -3, 4)$ and a linear subspace defined by the vectors

$$\mathbf{e}_1 = (1, 1, 1, 1),$$
$$\mathbf{e}_2 = (1, 2, 2, -1),$$
$$\mathbf{e}_3 = (1, 0, 0, 3).$$

12. Let $\mathcal{L}$ be a subspace defined by the system of linear equations

$$\begin{cases} 2x_1 + x_2 + x_3 + 3x_4 = 0, \\ 3x_1 + 2x_2 + 2x_3 + x_4 = 0, \\ x_1 + 2x_2 + 2x_3 - 9x_4 = 0. \end{cases}$$

   Consider the vector $\mathbf{x} = (7, -4, -1, 2)$. Find the orthogonal projection (call it $\mathbf{y}$) of $\mathbf{x}$ on $\mathcal{L}$. Find also $\mathbf{x} - \mathbf{y}$.

13. (a) Find the distance between the vector $\mathbf{x} = (4, 2, -5, 1)$ and the linear space $\mathcal{L}$ defined by the system

$$\begin{cases} 2x_1 - 2x_2 + x_3 + 2x_4 = 9, \\ 2x_1 - 4x_2 + 2x_3 + 3x_4 = 12. \end{cases}$$

   (b) Find the distance between the vector $\mathbf{x} = (2, 4, -4, 2)$ and the linear space $\mathcal{L}$ defined by the system

$$\begin{cases} x_1 + 2x_2 + x_3 - x_4 = 1, \\ x_1 + 3x_2 + x_3 - 3x_4 = 2. \end{cases}$$

14. Find the least square approximation for the function $y(x_1, x_2) = c_1 x_1 + c_2 x_2$, if its three measurements give $y(0, 1) = 3$, $y(1, 2) = 8$, $y(-1, 0) = 0$.

15. The market price (p) of some good in several dates are given in the following table:

Date	12 January	14 January	15 January	20 January
p	$13.2	$10.2	$9.8	$12.9

Using the least square approximation $p(x) = a + bx$ for the function p of the day x of the year, give your prognosis for the price p in January 27.

Linear Transformations

8

We will begin with a general definition of transformations and then study linear transformations.[1]

Definition 8.1. A *mapping* F from a set S to another set S' is a relation which, to every element x of S, associates an element y of S'. This mapping is denoted by $F : S \to S'$.

If x is an element of S, then $y = F(x)$ is called the *image* of x under F. Analogously, the set
$$B = \{F(x) \mid x \in S\}$$
is called the image of S (or full image). Naturally, $B \subseteq S'$. In the same way, for $W \subseteq S$, the set
$$F(W) = \{F(x) \mid x \in W\}$$
is called the image of W under F.

Example 8.1. (a) Let $S = S' = \mathbb{R}$ and $F(x) = x^2$. Then the image of S is the set of all non-negative numbers.
(b) Let $L : \mathbb{R}^3 \to \mathbb{R}$ be a mapping such that for all $\mathbf{x} \in \mathbb{R}^3$, $L(\mathbf{x}) = (\mathbf{a}, \mathbf{x})$, where $\mathbf{a} = (3, 2, 1)$.

Given any set S, the identity mapping I_S is defined such that $I_S(x) = x$ for all $x \in S$.

Definition 8.2. A mapping $F : S \to S'$ is said to have the inverse if there exists a mapping $G : S' \to S$ such that
$$G \cdot F = I_S \text{ and } F \cdot G = I_{S'},$$
that is, $G(F(x)) = x$ and $F(G(y)) = y$ for all $x \in S$, $y \in S'$.

---

[1] One may use the terms linear mapping, linear map, homomorphism and linear transformation interchangeably.

F. Aleskerov et al., *Linear Algebra for Economists*, Springer Texts in Business and Economics, DOI 10.1007/978-3-642-20570-5_8,
© Springer-Verlag Berlin Heidelberg 2011

Example 8.2. (a) Let $S = S' = \mathbb{R}_+$ (the nonnegative real line) and $f : S \to S'$ such that $f(x) = x^2$. Then the inverse mapping $g : S' \to S$ is given by $g(x) = \sqrt{x}$.
(b) Let R be a linear space, and $\mathbf{u}$ be a fixed element of R. Define $T_\mathbf{u} : R \to R$ such that $T_\mathbf{u}(x) = \mathbf{x} + \mathbf{u}$ for all $\mathbf{x} \in R$. The mapping $T_\mathbf{u}$ is called the *translation* by $\mathbf{u}$. For any $S \subset R$, it follows that

$$T_\mathbf{u}(S) = \{\mathbf{x} + \mathbf{u} \mid \mathbf{x} \in S\}.$$

Definition 8.3. A mapping A from one linear R space to another linear space S is called a *transformation*. If S is the same linear space as R, the transformation is also called an *operator*.

Definition 8.4. A transformation (or an operator) A from a linear space R to a linear space S is called *linear* if

$$A(\lambda_1 \mathbf{x}_1 + \lambda_2 \mathbf{x}_2) = \lambda_1 A(\mathbf{x}_1) + \lambda_2 A(\mathbf{x}_2)$$

for all $\mathbf{x}_1, \mathbf{x}_2 \in R$ and $\lambda_1, \lambda_2 \in \mathbb{R}$.

Very often we will write $A\mathbf{x}$ instead of $A(\mathbf{x})$.

Example 8.3. Some linear transformations:
(a) Rotation of vectors $\mathbb{R}^2$ around the origin by a fixed angle φ (see Fig. 2.10 and Example 2.12).
(b) For any $\mathbf{x} \in \mathbb{R}^2$ consider the projection of $\mathbf{x}$ on the line $y = Cx$, where $C \in \mathbb{R}$ (Fig. 8.1).
(c) Let $R = \mathbb{R}^n$ and $S = \mathbb{R}^m$. For each $\mathbf{x} = (\xi_1, \ldots, \xi_n) \in R$, consider the corresponding vector
$$\mathbf{y} = A\mathbf{x} = (\eta_1, \ldots, \eta_n),$$
where $A = \|a_{ik}\|_{m \times n}$ is a matrix, and

$$\eta_i = \sum_{k=1}^n a_{ik} \xi_k.$$

(Check that A is a linear transformation!)

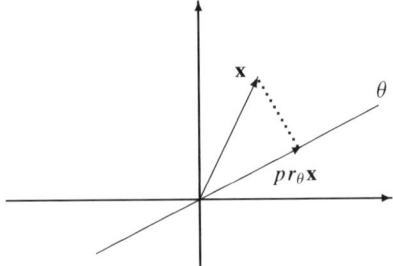

Fig. 8.1 A projection

(d) Let R be the linear space of all polynomials (over real numbers) having the degree at most $n - 1$. Define A such that

$$A(P(t)) = P'(t),$$

where $P'(t)$ is the derivative of polynomial $P(t)$ with respect to t. Then A is a linear operator on R, since

$$\begin{aligned} A(\lambda_1 P(t) + \lambda_2 Q(t)) &= (\lambda_1 P(t) + \lambda_2 Q(t))' \\ &= \lambda_1 P'(t) + \lambda_2 Q'(t) \\ &= \lambda_1 A(P(t)) + \lambda_2 A(Q(t)). \end{aligned}$$

for all $P, Q \in R$ and $\lambda_1, \lambda_2 \in \mathbb{R}$.

(e) Let R be the linear space of all continuous functions $f(t)$ on $[0, 1]$. Let

$$A(f(t)) = \int_0^1 f(r) dr.$$

Then A is a linear transformation from R to a one-dimensional vector space $\mathbb{R}$, because

$$A(\lambda_1 f_1 + \lambda_2 f_2) = \int_0^1 [\lambda_1 f_1(r) + \lambda_2 f_2(r)] dr = \lambda_1 A(f_1) + \lambda_2 A(f_2)$$

for all $f_1, f_2 \in R$ and $\lambda_1, \lambda_2 \in \mathbb{R}$.

Two special transformations are the following
- Identity operator (I): For all linear spaces R and $\mathbf{x} \in R$, $I\mathbf{x} = \mathbf{x}$.
- Null operator (Θ): For all linear spaces R and $\mathbf{x} \in R$, $\Theta\mathbf{x} = \mathbf{0}$.

Theorem 8.1. *Let R be an n-dimensional linear space, and let $\mathbf{e}_1, \ldots, \mathbf{e}_n$ be a basis for R. Then for any set of vectors $\mathbf{g}_1, \ldots, \mathbf{g}_n$ of another linear space S there exists a unique linear transformation A from R to S such that*

$$A\mathbf{e}_i = \mathbf{g}_i, \text{ for all } i = 1, \ldots, n.$$

Proof. Let us first prove that for any vector $\mathbf{g}_1, \ldots, \mathbf{g}_n$ there exists a linear transformation A such that

$$A\mathbf{e}_i = \mathbf{g}_i \text{ for all } i = 1, \ldots, n.$$

For each

$$\mathbf{x} = \xi_1 \mathbf{e}_1 + \cdots + \xi_n \mathbf{e}_n \in R$$

construct the corresponding vector

$$\mathbf{y} = \xi_1 \mathbf{g}_1 + \cdots + \xi_n \mathbf{g}_n \in S.$$

Since $\mathbf{x}$ is defined through $\mathbf{e}_1, \ldots, \mathbf{e}_n$ uniquely, $\mathbf{y}$ is well defined. Then let $A\mathbf{x} = \mathbf{y}$, hence A is a transformation. It is not hard to check that the transformation A is linear.

To show that A is uniquely defined by $\mathbf{g}_i$'s, suppose there exist two linear transformations A_1 and A_2 such that for some $\mathbf{x} = \xi_1 \mathbf{e}_1 + \cdots + \xi_n \mathbf{e}_n \in R$, $A_1 \mathbf{x} \neq A_2 \mathbf{x}$ while $A_1 \mathbf{e}_i = A_2 \mathbf{e}_i = \mathbf{g}_i$ for all $i = 1, \ldots, n$. But notice that

$$\xi_1 \mathbf{g}_1 + \cdots + \xi_n \mathbf{g}_n = \xi_1 A_1 \mathbf{e}_1 + \cdots + \xi_n A_1 \mathbf{e}_n$$
$$= \xi_1 A_2 \mathbf{e}_1 + \cdots + \xi_n A_2 \mathbf{e}_n$$

implying $A_1 \mathbf{x} = A_2 \mathbf{x}$, which is a contradiction. So, the linear transformation defined by $\mathbf{g}_i$'s must be unique. $\square$

Let $f_1, \ldots, f_m$ be a basis of the linear space S. Denote the coordinates of each $\mathbf{g}_k = A\mathbf{e}_k$ with respect to the given basis by $a_{1k}, a_{2k}, \ldots, a_{mk}$, i.e.,

$$\mathbf{g}_k = A\mathbf{e}_k = \sum_{i=1}^{m} a_{ik} \mathbf{f}_i, \quad k = 1, \ldots, n. \tag{8.1}$$

All these coordinates make a matrix $\|a_{ik}\|_{m \times n}$ which is called the matrix of linear transformation A with respect to the bases $\mathbf{e}_1, \ldots, \mathbf{e}_n$ and $f_1, \ldots, f_m$.

Then the following result holds.

Lemma 8.2. *Consider any two bases $\mathbf{e}_1, \ldots, \mathbf{e}_n$ and $\mathbf{f}_1, \ldots, \mathbf{f}_m$ for an n-dimensional linear space R and an m-dimensional linear space S, respectively. Then for any linear transformation A from R to S, there exists a unique transformation matrix $\|a_{ik}\|_{n \times m}$, and to each matrix $\|a_{ik}\|_{n \times m}$ corresponds a unique linear transformation A.*

In particular, one can assign to any linear operator in R a square matrix of order n.

Example 8.4. In what follows, we construct transformation matrices from the given linear transformations.

(a) Let $R = \mathbb{R}^3$ and A be a linear operator which maps each vector $\mathbf{a} \in R$ to its projection on the XY plane (Fig. 8.2).

Consider the canonical basis for $\mathbb{R}^3$:

$$\mathbf{e}_1 = (1, 0, 0)$$
$$\mathbf{e}_2 = (0, 1, 0)$$
$$\mathbf{e}_3 = (0, 0, 1)$$

8 Linear Transformations

Fig. 8.2 The operator of a projection on the plane XY

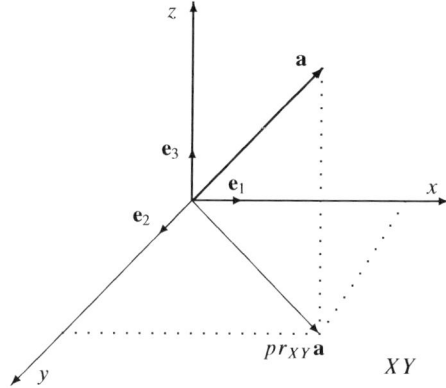

Then we have
$$Ae_1 = e_1$$
$$Ae_2 = e_2$$
$$Ae_3 = 0.$$

So, the transformation matrix A is

$$A = \begin{bmatrix} 1 & 0 & 0 \\ 0 & 1 & 0 \\ 0 & 0 & 0 \end{bmatrix}.$$

(b) Consider the identify operator I in an n dimensional space R, i.e., $Ie_i = e_i$ for any $i = 1, \ldots, n$. Then, obviously, the associated transformation matrix is

$$I = \begin{bmatrix} 1 & 0 & \ldots & 0 \\ 0 & 1 & \ldots & 0 \\ \cdot & \cdot & & \cdot \\ \cdot & \cdot & & \cdot \\ \cdot & \cdot & & \cdot \\ 0 & 0 & \ldots & 1 \end{bmatrix}.$$

Analogously, the transformation matrix for the null operator is given by $\Theta = 0$.

Let A be a linear transformation from R to S, let $e_1, \ldots, e_n$ be a basis for R, let $f_1, \ldots, f_m$ be a basis for S, and let $\|a_{ik}\|$ be the transformation matrix of A with respect to the given bases. Consider

$$x = \xi_1 e_1 + \cdots + \xi_n e_n,$$
$$Ax = \eta_1 f_1 + \cdots + \eta_m f_m.$$

Let us find how the coordinates of $A\mathbf{x}$ can be expressed through those of $\mathbf{x}$. Observe that

$$A\mathbf{x} = A(\xi_1 \mathbf{e}_1 + \cdots + \xi_n \mathbf{e}_n)$$
$$= \xi_1(a_{11}\mathbf{f}_1 + a_{21}\mathbf{f}_2 + \cdots + a_{m1}\mathbf{f}_m) + \xi_2(a_{12}\mathbf{f}_1 + a_{22}\mathbf{e}_2 + \cdots + a_{m2}\mathbf{f}_m)$$
$$+ \cdots + \xi_n(a_{1n}\mathbf{f}_1 + a_{2n}\mathbf{f}_2 + \cdots + a_{mn}\mathbf{f}_m)$$
$$= (a_{11}\xi_1 + a_{12}\xi_2 + \cdots + a_{1n}\xi_n)\mathbf{f}_1 + (a_{21}\xi_1 + a_{22}\xi_2 + \cdots + a_{2n}\xi_n)\mathbf{f}_2$$
$$+ \cdots + (a_{m1}\xi_1 + a_{m2}\xi_2 + \cdots + a_{mn}\xi_n)\mathbf{e}_m.$$

Then

$$\eta_i = \sum_{k=1}^{n} a_{ik}\xi_k, \quad i = 1, \ldots, n.$$

The last equality may be simply re-written in matrix from. Let ξ be the column vector of coordinates of the vector x and let η be the column vector of coordinates of the vector $A\mathbf{x}$ in the given bases, that is,

$$\xi = \begin{pmatrix} \xi_1 \\ \vdots \\ \xi_n \end{pmatrix} \text{ and } \eta = \begin{pmatrix} \eta_1 \\ \vdots \\ \eta_m \end{pmatrix}$$

Then

$$\eta = \|a_{ik}\| \xi.$$

Therefore, we have

Theorem 8.3. *Given a linear transformation A from a linear space R to another linear space S and some bases in R and S, let ξ be a column vector of coordinates of a vector $\mathbf{x} \in R$, let η be the column vector of coordinates of the vector $A\mathbf{x} \in S$, and let M denotes the matrix of A. Then $\eta = M\xi$.*

Example 8.5. Let A be a linear transformation from $\mathbb{R}^3$ to $\mathbb{R}^2$ which transforms the canonical basis vectors

$$\mathbf{e}_1 = (1, 0, 0),$$
$$\mathbf{e}_2 = (0, 1, 0),$$
$$\mathbf{e}_3 = (0, 0, 1)$$

to the vectors

$$A\mathbf{e}_1 = (1, 2),$$
$$A\mathbf{e}_2 = (-1, -2),$$
$$A\mathbf{e}_3 = (11, 22).$$

Then the matrix of A (in the canonical bases) is

$$M = \begin{pmatrix} 1 & -1 & 11 \\ 2 & -2 & 22 \end{pmatrix}$$

Let us calculate the value of the transformation of the vector, say, $x = (1, 2, 3)$. Note that, in the notation above, the vectors $\xi = x^T$ and $\eta = (Ax)^T$ are simply the vectors as $\mathbf{x}$ and $A\mathbf{x}$ in the column form, so that

$$\eta = M\xi = \begin{pmatrix} 1 & -1 & 11 \\ 2 & -2 & 22 \end{pmatrix} \begin{pmatrix} 1 \\ 2 \\ 3 \end{pmatrix} = \begin{pmatrix} 32 \\ 64 \end{pmatrix}$$

and $A\mathbf{x} = (32, 64)$.

Example 8.6 (Prices and Demand). Consider the following system of demand equations for n goods. Notice that the demand for each good is not only a function of its own price, but also the prices of other goods,

$$q_i = \sum_{i=1}^{n} \alpha_{ij} p_j, \quad j = 1, \ldots, n. \tag{8.2}$$

This can be written is a more compact form as

$$\mathbf{q} = A\mathbf{p}, \tag{8.3}$$

where $\mathbf{q}$ is an $n \times 1$ vector of quantities of goods demanded, $\mathbf{p}$ is the vector of prices of these goods and A is the matrix of coefficients given in (8.2).

Let P, Q denote the price and quantity spaces, respectively. Assuming that both variables are perfectly divisible, we can assume that the both are the subsets of the n-dimensional Euclidean space $\mathbb{R}^n$ (in order to make sense from an economic point of view, prices and quantities should be non-negative)

Question. Show that (8.3) is a linear transformation (isomorphism) from P space to Q space

Answer. Equation (8.3) can be written as

$$\mathbf{q} = f(\mathbf{p}) = A\mathbf{p}, \tag{8.4}$$

$\mathbf{p} \in P$ and $\mathbf{q} \in Q$, P and Q are Euclidean spaces. In order to show that the mapping

$$f : P \to Q$$

is a linear transformation it is sufficient to show that

$$f(\alpha \mathbf{p}^1 + \beta \mathbf{p}^2) = \alpha f(\mathbf{p}^1) + \beta f(\mathbf{p}^2).$$

From (8.4), by using matrix addition and multiplication of a matrix by a scalar, one can easily obtain that

$$A(\alpha \mathbf{p}^1 + \beta \mathbf{p}^2) = \alpha A(\mathbf{p}^1) + \beta A(\mathbf{p}^2).$$

Therefore, (8.4) is a linear transformation.

8.1 Addition and Multiplication of Linear Operators

Definition 8.5. Let R, S, T be 3 linear spaces and let $A : R \to S$, $B : S \to T$ and $C : R \to T$ be 3 linear transformations. The transformation C is the product of the transformations A and B if for all $\mathbf{x} \in R$

$$C\mathbf{x} = A(B\mathbf{x}).$$

The product transformation C in the above definition is also linear, since

$$\begin{aligned} C(\lambda_1 \mathbf{x}_1 + \lambda_2 \mathbf{x}_2) &= A[B(\lambda_1 \mathbf{x}_1 + \lambda_2 \mathbf{x}_2)] \\ &= A(B\lambda_1 \mathbf{x}_1 + B\lambda_2 \mathbf{x}_2) = \lambda_1 A B \mathbf{x}_1 + \lambda_2 A B \mathbf{x}_2 \\ &= \lambda_1 C \mathbf{x}_1 + \lambda_2 C \mathbf{x}_2 \end{aligned}$$

for all $\mathbf{x}_1, \mathbf{x}_2 \in R$ and $\lambda_1, \lambda_2 \in \mathbb{R}$.

For example, one can multiply any linear operators acting in the same linear space R. If I is the identity operator and A is an arbitrary operator in a linear space R, then

$$AI = IA = A.$$

Of a special interest, the powers of an operator A are defined as

$$A^2 = AA, \ A^3 = AA^2, \ \ldots, \ A^n = AA^{n-1}, \ \ldots$$

with $A^0 = I$.

Example 8.7. Let us denote the operator of rotation of the plane $\mathbb{R}^2$ by an angle α by A_α. It follows that the product of two such operators A_α and A_β is the rotation by the composite angle $\alpha + \beta$

$$A_\alpha A_\beta = A_{\alpha+\beta}.$$

Moreover, for every positive integer n the n-fold rotation by α is the rotation by $n\alpha$, that is,

$$A_\alpha^n = A_{n\alpha}.$$

8.1 Addition and Multiplication of Linear Operators

Let A and B be two linear transformations as above while $\widehat{A} = \|a_{ik}\|$ and $\widehat{B} = \|b_{ik}\|$ be the corresponding transformation matrices with respect to the basis $e_1, \ldots, e_n$ of R. Then, let us find the transformation matrix $\widehat{C}$ of the operator $C = AB$. By definition

$$C e_k = \sum_{i=1}^{n} c_{ik} e_i$$

and

$$AB e_k = A(\sum_{j=1}^{n} b_{jk} e_j) = \sum_{j=1}^{n} b_{jk} A e_j = \sum_{i=1}^{n} \sum_{j=1}^{n} b_{jk} a_{ij} e_i.$$

Comparing the coefficients we obtain that $C = AB$ if and only if

$$c_{ik} = \sum_{j=1}^{n} a_{ij} b_{jk}$$

for all $i, k = 1, \ldots, n$, that is, the matrix of a product of transformations is a product of the matrices of the multipliers, $\widehat{C} = \widehat{A}\widehat{B}$.

Example 8.8. Let $A : \mathbb{R}^2 \to \mathbb{R}^2$ and $B : \mathbb{R}^3 \to \mathbb{R}^2$ be two transformations with matrices (denoted by the same letters)

$$A = \begin{pmatrix} 0 & 1 \\ 1 & 1 \end{pmatrix} \text{ and } B = \begin{pmatrix} 0 & 1 & 10 \\ 0 & 2 & 20 \end{pmatrix}.$$

For example, for $\mathbf{x} = (1, 2, 3)$ we have (in the column vector form)

$$B\mathbf{x} = \begin{pmatrix} 0 & 1 & 10 \\ 0 & 2 & 20 \end{pmatrix} \begin{pmatrix} 1 \\ 2 \\ 3 \end{pmatrix} = \begin{pmatrix} 32 \\ 64 \end{pmatrix} \text{ and } A(B\mathbf{x}) = \begin{pmatrix} 0 & 1 \\ 1 & 1 \end{pmatrix} \begin{pmatrix} 32 \\ 64 \end{pmatrix} = \begin{pmatrix} 64 \\ 96 \end{pmatrix}.$$

By the other hand, the matrix of the product transformation $C = AB$ is

$$C = AB = \begin{pmatrix} 0 & 1 \\ 1 & 1 \end{pmatrix} \begin{pmatrix} 0 & 1 & 10 \\ 0 & 2 & 20 \end{pmatrix} = \begin{pmatrix} 0 & 2 & 20 \\ 0 & 3 & 30 \end{pmatrix},$$

so that

$$A(B\mathbf{x}) = C\mathbf{x} = \begin{pmatrix} 0 & 2 & 20 \\ 0 & 3 & 30 \end{pmatrix} \begin{pmatrix} 1 \\ 2 \\ 3 \end{pmatrix} = \begin{pmatrix} 64 \\ 96 \end{pmatrix},$$

as before.

The addition of any two linear transformations A and B from R to S is also a linear transformation acting in the same vector spaces, denoted by C, such that for all $\mathbf{x} \in R$, $C = (A + B)\mathbf{x} \equiv A\mathbf{x} + B\mathbf{x}$.

Let A and B be two linear transformations from R to S with $A = \|a_{ik}\|$ and $B = \|b_{ik}\|$ being the associated transformation matrices with respect to some bases $\mathbf{e}_1, \ldots, \mathbf{e}_n$ of R and $\mathbf{f}_1, \ldots, \mathbf{f}_m$ of S. Then, let us find the transformation matrix C of the operator $C = A + B$. We know that

$$A\mathbf{e}_k = \sum_{i=1}^{n} a_{ik}\mathbf{e}_i$$
$$B\mathbf{e}_k = \sum_{i=1}^{n} b_{ik}\mathbf{e}_i$$
$$C\mathbf{e}_k = \sum_{i=1}^{n} c_{ik}\mathbf{e}_i.$$

On the other hand,

$$(A + B)\mathbf{e}_k = \sum_{i=1}^{n}(a_{ik} + b_{ik})\mathbf{e}_i.$$

Hence $C = A + B$ if and only if

$$c_{ik} = a_{ik} + b_{ik}$$

for all $i, k = 1, \ldots, n$, that is, the matrix of a sum of transformations is again a sum of their matrices.

Example 8.9. If A and B are two linear operators of $\mathbb{R}^2$ with matrices $A = \begin{pmatrix} 0 & 1 \\ 2 & 3 \end{pmatrix}$ and $B = \begin{pmatrix} 10 & 20 \\ 30 & 40 \end{pmatrix}$ with respect to the canonical basis, then the sum $C = A + B$ is the operator with matrix $C = \begin{pmatrix} 10 & 21 \\ 32 & 43 \end{pmatrix}$.

Remark 8.1. Since there is a one-to-one correspondence between the transformations and transformation matrices, the properties of matrix multiplication and addition (such as commutativity of addition, associativity of addition and multiplication, distributive law, etc.) are pertained by the linear operators as well.

8.2 Inverse Transformation, Image and Kernel under a Transformation

The next definition is a particular case of Definition 8.2.

Definition 8.6. An operator B in a linear space R is said to be the *inverse* of an operator A in R if

$$AB = BA = I,$$

where I is the identity operator.

8.2 Inverse Transformation, Image and Kernel under a Transformation

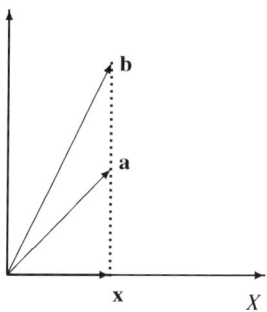

Fig. 8.3 x is a projection of both vectors **a** and **b**

The inverse of the operator A is denoted by A^{-1}. By definition we have

$$A^{-1}(A\mathbf{x}) = \mathbf{x},$$

for all $\mathbf{x} \in R$. That is, A transforms $\mathbf{x}$ to $A\mathbf{x}$, and A^{-1} transforms $A\mathbf{x}$ back to $\mathbf{x}$.

Example 8.10. Let $A = A_\phi$ be an operator of rotation of the plane $\mathbb{R}^2$ by an angle ϕ (see Example 8.3 a). Then the inverse operator A^{-1} is $A_{-\phi}$, a rotation by the opposite angle $-\phi$.

Does every operator have the inverse? To answer this question, consider an operator in $\mathbb{R}^2$, which maps every vector **t** to its projection on the X axis. It is easy to see that this operator does not have the inverse! This is best illustrated in Fig. 8.3: the vector **x** is the projection of **a** on X, but it is also the projection of **b** on X.

Given an operator A, the associated transformation matrix of the inverse operator A^{-1} is simply defined by

$$A^{-1}A = AA^{-1} = I,$$

where I is the identity matrix.

Definition 8.7. Given two linear spaces R and S and a transformation A from R to S, the set of vectors $A\mathbf{x}$, where $\mathbf{x} \in R$, is called the *image* of R under A, and denoted by $\text{Im}_R(A)$.

The image of R under a transformation A is often referred as *image of the transformation A* and denoted simply by Im A.

Example 8.11. Let $R = \mathbb{R}^2$ and A be the projection of each vector $\mathbf{z} \in R$ to X axis. Then the image of $\mathbb{R}^2$ under A is $\mathbb{R}$.

Lemma 8.4. *Let A be a linear transformation from R to S. Then $\text{Im}_R(A)$ is a subspace of S.*

Proof. Let $\mathbf{y}_1, \mathbf{y}_2 \in \text{Im}_R(A)$. Then there exists $\mathbf{x}_1, \mathbf{x}_2 \in R$ such that

$$\mathbf{y}_1 = A\mathbf{x}_1, \text{ and } \mathbf{y}_2 = A\mathbf{x}_2.$$

Take any $\lambda_1, \lambda_2 \in \mathbb{R}$. We have

$$\lambda_1 \mathbf{y}_1 + \lambda_2 \mathbf{y}_2 = \lambda_1 A \mathbf{x}_1 + \lambda_2 A \mathbf{x}_2$$
$$= A(\lambda_1 \mathbf{x}_1 + \lambda_2 \mathbf{x}_2) \in \operatorname{Im}_R(A)$$

for $\lambda_1 \mathbf{x}_1 + \lambda_2 \mathbf{x}_2 \in R$. □

Lemma 8.5. *Let R be a linear space and A be a linear operator in R. If A has an inverse, then $\operatorname{Im}_R(A) = R$.*

Proof. Exercise. □

Definition 8.8. Given two spaces R and S and a transformation A from R to S, the set of vectors $\mathbf{x}$, satisfying $A\mathbf{x} = \mathbf{0}$, is called the *kernel* of A in R, and is denoted by $\operatorname{Ker}_R(A)$.

Lemma 8.6. *Let R be a linear space and A be a linear transformation in R. Then $\operatorname{Ker}_R(A)$ is a subspace of R.*

Proof. Exercise. □

Theorem 8.7. *Let R and S be two linear spaces and let A be a linear transformation from R to S. Then the sum of the dimensions of the kernel and image of R under A is equal to the dimension of R, that is*

$$\dim(\operatorname{Im}_R(A)) + \dim(\operatorname{Ker}_R(A)) = \dim R$$

Proof. Let $\dim R = n$ and $\dim(\operatorname{Ker}_R(A)) = k \leq n$. Select for $\operatorname{Ker}_R(A)$, a basis $\mathbf{e}_1, \ldots, \mathbf{e}_k$ and complete it to a basis for R by the vectors $\mathbf{e}_{k+1}, \ldots, \mathbf{e}_n$.

Consider the vectors $A\mathbf{e}_{k+1}, \ldots, A\mathbf{e}_n$. The set of all linear combinations of these vectors coincides with $\operatorname{Im}_R(A)$.

Now, take any $\mathbf{y} \in \operatorname{Im}_R(A)$. By definition, there exists $\mathbf{x} \in R$ such that $\mathbf{y} = A\mathbf{x}$. Since $\mathbf{e}_1, \ldots, \mathbf{e}_n$ is a basis for R, we have

$$\mathbf{x} = \gamma_1 \mathbf{e}_1 + \cdots + \gamma_n \mathbf{e}_n$$

for some γ_i's. Notice that $A\mathbf{e}_1 = A\mathbf{e}_2 = \cdots = A\mathbf{e}_k = \mathbf{0}$ since $\mathbf{e}_1, \ldots, \mathbf{e}_k$ is a basis for $\operatorname{Ker}_R(A)$. Hence, we have

$$\mathbf{y} = A\mathbf{x} = \gamma_{k+1} A\mathbf{e}_{k+1} + \cdots + \gamma_n A\mathbf{e}_n.$$

Now let us show that $n - k$ vectors given by $A\mathbf{e}_{k+1}, \ldots, A\mathbf{e}_n$ are linearly independent. Suppose on the contrary that there exist $\alpha_1, \ldots, \alpha_{n-k}$, not all being equal to zero such that

$$\alpha_1 A\mathbf{e}_{k+1} + \cdots + \alpha_{n-k} A\mathbf{e}_n = \mathbf{0}.$$

8.3 Linear Transformation Matrices with Respect to Different Bases

Consider now the vector $\mathbf{x} = \alpha_1 \mathbf{e}_{k+1} + \cdots + \alpha_{n-k} \mathbf{e}_n$. We have

$$A\mathbf{x} = A(\alpha_1 \mathbf{e}_{k+1} + \cdots + \alpha_{n-k} \mathbf{e}_n) = \alpha_1 A\mathbf{e}_{k+1} + \cdots + \alpha_{n-k} A\mathbf{e}_n = \mathbf{0}$$

implying $\mathbf{x} \in \text{Ker}_R(A)$. Thus we have obtained a contradiction, since $\mathbf{x}$ as an element of the kernel can be represented through $\mathbf{e}_1, \ldots, \mathbf{e}_k$ as well as the basis $\mathbf{e}_{k+1}, \ldots, \mathbf{e}_n$ for the image, which is impossible as the vector $\mathbf{x}$ must have a unique representation with respect to the basis $\mathbf{e}_1, \ldots, \mathbf{e}_n$.

Hence, $n - k$ vectors $A\mathbf{e}_{k+1} + \cdots + A\mathbf{e}_n$ are linearly independent, and therefore any vector in $\text{Im}_R(A)$ can be represented as a linear combination of $n - k$ vectors, i.e. $\dim(\text{Im}_R(A)) = n - k$. □

Example 8.12. Let A be an operator in $R = \mathbb{R}^3$ with the matrix (in the canonical basis)

$$A = \begin{bmatrix} 0 & 1 & 1 \\ 1 & 0 & 1 \\ 0 & 1 & 1 \end{bmatrix}$$

The vector space $\text{Im}_R(A)$ by the vectors $A\mathbf{e}_1$, $A\mathbf{e}_2$ and $A\mathbf{e}_3$, that is, by the columns of the above matrix A. The maximal number of linearly independent vectors among these three is 2 (because the third column is the sum of the first and second columns), hence $\dim \text{Im}_R(A) = 2$. The kernel $\text{Ker}_R(A)$ consists of all vectors $\mathbf{x}$ which satisfy the equation $A\mathbf{x} = \mathbf{0}$, that is,

$$\begin{bmatrix} 0 & 1 & 1 \\ 1 & 0 & 1 \\ 0 & 1 & 1 \end{bmatrix} \mathbf{x}^T = \begin{bmatrix} 0 \\ 0 \\ 0 \end{bmatrix}$$

The solutions of this equation are $x = \lambda(1, 1, -1)$, so that $\text{Ker}_R(A)$ is a one-dimensional subspace spanned by the vector $(1, 1, -1)$.

8.3 Linear Transformation Matrices with Respect to Different Bases

Let R be an n-dimensional linear space with two bases $\mathbf{e}_1, \ldots, \mathbf{e}_n$ and $\mathbf{f}_1, \ldots, \mathbf{f}_n$. Let A be a linear operator in R, and C denote a matrix of transformation from $\mathbf{e}_1, \ldots, \mathbf{e}_n$ to $\mathbf{f}_1, \ldots, \mathbf{f}_n$, i.e.,

$$\mathbf{f}_1 = c_{11}\mathbf{e}_1 + c_{21}\mathbf{e}_2 + \cdots + c_{n1}\mathbf{e}_n,$$
$$\mathbf{f}_2 = c_{12}\mathbf{e}_1 + c_{22}\mathbf{e}_2 + \cdots + c_{n2}\mathbf{e}_n,$$
$$\vdots$$
$$\mathbf{f}_n = c_{1n}\mathbf{e}_1 + c_{2n}\mathbf{e}_2 + \cdots + c_{nn}\mathbf{e}_n.$$

Define a linear operator C as

$$Ce_i = f_i, \ i = 1, \ldots, n.$$

Let $A = \|a_{ik}\|$ and $B = \|b_{ik}\|$ be the transformation matrices of A with respect to the bases $e_1, \ldots, e_n$ and $f_1, \ldots, f_n$, respectively. We have

$$Ae_k = \sum_{i=1}^{n} a_{ik} e_i,$$
$$Af_k = \sum_{i=1}^{n} b_{ik} f_i.$$

It then follows that

$$AC(e_k) = \sum_{i=1}^{n} b_{ik} C e_i.$$

Multiply now both sides of the last equality by the inverse operator C^{-1}, which exists since $f_1, \ldots, f_n$ are linearly independent, to get

$$C^{-1} A C e_k = \sum_{i=1}^{n} b_{ik} C^{-1} C e_i = \sum_{i=1}^{n} b_{ik} e_i.$$

Then B is a transformation matrix for the operator $C^{-1} AC$ with respect to the basis $e_1, \ldots, e_n$. Hence

$$B = C^{-1} AC. \tag{8.5}$$

Example 8.13. Problem. A linear operator has the following matrix with respect to some basis e_1, e_2

$$A = \begin{bmatrix} 2 & -1 \\ 3 & 8 \end{bmatrix}.$$

Find its transformation matrix with respect to the basis $f_1 = 2e_1 + e_2, f_2 = 3e_1 + 2e_2$.

Solution. From the above definition of f_1 and f_2, the matrix C of the transformation of coordinates has the form

$$C = \begin{bmatrix} 2 & 3 \\ 1 & 2 \end{bmatrix}.$$

Then the matrix of the given linear operator with respect to the basis $\{f_1, f_2\}$ is

$$B = C^{-1} AC = \begin{bmatrix} 2 & -3 \\ -1 & 2 \end{bmatrix} \begin{bmatrix} 2 & -1 \\ 3 & 8 \end{bmatrix} \begin{bmatrix} 2 & 3 \\ 1 & 2 \end{bmatrix} = \begin{bmatrix} -36 & -77 \\ 33 & 46 \end{bmatrix}.$$

Matrices A and B related by (8.5) are called *similar*. The matrix B is also said to be obtained from A by a *conjugation* by C.

Example 8.14. The matrices

$$A = \begin{bmatrix} 0 & 1 \\ 0 & 0 \end{bmatrix} \text{ and } B = \begin{bmatrix} 0 & 0 \\ 1 & 0 \end{bmatrix}$$

are similar under conjugation by

$$C = \begin{bmatrix} 0 & 1 \\ 1 & 0 \end{bmatrix}.$$

Remark 8.2. If we consider transformation of one linear space to another one, the above formula becomes a bit more complicated. Let R and S be two linear spaces, let E and E' be two bases in R, let F and F' be two bases in S, and let C and D be the matrices of basis transformations (as above) from E to E' and from F to F', respectively. Suppose that A and A' are the matrices of the same linear transformation from R to S with respect to the bases E and F and the bases E' and F', respectively. Then

$$A' = D^{-1}AC.$$

8.4 Problems

1. Let $T_\mathbf{u}$ be a translation. Prove that

$$T_\mathbf{u}(\mathbf{x} + \mathbf{y}) = T_\mathbf{u}\mathbf{x} + T_\mathbf{u}\mathbf{y}.$$

2. Let $\mathbf{a} = (2, 4, -3)$. Define $F : \mathbb{R}^3 \to \mathbb{R}$ such that for all $\mathbf{x} \in \mathbb{R}^3$, $F(\mathbf{x}) = 3(\mathbf{a}, \mathbf{x}) + 1$. What is the value of $F(\mathbf{x})$ if
   (a) $\mathbf{x} = (3, 2, 1)$;   (b) $\mathbf{x} = (1, 2, 3)$.
3. Determine which of the following mappings are linear:
   (a) $F : \mathbb{R}^3 \to \mathbb{R}^2$, $F(x, y, z) = (x, z)$.
   (b) $F : \mathbb{R}^4 \to \mathbb{R}^4$, $F(\mathbf{x}) = -\mathbf{x}$.
   (c) $F : \mathbb{R}^3 \to \mathbb{R}^3$, $F(\mathbf{x}) = \mathbf{x} + (0, -1, 0)$.
   (d) $F : \mathbb{R}^2 \to \mathbb{R}^2$, $F(x, y) = (2x + 4, y)$.
   (e) $F : \mathbb{R}^2 \to \mathbb{R}$, $F(x, y) = xy$.
4. Prove that the rotation of the XY plane by an angle α around the origin is a linear transformation.
5. Let R be a three dimensional space with a basis $\mathbf{e}_1, \mathbf{e}_2, \mathbf{e}_3$. Consider the operator which maps any $\mathbf{x} \in R$ to its projection on the space spanned by $\mathbf{e}_1$. Prove that this operator is linear. Find the transformation matrix with respect to the canonical basis of $R = \mathbb{R}^3$.
6. Let F be the orthogonal projection of the vectors $\mathbf{x} \in \mathbb{R}^3$ on the XY plane. Prove that F is a linear transformation. Find its transformation matrix with respect to the canonical basis.

7. Determine which of the following mappings $F : \mathbb{R}^3 \to \mathbb{R}^3$ are linear, and then find the associated transformation matrix of F with respect to the same basis through which $\mathbf{x}$ and $F(\mathbf{x})$ are represented.
   (a) $F(\mathbf{x}) = (x_2 + x_3, 2x_1 + x_3, 3x_1 - x_2 + x_3)$.
   (b) $F(\mathbf{x}) = (x_1, x_2 + 1, x_3 + 2)$.
   (c) $F(\mathbf{x}) = (2x_1 + x_2, x_1 + x_3, x_3^2)$.
   (d) $F(\mathbf{x}) = (x_1 - x_2 + x_3, x_3, x_2)$.

8. Consider the following operator which maps each a_i to b_i:

$$\mathbf{a}_1 = (2, 3, 5) \to \mathbf{b}_1 = (1, 1, 1),$$
$$\mathbf{a}_2 = (0, 1, 2) \to \mathbf{b}_2 = (1, 1, -1),$$
$$\mathbf{a}_3 = (1, 0, 0) \to \mathbf{b}_3 = (2, 1, 2).$$

Find the matrix of this operator with respect to the basis in which coordinates of the vectors are given.

9. Let $F : \mathbb{R}^3 \to \mathbb{R}^3$ be such that $F(\mathbf{x}) = (\mathbf{x}, \mathbf{a})\mathbf{a}$, where $\mathbf{a} = (1, 2, 3)$. Prove that F is a linear operator, and find its transformation matrix with respect to the canonical basis for $\mathbb{R}^3$, and also with respect to the basis:

$$\mathbf{b}_1 = (1, 0, 1),$$
$$\mathbf{b}_2 = (2, 0, -1),$$
$$\mathbf{b}_3 = (1, 1, 0).$$

10. Assume a linear operator has the following matrix with respect to some basis $\mathbf{e}_1, \mathbf{e}_2, \mathbf{e}_3, \mathbf{e}_4$

$$\begin{bmatrix} 1 & 2 & 0 & 1 \\ 3 & 0 & -1 & 2 \\ 2 & 5 & 3 & 1 \\ 1 & 2 & 1 & 3 \end{bmatrix}$$

Find the matrix of this operator with respect to the bases:
   (a) $\mathbf{e}_2, \mathbf{e}_3, \mathbf{e}_4, \mathbf{e}_1$
   (b) $\mathbf{e}_1, \mathbf{e}_1 + \mathbf{e}_2, \mathbf{e}_1 + \mathbf{e}_2 + \mathbf{e}_3, \mathbf{e}_1 + \mathbf{e}_2 + \mathbf{e}_3 + \mathbf{e}_4$

11. A linear operator has the following matrix with respect to some basis $\mathbf{e}_1, \mathbf{e}_2, \mathbf{e}_3$

$$\begin{bmatrix} 15 & -11 & 5 \\ 20 & -15 & 8 \\ 8 & -7 & 6 \end{bmatrix}$$

Find its transformation matrix with respect to the basis $\mathbf{f}_1 = 2\mathbf{e}_1 + 3\mathbf{e}_2 + \mathbf{e}_3$, $\mathbf{f}_2 = 3\mathbf{e}_1 + 4\mathbf{e}_2 + \mathbf{e}_3$, $\mathbf{f}_3 = \mathbf{e}_1 + \mathbf{e}_2 + \mathbf{e}_3$.

8.4 Problems

12. A linear operator has the following transformation matrix

$$\begin{bmatrix} 1 & -18 & 15 \\ -1 & -22 & 15 \\ 1 & -25 & 22 \end{bmatrix}$$

with respect to the basis

$$\mathbf{a}_1 = (8, -6, 7)$$
$$\mathbf{a}_2 = (-16, 7, -13)$$
$$\mathbf{a}_3 = (9, -3, 7)$$

Find the operator matrix with respect to the basis

$$\mathbf{b}_1 = (1, -2, 1)$$
$$\mathbf{b}_2 = (3, -1, 2)$$
$$\mathbf{b}_3 = (2, 1, 2)$$

13. Let $F_1 : \mathbb{R}^2 \to \mathbb{R}^2$ has the transformation matrix

$$\begin{bmatrix} 2 & -1 \\ 5 & -3 \end{bmatrix}$$

with respect to the basis $\mathbf{a}_1 = (-3, 7)$, $\mathbf{a}_2 = (1, -2)$ and let $F_2 : \mathbb{R}^2 \to \mathbb{R}^2$ have the transformation matrix

$$\begin{bmatrix} 1 & 0 \\ 2 & -1 \end{bmatrix}$$

with respect to the basis $\mathbf{b}_1 = (6, -7)$, $\mathbf{b}_2 = (-5, 6)$. Find the transformation matrix of the product $F_1 F_2$ with respect to the bases:
(a) $\mathbf{a}_1, \mathbf{a}_2$.
(b) $\mathbf{b}_1, \mathbf{b}_2$.
(c) The canonical basis of $\mathbb{R}^2$.

14. Prove that any linear operator $F : \mathbb{R} \to \mathbb{R}$ can be represented as $F(\mathbf{x}) = \alpha \mathbf{x}$, where α is a real number.

15. Let $T_{\mathbf{u}_1}$ and $T_{\mathbf{u}_2}$ be two translations. Prove that

$$T_{\mathbf{u}_1} T_{\mathbf{u}_2} = T_{\mathbf{u}_1 + \mathbf{u}_2}.$$

16. Let $A : R \to S$ and $B : S \to T$ be two linear transformations such that Ker $B =$ Im A. Prove that

$$\dim (\text{Ker } A) + \dim (\text{Im } B) = \dim R - \dim S + \dim T.$$

17. (a) Show that if the matrices A and B are conjugated, then the matrices B and A (the order is opposite to the given one) are conjugated as well.
    (b) Show that if the matrix A is conjugated to B and B is conjugated to some C, then A is conjugated to C.
18. Let A and B be two conjugated matrices of order n. Prove that:
    (a) $\det A = \det B$.
    (b) $\operatorname{rank} A = \operatorname{rank} B$.
    (c) $\operatorname{Tr} A = \operatorname{Tr} B$.

Eigenvectors and Eigenvalues

Let $\mathcal{L}$ be a linear space, $\mathcal{L}_1$ be a linear subspace of $\mathcal{L}$ and A be a linear operator in $\mathcal{L}$. In general, for any vector $\mathbf{x} \in \mathcal{L}_1$, $A\mathbf{x}$ may not belong to $\mathcal{L}_1$. The following example makes this claim clear. Let $\mathcal{L} = \mathbb{R}^2$ and $\mathcal{L}_1 = \{\mathbf{x} = (\xi_1, \xi_2) \in \mathbb{R}^2 \mid \xi_1 = \xi_2\}$, i.e., $\mathcal{L}_1$ contains all vectors on the bisectrice of the Euclidean plane. Let A be a linear operator in $\mathcal{L}$ such that for all $\mathbf{x} \in \mathcal{L}_1$, $A\mathbf{x} = (\xi_1, 2\xi_2)$. So, for all nonzero $\mathbf{x} \in \mathcal{L}_1$ we have $A\mathbf{x} \notin \mathcal{L}_1$ (Fig. 9.1).

Definition 9.1. Let A be a linear operator in a linear space $\mathcal{L}$. Then the linear subspace $\mathcal{L}_1$ of $\mathcal{L}$ is said to be *invariant* with respect to A if for all $\mathbf{x} \in \mathcal{L}_1$, $A\mathbf{x} \in \mathcal{L}_1$.

Example 9.1. (a) Trivial invariant subspaces of any linear space $\mathcal{L}$ are $\{\mathbf{0}\}$ and $\mathcal{L}$ itself.
(b) Let $\mathcal{L} = \mathbb{R}^2$. Given any $\mathbf{x} = \xi_1 \mathbf{e}_1 + \xi_2 \mathbf{e}_2 \in \mathbb{R}^2$, where $\mathbf{e}_1 = (1, 0)$ and $\mathbf{e}_2 = (0, 1)$, we define
$$A\mathbf{x} = \lambda_1 \xi_1 \mathbf{e}_1 + \lambda_2 \xi_2 \mathbf{e}_2,$$
where $\lambda_1, \lambda_2 \in \mathbb{R}$. Then, the axes of the Cartesian plane are invariant subspaces under this operator.
(c) Let $\mathcal{L}$ be the set of all polynomials of degree at most $n-1$. Let $A : P(t) \to P'(t)$ map each polynomial to its derivative. Then the set of all polynomials of degree at most k, where $0 \leq k \leq n-1$ is an invariant subspace of $\mathcal{L}$.
(d) Let $\mathcal{L}$ be any linear space and A be any linear operator in $\mathcal{L}$. Consider $\text{Ker}_\mathcal{L}(A)$ and $\text{Im}_\mathcal{L}(A)$. Both of these subspaces are invariant with respect to A.
Let $\mathcal{L}_1$ be a one-dimensional subspace of the linear space $\mathcal{L}$ such that $\mathcal{L}_1 = \{\alpha \mathbf{y} \mid \alpha \in \mathbb{R}\}$ for some nonzero $\mathbf{y} \in \mathcal{L}$. Now $\mathcal{L}_1$ is invariant with respect to A if and only if for any $\mathbf{x} \in \mathcal{L}_1$ we have $A\mathbf{x} \in \mathcal{L}_1$, i.e., $A\mathbf{x} = \lambda \mathbf{x}$ for some $\lambda \in \mathbb{R}$.

Definition 9.2. The non-zero vector $\mathbf{x} \in \mathcal{L}$ satisfying

$$A\mathbf{x} = \lambda \mathbf{x}$$

Fig. 9.1 $A\mathbf{x} \notin \mathcal{L}_1$ for all nonzero $\mathbf{x} \in \mathcal{L}_1$

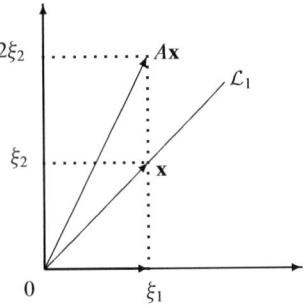

for some $\lambda \in \mathbb{R}$, is called the *eigenvector*, while the corresponding λ is called the *eigenvalue*, of the linear operator A.

Eigenvectors are also called *characteristic vectors*, and eigenvalues are also called *characteristic values* of A.

Example 9.2. Let $A = \begin{bmatrix} 1 & 2 \\ 0 & 3 \end{bmatrix}$ be a linear operator. Let us find its eigenvectors $\mathbf{x}$. The equation $A\mathbf{x} = \lambda\mathbf{x}$ gives

$$\begin{bmatrix} 1 & 2 \\ 0 & 3 \end{bmatrix} \begin{bmatrix} x_1 \\ x_2 \end{bmatrix} = \lambda \begin{bmatrix} x_1 \\ x_2 \end{bmatrix},$$

or

$$\begin{cases} (1-\lambda)x_1 + 2x_2 = 0, \\ (3-\lambda)x_2 = 0. \end{cases}$$

It follows that there are two eigenvalues: $\lambda = 3$ and $\lambda = 1$. The eigenvectors are $\mathbf{x} = c(1, 1)$ (for $\lambda = 3$) and $\mathbf{x} = c(1, 0)$ (for $\lambda = 1$), where c is an arbitrary nonzero real number.

Example 9.3. (a) Let $\mathcal{L}$ be the set of all differentiable functions, and $\mathcal{L}_1$ be the set of all infinitely many times differentiable functions. Then the function

$$f(t) = e^{\lambda t}$$

with $\lambda \in \mathbb{R}$, is an eigenvector of the linear operator $A = d/dt$ since $df(t)/dt = \lambda f(t)$.

(b) Let

$$\begin{bmatrix} a_1 & & 0 \\ & \ddots & \\ 0 & & a_n \end{bmatrix}$$

be the transformation matrix of a linear operator A. Then for all $i = 1, \ldots, n$ the unit vector

9 Eigenvectors and Eigenvalues

$$\mathbf{e}^i = \begin{bmatrix} 0 \\ \vdots \\ 1 \\ \vdots \\ 0 \end{bmatrix} \leftarrow i\text{th row}$$

is an eigenvector of A. Indeed, $A\mathbf{e}^i = a_i \mathbf{e}^i$. So, a_i is the eigenvalue associated with the eigenvector $\mathbf{e}^i$.

Example 9.4. Consider an industry involving three firms which share the market for a certain commodity. Let the market shares of firms in a given year is denoted by the vector $\mathbf{s}$, with its ith row being the market share of firm i.

Assume that the transition matrix of the economy is given by $T = |t_{ij}|$, where t_{ij} denotes the share of customers of firm i which go to firm j next year. Specifically, we are given

$$T = \begin{bmatrix} 0.85 & 0.10 & 0.10 \\ 0.05 & 0.55 & 0.05 \\ 0.10 & 0.35 & 0.85 \end{bmatrix}.$$

We are interested in finding whether there exists a vector $\mathbf{v}$ of current market shares which will, under the transition operator T, remain unchanged next year, i.e.,

$$T\mathbf{v} = \mathbf{v}.$$

Here, we can consider $\mathbf{v}$ as an eigenvector if 1 is an eigenvalue of T. One can check that

$$\mathbf{v} = \begin{bmatrix} 0.4 \\ 0.1 \\ 0.5 \end{bmatrix}$$

is a solution to this problem, i.e., it is an eigenvector of T.

Remark 9.1. Let A be a linear operator and let $\mathbf{x}$ be an eigenvector of A. Then $\alpha \mathbf{x}$ is also an eigenvector of A.

Theorem 9.1. *Let $\mathcal{L}$ be a linear space and A be a linear operator in $\mathcal{L}$. Assume that $\mathbf{q}_1, \ldots, \mathbf{q}_m$ are eigenvectors of A with eigenvalues $\lambda_1, \ldots, \lambda_m$, respectively. Assume that for all $i \neq j$, $\lambda_i \neq \lambda_j$. Then $\mathbf{q}_1, \ldots, \mathbf{q}_m$ are linearly independent.*

Proof (by induction). For $m = 1$, $\mathbf{q}_1 \neq \mathbf{0}$ is linearly independent (since an eigenvector is not a null vector, by definition). Assume $m > 1$, and that $\mathbf{q}_1, \ldots, \mathbf{q}_m$ are linearly dependent, i.e., there exist $c_1, \ldots, c_m \in \mathbb{R}$ not all being equal to zero such that

$$c_1 \mathbf{q}_1 + \cdots + c_m \mathbf{q}_m = \mathbf{0}. \tag{9.1}$$

Multiply (9.1) by λ_1 to get

$$c_1\lambda_1\mathbf{q}_1 + \cdots + c_m\lambda_1\mathbf{q}_m = \mathbf{0}. \tag{9.2}$$

Operating A on (9.1) yields

$$c_1\lambda_1\mathbf{q}_1 + \cdots + c_m\lambda_m\mathbf{q}_m = \mathbf{0}. \tag{9.3}$$

Subtracting (9.2) from (9.3) we obtain

$$c_2(\lambda_2 - \lambda_1)\mathbf{q}_2 + \cdots + c_m(\lambda_m - \lambda_1)\mathbf{q}_m = \mathbf{0},$$

implying now that the vectors $\mathbf{q}_2, \ldots, \mathbf{q}_m$ are linearly dependent. Repeating the above steps to remove the vectors $\mathbf{q}_2, \ldots, \mathbf{q}_{m-1}$, one by one, from the above linear equation, yields in the end

$$c_m\mathbf{q}_m = 0,$$

implying $c_m = 0$. Then moving in the backwards direction, we get also $c_2 = \cdots = c_m = 0$, and finally $c_1 = 0$, which is a contradiction, since we assumed that $c_1, \ldots, c_m$ are not all zero. Therefore, $\mathbf{q}_1, \ldots, \mathbf{q}_m$ are linearly independent.

Corollary 9.2. *Let $\mathcal{L}$ be an n-dimensional linear space, and A be a linear operator in $\mathcal{L}$ with eigenvectors $\mathbf{e}_1, \ldots, \mathbf{e}_n$ and with distinct eigenvalues $\lambda_1, \ldots, \lambda_n$. Then $\mathbf{e}_1, \ldots, \mathbf{e}_n$ is a basis for $\mathcal{L}$.*

How to find the eigenvalues and eigenvectors? Let us discuss conditions for an arbitrary vector to be an eigenvector.

Let A is a linear operator in an n-dimensional linear space $\mathcal{L}$, and let $\mathbf{e}_1, \ldots, \mathbf{e}_n$ be a basis for $\mathcal{L}$. Let the linear operator A in $\mathcal{L}$ corresponds the transformation matrix $\|a_{ik}\|$. Pick an arbitrary vector

$$\mathbf{x} = \xi_1\mathbf{e}_1 + \cdots + \xi_n\mathbf{e}_n$$

in $\mathcal{L}$. Then the coordinates $\eta_1, \ldots, \eta_n$ of the vector $A\mathbf{x}$ can be represented as

$$\begin{cases} \eta_1 = a_{11}\xi_1 + a_{12}\xi_2 + \cdots + a_{1n}\xi_n, \\ \quad \cdots\cdots\cdots \\ \eta_n = a_{n1}\xi_1 + a_{n2}\xi_2 + \cdots + a_{nn}\xi_n. \end{cases}$$

If $\mathbf{x}$ is an eigenvector then we have

$$\begin{cases} a_{11}\xi_1 + a_{12}\xi_2 + \cdots + a_{1n}\xi_n = \lambda\xi_1, \\ \quad \cdots\cdots\cdots \\ a_{n1}\xi_1 + a_{n2}\xi_2 + \cdots + a_{nn}\xi_n = \lambda\xi_n, \end{cases}$$

or rearranging

9 Eigenvectors and Eigenvalues

$$\begin{cases} (a_{11} - \lambda)\xi_1 + a_{12}\xi_2 + \cdots + a_{1n}\xi_n = 0, \\ a_{21}\xi_1 + (a_{22} - \lambda)\xi_2 + \cdots + a_{2n}\xi_n = 0, \\ \cdots\cdots\cdots \\ a_{n1}\xi_1 + a_{22}\xi_2 + \cdots + (a_{nn} - \lambda)\xi_n = 0. \end{cases} \quad (9.4)$$

We have a non-zero solution of (9.4) if and only if $\det A_\lambda = 0$, where

$$A_\lambda = \begin{bmatrix} a_{11} - \lambda & a_{12} & \cdots & a_{1n} \\ a_{21} & a_{22} - \lambda & \cdots & a_{2n} \\ \cdots & \cdots & \ddots & \cdots \\ a_{n1} & a_{n2} & \cdots & a_{nn} - \lambda \end{bmatrix} = A - \lambda I.$$

The determinant $\det A_\lambda$ is a polynomial in λ of degree n.

Suppose that λ_0 is a root of this polynomials. For $\lambda = \lambda_0$, the matrix A_{λ_0} of the system (9.4) of linear equations is degenerate, so, there exists a non-zero solution $\xi_1^0, \ldots, \xi_n^0$ of (9.4). Then

$$\mathbf{x} = \xi_1^0 \mathbf{e}_1 + \cdots + \xi_n^0 \mathbf{e}_n$$

is an eigenvector, while λ_0 is an eigenvalue, of A, since $A\mathbf{x}^0 = \lambda_0 \mathbf{x}^0$.

In contrast, if λ is not a root of the above polynomial, then the system (9.4) has non-degenerate matrix, so, the zero solution of the system is unique, therefore, λ is not an eigenvalue.

The polynomial

$$\chi_A(\lambda) = \det(A - \lambda I)$$

is called the *characteristic polynomial*.

Example 9.5. Let

$$A = \begin{bmatrix} a & b \\ c & d \end{bmatrix}$$

be an arbitrary 2×2 matrix. Then

$$\chi_A(\lambda) = \det(A - \lambda I) = \begin{vmatrix} a - \lambda & b \\ c & d - \lambda \end{vmatrix}$$

$$= \lambda^2 - (a + d)\lambda + ad - bc = \lambda^2 - (\text{Tr } A)\lambda + \det A.$$

Proposition 9.3. *The coefficients of the characteristic polynomial $\chi_A(\lambda)$ do not depend on the chosen basis for $\mathcal{L}$.*

Proof. Let A and A' be the matrices of the same linear operator in the bases E and E', respectively, and let C be a basis transformation matrix from E to E'. Then

$$\chi'_A(\lambda) = \det(A' - \lambda I) = \det(C^{-1}AC - \lambda I) = \det(C^{-1}AC - \lambda C^{-1}C)$$
$$= \det\left(C^{-1}(A - \lambda I)C\right) = \det C^{-1} \det(A - \lambda I) \det C$$
$$= \det(A - \lambda I) = \chi_A(\lambda).$$

$\square$

Hence, it is called the characteristic polynomial of the linear operator A, but not that of the transformation matrix A.

Thus we obtain

Theorem 9.4. *Let A be a linear operator in an n-dimensional linear space. Then each eigenvalue of A is a root of the characteristic polynomial $\chi_A(\lambda)$, and each root of this polynomial is an eigenvalue.*

To describe the eigenvectors $\mathbf{x}$ corresponding to a given root λ_0, one should solve the system (9.4) of linear equations, that is, the system

$$(A - \lambda_0 I)\mathbf{x} = \mathbf{0}.$$

Until now we have considered linear spaces over real numbers. In fact, our results remain valid if we consider linear spaces over complex numbers (we discuss complex numbers in Appendix C). We formulate this as our next result.

Corollary 9.5. *For every complex vector space $\mathcal{L}$, any linear operator A in $\mathcal{L}$ has at least one eigenvector.*

Proof. The polynomial $\chi_A(\lambda)$ is always of positive degree. Such a polynomial is known to have at least one (complex) root λ_0 (the fundamental theorem of algebra, see Theorem C.1). Then it is a (complex) eigenvalue, and the corresponding eigenvector $\mathbf{x}$ (probably, with complex coefficients) may be obtained as a solution of (9.4). $\square$

Now let A be a linear operator in $\mathcal{L}$ and $\mathbf{e}_1, \ldots, \mathbf{e}_n$ be its linearly independent eigenvectors, i.e.

$$A\mathbf{e}_i = \lambda_i \mathbf{e}_i, \quad i = 1, \ldots, n. \tag{9.5}$$

Let us choose $\mathbf{e}_1, \ldots, \mathbf{e}_n$ as a basis for $\mathcal{L}$. Then the above equalities imply that the transformation matrix of A with respect to this basis is

$$\begin{bmatrix} \lambda_1 & 0 & \ldots & 0 \\ 0 & \lambda_2 & \ldots & 0 \\ \vdots & \vdots & \ddots & \vdots \\ 0 & 0 & \ldots & \lambda_n \end{bmatrix},$$

which is called the *diagonal matrix*. Hence, we prove the following

Theorem 9.6. *If a linear operator $A \in \mathcal{L}$ has n linearly independent eigenvectors, then choosing them as a basis for $\mathcal{L}$, we can represent the transformation matrix of*

A in diagonal form. Conversely, if with respect to some basis in $\mathcal{L}$ the transformation matrix of a linear operator A is diagonal, then all vectors of this basis are the eigenvectors of A.

By the above theorem, for any complex linear space and any linear operator in that space, there exists an invariant one-dimensional linear subspace. But in the case of real linear spaces this assertion is not correct.

Example 9.6. Consider the rotation of $\mathbb{R}^2$ by some angle $\gamma \neq k\pi$ for any integer k (see Example 8.3a). Observe that with respect to this linear operator, no one-dimensional linear subspace of $\mathbb{R}^2$ is invariant.

Theorem 9.7. *For any linear operator A in an n-dimensional linear space $\mathcal{L}$ over the reals ($n \geq 2$), there exist either one- or two-dimensional invariant subspaces.*

Proof. Let $\mathbf{e}_1, \ldots, \mathbf{e}_n$ be a basis for $\mathcal{L}$ and $\|a_{ik}\|$ be the transformation matrix of the linear operator A with respect to this basis. Consider the equation

$$A\mathbf{x} = \lambda \mathbf{x} \tag{9.6}$$

in the unknowns $\mathbf{x}$ and λ.

This matrix equation has a non-zero solution if and only if $\det(A - \lambda I) = 0$. Let λ_0 be a root of the characteristic polynomial. We have two cases to consider:

Case 1: λ_0 is a real root. Let the corresponding eigenvector be $\mathbf{x}$, which generates a one-dimensional linear subspace of $\mathcal{L}$ (recall that any multiple of $\mathbf{x}$ is an eigenvector of A).

Case 2: $\lambda_0 = \alpha + i\beta$ is a complex root. Let $\mathbf{x} + i\mathbf{y}$ be a solution of the (9.6). Insert this solution and λ_0 into (9.6) and separate the real and imaginary parts to get

$$\begin{aligned} A\mathbf{x} &= \alpha\mathbf{x} - \beta\mathbf{y} \\ A\mathbf{y} &= \alpha\mathbf{y} + \beta\mathbf{x}. \end{aligned} \tag{9.7}$$

Notice that $\mathbf{x}$ and $\mathbf{y}$ generate a two-dimensional subspace of $\mathcal{L}$ which is invariant with respect to A. □

Example 9.7. Consider a operator A in $\mathbb{R}^2$ given by the transformation matrix

$$A = \begin{bmatrix} \lambda_0 & 0 \\ \mu & \lambda_0 \end{bmatrix}$$

where μ is an arbitrary nonzero real number.

The characteristic polynomial of A is $(\lambda_0 - \lambda)^2$, and it has the repeated root λ_0. Then to find the eigenvector(s) $\mathbf{x} = (\xi_1, \xi_2)$, we need to solve the system of equations

$$\begin{cases} 0\xi_1 + 0\xi_2 = 0 \\ \mu\xi_1 + 0\xi_2 = 0. \end{cases}$$

The above system adopts the unique non-zero solution $\xi_1 = 0$, $\xi_2 = 1$ (up to a constant multiple). Then the corresponding invariant subspace of $\mathbb{R}^2$ is of dimension one, which is less than the number of roots of the characteristic polynomial.

9.1 Macroeconomic Example: Growth and Consumption

9.1.1 The Model

Consider an n-sector economy that produces n goods denoted by the vector **x**. All goods are both used as inputs in production and consumed by workers. The commodity input requirements of production is given by $n \times n$ input/output coefficients matrix A and labor requirements by labor/output coefficients row vector **a**. (Notice that all the elements of A and **a** are non-negative). It is assumed that the consumption basket of workers is given by the column vector **c**. (This can be interpreted as a wage basket that reflects social minimum wage, determined by social and historical conditions of the society in question.)

This system of production and its use can be summarized as follows:

Output = input requirements of production+ consumption (of workers), or

$$\mathbf{x} = A\mathbf{x} + \mathbf{c}a\mathbf{x}.$$

This economy uses all its output for current production and consumption, therefore it is incapable of growing. Now let us introduce growth into this picture. For the sake of simplicity suppose that policy makers aim at growing all sectors at the same rate, i.e. on a 'balanced growth path'. Therefore, the extra amount of goods necessary for achieving this growth performance is given by

$$\mathbf{s} = g[A\mathbf{x} + \mathbf{c}a\mathbf{x}], \tag{9.8}$$

where g is the balanced growth rate (a scalar) of the economy. When (9.8) is taken into account, the production system becomes

$$\mathbf{x} = A\mathbf{x} + \mathbf{c}a\mathbf{x} + g[A\mathbf{x} + \mathbf{c}a\mathbf{x}], \tag{9.9}$$

where the first term in the right hand corresponds to the input requirements for current consumption and the second term stands for the input requirements of growth. Let us define

$$B = A + \mathbf{c}a$$

9.1 Macroeconomic Example: Growth and Consumption

As the 'augmented input matrix', i.e., that takes into account the consumption needs of the labor, and treats the latter as an input of production. Therefore, (9.9) can be rewritten as

$$\mathbf{x} = (1+g)B\mathbf{x}. \tag{9.10}$$

Question 1. Suppose that the policy makers want to know the maximum feasible rate of growth of such economy, given the technology $(A, \mathbf{a})$ and exogenously given consumption basket of the workers, $\mathbf{c}$.

Answer. Notice that (9.10) can be rewritten as

$$[\lambda I - B]\mathbf{x} = \mathbf{0}, \tag{9.11}$$

where $\lambda = (1+g) - 1$, I is the $n \times n$ identity matrix and $\mathbf{0}$ is the $n \times 1$ zero vector. From (9.11) it is clear that the required answer can be obtained by the eigenvalues and corresponding eigenvectors of the matrix B and picking the one that allows maximum rate of growth. Obviously in this context a production system is feasible if at least it can sustain itself, i.e., g is a non-negative scalar and corresponding eigenvector is also non-negative.

9.1.2 Numerical Example

Suppose the technology of the economy is given by the following input/output matrix A, and labor/output vector $\mathbf{a}$

$$A = \begin{bmatrix} 0.120 & 0.170 & 0.120 \\ 0.140 & 0.110 & 0.140 \\ 0.110 & 0.130 & 0.110 \end{bmatrix},$$

$$\mathbf{a} = [0.36, 0.37, 0.40].$$

Let the consumption vector be

$$\mathbf{c} = \begin{bmatrix} 0.6 \\ 0.45 \\ 0.45 \end{bmatrix}.$$

Using this information, the augmented input coefficients matrix can be calculated as

$$B = A + \mathbf{ca} = \begin{bmatrix} 0.336 & 0.392 & 0.360 \\ 0.253 & 0.277 & 0.320 \\ 0.272 & 0.297 & 0.290 \end{bmatrix}.$$

The eigenvalues of the matrix B are

$$\lambda_1 = 0{,}9458, \lambda_2 = -0{,}0003, \lambda_3 = -0{,}0425.$$

Only the first eigenvalue is positive. Therefore, the others have no economic meaning. The eigenvector that corresponds to λ_1 is $(0.6574, 0.5455, 0.5197)$. The technically maximum feasible growth rate for this economy is

$$g = (1-\lambda)/\lambda = 5{,}7$$

Question 1. Can this economy afford a 10% increase in the consumption good vector of workers?

(*Hint*: Change the consumption vector by increasing its each component by 10%. Then calculate the eigenvalues of the augmented matrix and check if the corresponding maximum feasible growth rate is non-negative or not.)

9.2 Self-Adjoint Operators

A linear operator A in a real Euclidean space $\mathcal{E}$ is called *self-adjoint* if for all $\mathbf{x}, \mathbf{y} \in \mathcal{E}$

$$(A\mathbf{x}, \mathbf{y}) = (\mathbf{x}, A\mathbf{y}). \tag{9.12}$$

Example 9.8. Let us show that the operator $A = \begin{bmatrix} -1 & 4 \\ 4 & 5 \end{bmatrix}$ is self-adjoint. Indeed, for $\mathbf{x} = (x_1, x_2)$ and $\mathbf{y} = (y_1, y_2)$ we have $A\mathbf{x} = (-x_1 + 4x_2, 4x_1 + 5x_2)$ and $A\mathbf{y} = (-y_1 + 4y_2, 4y_1 + 5y_2)$, so that

$$(A\mathbf{x}, \mathbf{y}) = -x_1 y_1 + 4x_2 y_1 + 4x_1 y_2 + 5x_2 y_2 = (\mathbf{x}, A\mathbf{y}).$$

Theorem 9.8. *Any linear operator A in a real Euclidean space $\mathcal{E}$ is self-adjoint if and only if for any orthogonal basis of $\mathcal{E}$, the transformation matrix of A is symmetric.*

Proof. Let $\mathbf{e}_1, \ldots, \mathbf{e}_n$ be an orthogonal basis for $\mathcal{E}$. Pick any two vectors

$$\mathbf{x} = \xi_1 \mathbf{e}_1 + \cdots + \xi_n \mathbf{e}_n$$
$$\mathbf{y} = \eta_1 \mathbf{e}_1 + \cdots + \eta_n \mathbf{e}_n$$

in $\mathcal{E}$. Let $\zeta_1, \ldots, \zeta_n$ be coordinates of the vector $\mathbf{z} = A\mathbf{x}$, i.e.

$$\zeta_i = \sum_{k=1}^{n} a_{ik} \xi_k,$$

9.2 Self-Adjoint Operators

where $\|a_{ik}\|$ is the transformation matrix of A with respect to $\mathbf{e}_1, \ldots, \mathbf{e}_n$. Then

$$(A\mathbf{x}, \mathbf{y}) = (\mathbf{z}, \mathbf{y}) = \sum_{i=1}^{n} \zeta_i \eta_i = \sum_{k=1}^{n}\sum_{i=1}^{n} a_{ik} \xi_k \eta_i.$$

On the other hand, we have

$$(\mathbf{x}, A\mathbf{y}) = \sum_{k=1}^{n}\sum_{i=1}^{n} \xi_i a_{ik} \eta_k.$$

We see that $(A\mathbf{x}, \mathbf{y}) = (\mathbf{x}, A\mathbf{y})$ provided $a_{ik} = a_{ki}$ for all $i, k = 1, \ldots, n$, that is, the operator with symmetric matrix is self-adjoint.

Now, suppose that the operator is self-adjoint, that is, $(A\mathbf{x}, \mathbf{y}) = (\mathbf{x}, A\mathbf{y})$ for all $\mathbf{x}, \mathbf{y} \in \mathcal{E}$. Let us put $\mathbf{x} = \mathbf{e}_i$ and $\mathbf{y} = \mathbf{e}_k$ be two arbitrary elements of the basis. Then $(\mathbf{e}_i, A\mathbf{e}_k) = (A\mathbf{e}_i, \mathbf{e}_k)$, that is, $a_{ik} = a_{ki}$ for arbitrary $i, k = 1, \ldots, n$. Thus the matrix $\|a_{ij}\|$ is symmetric. □

Now we will show that for any self-adjoint operator A in a real Euclidean space $\mathcal{E}$, there exists an orthogonal basis for $\mathcal{E}$, with respect to which the transformation matrix of A is diagonal. But let us first prove some useful lemmas.

Lemma 9.9. *Any self-adjoint operator in a real Euclidean space has a one-dimensional invariant subspace.*

Proof. By Theorem 9.7 any linear operator has a one dimensional invariant subspace if the root λ of the characteristic polynomial is real and two-dimensional invariant subspace if λ is complex. So, consider the second case. Suppose that λ is complex, i.e. $\lambda = \alpha + i\beta$, for some $\alpha, \beta \in \mathbb{R}$. Let $\mathbf{x} + i\mathbf{y}$ be an eigenvector associated with λ. By the proof of Theorem 9.7 it follows that

$$A\mathbf{x} = \alpha\mathbf{x} - \beta\mathbf{y}$$
$$A\mathbf{y} = \beta\mathbf{x} + \alpha\mathbf{y}.$$

Hence

$$(A\mathbf{x}, \mathbf{y}) = \alpha(\mathbf{x}, \mathbf{y}) - \beta(\mathbf{y}, \mathbf{y})$$
$$(\mathbf{x}, A\mathbf{y}) = \beta(\mathbf{x}, \mathbf{x}) + \alpha(\mathbf{x}, \mathbf{y}).$$

Subtracting the first equation from the second yields

$$\beta\left[(\mathbf{x}, \mathbf{x}) + (\mathbf{y}, \mathbf{y})\right] = 0$$

since $(A\mathbf{x}, \mathbf{y}) = (\mathbf{x}, A\mathbf{y})$. Using $(\mathbf{x}, \mathbf{x}) + (\mathbf{y}, \mathbf{y}) \neq 0$, we conclude that $\beta = 0$, i.e. λ is real. Then the eigenvector associated with λ is also real and creates a one-dimensional invariant subspace. □

Lemma 9.10. *Let A be a self-adjoint operator in an n-dimensional real Euclidean space $\mathcal{E}$ and $\mathbf{e}$ be an eigenvector of A. Then the set $\mathcal{E}'$ of all vectors which are orthogonal to $\mathbf{e}$ is an $(n-1)$-dimensional invariant subspace of $\mathcal{E}$.*

Proof. The fact that $\mathcal{E}'$ is a subspace of $\mathcal{E}$ is obvious. To show that $\mathcal{E}'$ is invariant with respect to A, pick any $\mathbf{x} \in \mathcal{E}'$ such that $(\mathbf{x}, \mathbf{e}) = 0$. Then $(A\mathbf{x}, \mathbf{e}) = (\mathbf{x}, A\mathbf{e}) = (\mathbf{x}, \lambda \mathbf{e}) = \lambda(\mathbf{x}, \mathbf{e}) = 0$, implying $A\mathbf{x} \in \mathcal{E}'$. □

Theorem 9.11. *Let $\mathcal{E}$ be an n-dimensional real Euclidean space and A be a self-adjoint operator in $\mathcal{E}$. Then there exists orthonormal basis for $\mathcal{E}$ with respect to which the transformation matrix of A is diagonal.*

Proof. By Lemma 9.9, A has at least one eigenvector $\mathbf{e}_1$. Let $\mathcal{E}'$ be a space containing vectors which are orthogonal to $\mathbf{e}_1$. By Lemma 9.10, $\mathcal{E}'$ is an invariant subspace $\mathcal{E}$. Then, again by Lemma 9.9, there exists an eigenvector $\mathbf{e}_2 \in \mathcal{E}'$. Repeating this procedure eventually yields the set of n eigenvectors $\mathbf{e}_1, \ldots, \mathbf{e}_n$ which are pairwise orthogonal. Choose this set as a basis for $\mathcal{E}$. Note that

$$A\mathbf{e}_i = \lambda_i \mathbf{e}_i,$$

for some $\lambda_i \in \mathbb{R}$ and for all $i = 1, \ldots, n$. So, the transformation matrix of A with respect to the above basis is equal to

$$\begin{bmatrix} \lambda_1 & & 0 \\ & \ddots & \\ 0 & & \lambda_n \end{bmatrix}.$$

□

Theorem 9.11 can be reformulated as so-called Spectral Theorem of Algebra.

Spectral Theorem. *Let $\mathcal{L}$ be a n-dimensional Euclidean space with $n > 1$. Let A be a linear operator symmetric with respect to an inner-product defined in $\mathcal{L}$. Then there exists an orthogonal basis for $\mathcal{L}$ consisting of the eigenvectors of A.*

Example 9.9. Consider the self-adjoint operator $A = \begin{bmatrix} -1 & 4 \\ 4 & 5 \end{bmatrix}$ from Example 9.8. Its characteristic polynomial has the form

$$\chi_A(\lambda) = \begin{vmatrix} -1-\lambda & 4 \\ 4 & 5-\lambda \end{vmatrix} = \lambda^2 - 4\lambda - 21.$$

It has two roots (eigenvalues) $\lambda_1 = -3$ and $\lambda_2 = 7$. Pick any two eigenvectors corresponding to these eigenvalues (that is, solutions of the corresponding versions of the system (9.4)), say, $\mathbf{v}_1 = (2, -1)$ and $\mathbf{v}_2 = (1, 2)$. We have $(\mathbf{v}_1, \mathbf{v}_2) = 0$, so the vectors v_1 and v_2 form an orthogonal basis of $\mathbb{R}^2$. The matrix of the operator A with respect to this basis has the form

$$\begin{bmatrix} -3 & 0 \\ 0 & 7 \end{bmatrix}$$

9.3 Orthogonal Operators

Definition 9.3. A linear operator A in a real n-dimensional Euclidean space $\mathcal{E}$ is called *orthogonal operator* if for all $\mathbf{x}, \mathbf{y} \in \mathcal{E}$

$$(A\mathbf{x}, A\mathbf{y}) = (\mathbf{x}, \mathbf{y}). \qquad (9.13)$$

Example 9.10. The matrix $A = \begin{bmatrix} 1/2 & \sqrt{3}/2 \\ -\sqrt{3}/2 & 1/2 \end{bmatrix}$ is orthogonal.

Note that orthogonal operators preserves the inner-product. To see a special implication of this observation, insert $\mathbf{x} = \mathbf{y}$ into (9.13) to obtain

$$|A\mathbf{x}|^2 = |\mathbf{x}|^2.$$

Hence orthogonal operator preserves the length of a vector.

Now pick any $\mathbf{e}_1, \ldots, \mathbf{e}_n$ orthogonal basis for $\mathcal{E}$. Then we have

$$(A\mathbf{e}_i, A\mathbf{e}_k) = \begin{cases} 1, & \text{if } i = k, \\ 0, & \text{otherwise}. \end{cases} \qquad (9.14)$$

Note that, on the other side, any operator A which satisfies (9.14) is orthogonal.

Let $\|a_{ik}\|$ be the transformation matrix of A with respect to this basis. The ith column of this matrix contains the coordinates of $A\mathbf{e}_i$, hence (9.14) can be rewritten as

$$\sum_{j=1}^{n} a_{ji} a_{jk} = \begin{cases} 1, & \text{if } i = k, \\ 0, & \text{otherwise}. \end{cases}$$

But $\sum_{j=1}^{n} a_{ji} a_{jk}$ is the ik-th element of the product $A^T A$, hence it is also true that

$$A^T A = I.$$

Thus, we have

$$\det(A) = \pm 1,$$

i.e., the determinant of the orthogonal transformation matrix is either $+1$ or -1. An orthogonal operator A is called *non-singular* if $\det(A) = +1$ and *singular* if $\det(A) = -1$.

(Do not confuse with singular matrix A for which $\det A = 0$!)

Lemma 9.12. *Let $\mathcal{L}$ be a linear space, A be an orthogonal linear operator in $\mathcal{L}$, and $\mathcal{L}_1$ be a subspace of $\mathcal{L}$ invariant with respect to A. Then orthogonal complement $\mathcal{L}_2$ of $\mathcal{L}_1$ (the set of all vectors $\mathbf{y} \in \mathcal{L}$ orthogonal to each $\mathbf{x} \in \mathcal{L}_1$) is also an invariant subspace.*

Proof. Pick any $\mathbf{y} \in \mathcal{L}_2$. Then for all $\mathbf{x} \in \mathcal{L}_1$, $(\mathbf{x}, \mathbf{y}) = 0$. Since A is orthogonal, we have $\det(A) \neq 0$, and hence the inverse of A exists. So, $\text{Im}_{\mathcal{L}_1}(A) = \mathcal{L}_1$ by Lemma 9.3. Then any $\mathbf{x} \in \mathcal{L}_1$ is equal to

$$\mathbf{x} = A\mathbf{z}$$

for some $\mathbf{z} \in \mathcal{L}_1$. Thus

$$(\mathbf{x}, A\mathbf{y}) = (A\mathbf{z}, A\mathbf{y}) = (\mathbf{z}, \mathbf{y}) = 0,$$

implying that $A\mathbf{y} \in \mathcal{L}_2$. Therefore, $\mathcal{L}_2$ is an invariant subspace of $\mathcal{L}$. $\square$

We can now study orthogonal operators in one and two dimensional linear spaces. Let $\mathbf{e}$ be a vector generating a one-dimensional linear space, and A be an orthogonal operator in this space. Then $A\mathbf{e} = \lambda \mathbf{e}$, for some $\lambda \in \mathbb{R}$. By the fact

$$(A\mathbf{e}, A\mathbf{e}) = (\mathbf{e}, \mathbf{e}),$$

it follows that

$$\lambda^2 (\mathbf{e}, \mathbf{e}) = (\mathbf{e}, \mathbf{e}),$$

implying $\lambda = \pm 1$.

In other words, A is an orthogonal operator in a one-dimensional linear space if and only if for all $\mathbf{x}$ in this space we have either $A\mathbf{x} = \mathbf{x}$ or $A\mathbf{x} = -\mathbf{x}$.

Consider now a two-dimensional linear space $\mathcal{L}$. Let $\mathbf{e}_1, \mathbf{e}_2$ be an orthogonal basis for $\mathcal{L}$ and A be an orthogonal transformation in $\mathcal{L}$. Then the transformation matrix of A can be written as

$$A = \begin{bmatrix} \alpha & \beta \\ \gamma & \delta \end{bmatrix}.$$

Since $A^T A = I$, we must have

$$A^{-1} = \begin{bmatrix} \alpha & \beta \\ \gamma & \delta \end{bmatrix}^{-1} = \begin{bmatrix} \alpha & \gamma \\ \beta & \delta \end{bmatrix} = A^T.$$

First consider the case $\det(A) = -1$, i.e., $\alpha\delta - \gamma\beta = -1$. Then we have

$$\begin{bmatrix} \alpha & \beta \\ \gamma & \delta \end{bmatrix}^{-1} = \begin{bmatrix} -\delta & \beta \\ \gamma & -\alpha \end{bmatrix}.$$

Hence we get

$$A = \begin{bmatrix} \alpha & \beta \\ \beta & -\alpha \end{bmatrix}.$$

Note that for any eigenvector $\mathbf{e}$ of A, it must be true that $A\mathbf{e} = \lambda \mathbf{e}$. The orthogonality of A implies $\lambda = \pm 1$. One can check that the characteristic polynomial for the above matrix A is $\lambda^2 - 1$, which adopts the roots ± 1.

9.3 Orthogonal Operators

Using the assumption $-\alpha^2 - \beta^2 = -1$, we further get $\alpha, \beta \in [-1, 1]$, so one can introduce another parameter $\varphi \in [0, 2\pi]$ such that $\alpha = \cos\varphi$ and $\beta = \sin\varphi$. Thus, we can rewrite A as

$$A = \begin{bmatrix} \cos\varphi & \sin\varphi \\ \sin\varphi & -\cos\varphi \end{bmatrix}.$$

In other words, any non-singular orthogonal operator with respect to an orthogonal basis is a reflection with respect to the bisector of the angle (equal to φ) between the vectors e_1 and Ae_1.

Consider now the case of $\det(A) = \alpha\delta - \beta\gamma = +1$. Then we have

$$A^{-1} = \begin{bmatrix} \delta & -\beta \\ -\gamma & \alpha \end{bmatrix}.$$

From the equation $A^{-1} = A^T$ we get

$$A = \begin{bmatrix} \alpha & \beta \\ -\beta & \alpha \end{bmatrix}.$$

Using $\alpha^2 + \beta^2 = 1$ yields

$$A = \begin{bmatrix} \cos\varphi & -\sin\varphi \\ \sin\varphi & \cos\varphi \end{bmatrix}$$

for some $\varphi \in [0, 2\pi]$ such that $\alpha = \cos\varphi$ and $\beta = \sin\varphi$, that is, A is a rotation by the angle φ.

It is possible to show that any orthogonal operator in the n-dimensional case can be represented as the composition of these simple cases, that is, a composition of reflections in some hyperplanes and rotation over some axes which are orthogonal to the hyperplanes and to each other, see [33, Theorem 6.3.3]. In particular, there is a famous theorem of Gauss saying that any orthogonal operator in the 3-dimensional space $\mathbb{R}^3$ is either a rotation (see Fig. 9.2) over some axis or a composition of the rotation and a reflection in a plane orthogonal to the axis, see Fig. 9.3.

Theorem 9.13. *Let A be a symmetric matrix with different eigenvalues. Then the matrix Q of eigenvectors of length one, which we call the* modal *matrix of A, is orthogonal.*

Proof. We already proved that if A is a symmetric matrix and if its eigenvalues are distinct, then the corresponding eigenvectors are orthogonal (Theorem 9.11). Then, consider the matrix Q containing these eigenvectors. Since the columns of Q are pairwise orthogonal and have unit length, the conditions (9.14) hold. □

Example 9.11. Let us construct the modal matrix for the self-adjoint operator $A = \begin{bmatrix} -1 & 4 \\ 4 & 5 \end{bmatrix}$ from Examples 9.8 and 9.9. By Example 9.9, there are two orthogonal

Fig. 9.2 A rotation over the axis l

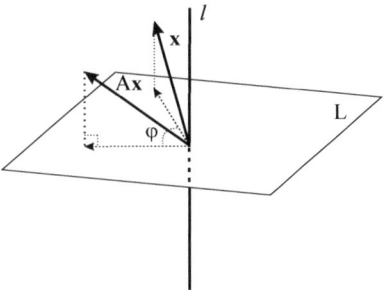

Fig. 9.3 A composition of a rotation over l and a reflection

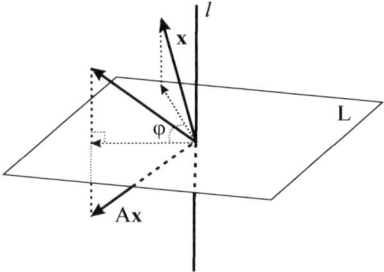

eigenvectors $\mathbf{v}_1 = (2, -1)$ and $\mathbf{v}_2 = (1, 2)$. To get eigenvectors of length 1, take $\mathbf{e}_1 = |\mathbf{v}_1|^{-1}\mathbf{v}_1 = (2/\sqrt{5}, -1/\sqrt{5})$ and $\mathbf{e}_2 = |\mathbf{v}_2|^{-1}\mathbf{v}_2 = (1/\sqrt{5}, 2/\sqrt{5})$. Then the modal matrix Q has the columns $\mathbf{e}_1^T$ and $\mathbf{e}_2^T$, that is,

$$Q = \begin{bmatrix} 2/\sqrt{5} & 1/\sqrt{5} \\ -1/\sqrt{5} & 2/\sqrt{5} \end{bmatrix}.$$

This matrix has the form

$$\begin{bmatrix} \cos\varphi & -\sin\varphi \\ \sin\varphi & \cos\varphi \end{bmatrix}$$

with $\phi = 2\pi - \arcsin(1/\sqrt{5})$, that is, Q is a rotation by the angle φ.

9.4 Quadratic Forms

Consider the matrix

$$A = \begin{bmatrix} 3 & 5 \\ 4 & 1 \end{bmatrix}$$

and any column vector $\mathbf{x} = [x_1\ x_2]$. The *quadratic form* of A is a function

$$q(\mathbf{x}) = \mathbf{x}^T A \mathbf{x} = 3x_1^2 + 5x_1x_2 + 4x_1x_2 + 1x_2^2$$
$$= 3x_1^2 + 9x_1x_2 + x_2^2.$$

9.4 Quadratic Forms

In general, for any matrix $A = \|a_{ij}\|_{n \times n}$ and any row vector $\mathbf{x} = [x_1, \ldots, x_n]$ of variables, the quadratic form is defined as

$$q(\mathbf{x}) = \sum_{i=1}^{n} \sum_{j=1}^{n} a_{ij} x_i x_j.$$

In the matrix form, this can be re-written as $q(\mathbf{x}) = \mathbf{x} A \mathbf{x}^T$.

Quadratic forms are heavily used in calculus to check the second order conditions in optimization problems. They have a particular use in econometrics, as well.

Before we consider some special issues on quadratic forms, let us introduce a more general notion, namely, bilinear forms. Given the variables $x_1, \ldots, x_n$ and $y_1, \ldots, y_n$, *bilinear form* is defined as the sum

$$\Omega = \sum_{i=1}^{n} \sum_{j=1}^{n} a_{ij} x_i y_j,$$

where a_{ij} are arbitrary numbers. Bilinear form is called symmetric if $a_{ij} = a_{ji}$ for all $i, j = 1, \ldots, n$. The dot product of two vectors is obviously a bilinear form.

Bilinear form is reduced to quadratic form if $x_i = y_i$ for all $i = 1, \ldots, n$. We denote this form as Φ. Then

$$\Phi = \sum_{i=1}^{n} \sum_{j=1}^{n} a_{ij} x_i x_j$$

or

$$\Phi = \sum_{i=1}^{n} x_i \phi_i,$$

where

$$\phi_i = \sum_{j=1}^{n} a_{ij} x_j$$

are linear combinations of variables. Here one can take

$$\phi_i = \frac{1}{2} \frac{\partial \Phi}{\partial x_i}.$$

Conversely, for every function of the form $\Phi(x_1, \ldots, x_n) = \sum_{i=1}^{n} \sum_{j=i}^{n} b_{ij} x_i x_j$ one can define a symmetric bilinear form $\Omega(\mathbf{x}, \mathbf{y})$ and a symmetric matrix B such that

$$\Phi(\mathbf{x}) = \Omega(\mathbf{x}, \mathbf{x}) = \mathbf{x} A \mathbf{x}^T.$$

To do this, one can take the coefficients of the above linear combinations[1] $\phi_i = \frac{1}{2}\frac{\partial \Phi}{\partial x_i}$ as the elements of the i-th row of the matrix A and then define $\Omega(\mathbf{x}, \mathbf{y}) = \mathbf{x}A\mathbf{y}$. Such symmetric matrix A is said to correspond to the quadratic form Φ.

Example 9.12. For a quadratic form $\Phi(x_1, x_2) = x_1^2 + 2x_1x_2 + 3x_2^2$ one has $\phi_1 = \frac{1}{2}\frac{\partial \Phi}{\partial x_1} = x_1 + x_2$ and $\phi_2 = \frac{1}{2}\frac{\partial \Phi}{\partial x_2} = x_1 + 3x_2$. Then

$$A = \begin{bmatrix} 1 & 1 \\ 1 & 3 \end{bmatrix},$$

so that $\Omega(x_1, x_2, y_1, y_2) = [x_1, x_2]A[y_1, y_2]^T = x_1y_1 + x_1y_2 + x_2y_1 + 3x_2y_2$.

While studying the properties of inner-products, we informally defined the positive definiteness of matrices using the quadratic forms. Let us formalize it.

Definition 9.4. Let A be an $n \times n$ matrix. A is said to be *positive definite* (*positive semi-definite*) if $q(\mathbf{x}) = \mathbf{x}^T A\mathbf{x} > 0$ ($q(\mathbf{x}) = \mathbf{x}^T A\mathbf{x} \geq 0$) for all nonzero column vectors $\mathbf{x}$ of size n. Analogously, A is said to be *negative definite* (*negative semi-definite*) if $q(\mathbf{x}) = \mathbf{x}^T A\mathbf{x} \geq 0$ ($q(\mathbf{x}) = \mathbf{x}^T A\mathbf{x} \leq 0$).

Theorem 9.14. *Let A be a symmetric matrix. Then A is positive definite (negative definite) if and only if all its eigenvalues are positive (negative). Moreover, A is positive semi-definite (negative semi-definite) if and only if all its eigenvalues are nonnegative (nonpositive).*

Proof. Let Q be the orthogonal matrix containing eigenvectors of A. Then by Theorems 9.11 and 9.13.

$$Q^T A Q = \Lambda,$$

where Λ is a diagonal matrix of eigenvalues. Consider an arbitrary vector $\mathbf{x}$. Then $\mathbf{x} = Q\mathbf{y}$ for some $\mathbf{y}$. It follows that

$$q(\mathbf{x}) = \mathbf{x}^T A \mathbf{x} = \mathbf{y}^T Q^T A Q \mathbf{y}$$
$$= \sum_{i=1}^{n} \lambda_i y_i^2.$$

Then, $\mathbf{x}^T A\mathbf{x}^T > 0$ ($\mathbf{x}^T A\mathbf{x}^T < 0$) if and only if $\lambda_i > 0$ ($\lambda_i < 0$) for all $i = 1, \ldots, n$. Analogously, $\mathbf{x}^T A\mathbf{x}^T \geq 0$ ($\mathbf{x}^T A\mathbf{x}^T \leq 0$) if and only if $\lambda_i \geq 0$ ($\lambda_i \leq 0$) for all $i = 1, \ldots, n$. □

---

[1]Let us note for a reader who is not familiar with partial derivatives that the coefficients of the linear combinations ϕ_i are calculated by the following rule: the coefficient of x_1 in ϕ_i is equal to the coefficient of x_i^2 in Φ while the coefficient of x_j for $j \neq i$ is equal to the coefficient of x_ix_j in Φ divided by two.

9.4 Quadratic Forms

It is sometimes hard to apply the above theorem to check if a quadratic form is positive (or negative) definite, because there is no simple general method to calculate the eigenvalues. The next criterion is more useful.

A k-th *principal minor* of a square matrix A is the determinant of its submatrix $A_{(k)}$ formed by the intersection of the first k rows and the first k columns of A, that is, the upper left $k \times k$ corner of A. It is denoted by $\Delta_k = \det A_{(k)}$.

Theorem 9.15 (Sylvester[2] criterion). *A symmetric matrix A is positive definite if and only if all its principal minors Δ_k are positive. It is negative definite if and only if all the numbers $(-1)^k \Delta_k$ are positive, that is, each principal minor Δ_k is positive for all even k and negative for all odd k.*

We begin with the following

Lemma 9.16. *Let A be an arbitrary square matrix and let L and U be a lower triangular matrix and an upper triangular matrix, all of the same order n. Then for all $k = 1, \ldots, n$ we have*

$$(LA)_{(k)} = L_{(k)} A_{(k)} \text{ and } (AU)_{(k)} = A_{(k)} U_{(k)}.$$

Proof. Left as an exercise. □

Proof of Theorem 9.15. Let us first consider the criterion for positive definite matrix. Let $q(x_1, \ldots, x_n)$ be the quadratic form corresponding to the matrix $A = \|a_{ij}\|_{n \times n}$. If $n = 1$, the criterion is obvious. By the induction argument, we assume that $n \geq 2$ and the criterion is true for all symmetric matrices of lower order.

Suppose that A is positive definite. For every $k \leq n - 1$, the quadratic form $q_k(x_1, \ldots, x_k) = q(x_1, \ldots, x_k, 0, \ldots, 0)$ is positive definite as well. By the induction assumption, it follows that the corresponding determinant Δ_k is positive. In order to show that $\Delta_n = \det A > 0$, recall from (4.3) that $A^* A = \Delta_n E$, where A^* is the transpose of the matrix of cofactors. Multiplying the both sides by A^*, we obtain $A^* A A^* = \Delta_n A^*$. Let A_n^* be the n-th row of A^*. Since A^* is symmetric, we have

$$A_n^* A (A_n^*)^T = \Delta_n A_{nn}.$$

It follows that $\Delta_n A_{nn} = \Delta_n \Delta_{n-1} = q(A_n^*) > 0$ (here A_n^* is a nonzero vector because its n-th component is $A_{nn} = \Delta_{n-1} > 0$). By the induction assumption, here $\Delta_{n-1} > 0$, hence $\Delta_n > 0$.

To prove the criterion in the opposite direction, suppose that $\Delta_1 > 0, \ldots, \Delta_n > 0$. The determinant Δ_k of matrix $A_{(k)}$ for each $k = 1..n$ is nonzero, so that the system of linear equations

---

[2] James Joseph Sylvester (1814–1897) was famous English and American mathematician of nineteenth century.

$$A_{(k)}\mathbf{X} = \begin{bmatrix} 0 \\ \vdots \\ 0 \\ 1 \end{bmatrix} \in \mathbb{R}^k$$

has a unique solution $\mathbf{X}_{(k)}$. Let B be the upper triangular matrix such that for each k its k-th column has $\mathbf{X}_{(k)}$ as the first k elements and zeroes on the other $n-k$ places. By Lemma 9.16, for the lower triangular matrix

$$C = AB = \begin{bmatrix} 1 & 0 & \ldots & 0 \\ * & 1 & \ldots & 0 \\ \ldots & \ldots & & \\ * & * & \ldots & 1 \end{bmatrix}$$

(where stars denote arbitrary numbers) we have $C_{(k)} = A_{(k)}B_{(k)}$ for each $k = 1, \ldots, n$. The matrix B^T is lower triangular, hence the product $Q = B^T C = B^T AB$ is lower triangular too,

$$Q = \begin{bmatrix} \alpha_1 & 0 & \ldots & 0 \\ * & \alpha_2 & \ldots & 0 \\ & & \ldots & \\ * & * & \ldots & \alpha_n \end{bmatrix}$$

for some $\alpha_1, \ldots, \alpha_n$. But $Q^T = (B^T AB)^T = B^T A(B^T)^T = Q$, so that Q is symmetric,

$$Q = \begin{bmatrix} \alpha_1 & 0 & \ldots & 0 \\ 0 & \alpha_2 & \ldots & 0 \\ & & \ldots & \\ 0 & 0 & \ldots & \alpha_n \end{bmatrix}$$

By Lemma 9.16, we have

$$\alpha_1 \ldots \alpha_k = \det Q_{(k)} = \det(B_{(k)}^T C_{(k)}) = \det(B_{(k)}^T A_{(k)} B_{(k)})$$
$$= \det B_{(k)}^T \det A_{(k)} \det B_{(k)} = \Delta_k (\det B_k)^2 > 0.$$

Then $\alpha_1 = \det Q_{(1)} > 0$ and $\alpha_k = \det Q_{(k)} / \det Q_{(k-1)} > 0$ for all $k > 1$. Note that the matrix B is non-singular because $1 = \det C = \det A \det B$, hence $\det B = 1/\Delta_n \neq 0$. Thus, for every nonzero (column) vector $\mathbf{v} \in \mathbb{R}^n$ we have

$$q(\mathbf{v}) = \mathbf{v}^T A \mathbf{v} = \mathbf{v}^T (B^T)^{-1} (B^T AB) B^{-1} \mathbf{v} = \mathbf{y}^T Q \mathbf{y} = \alpha_1 y_1^2 + \cdots + \alpha_n y_n^2 > 0,$$

where $\mathbf{y} = B^{-1}\mathbf{v}$. This means that the matrix A is positive definite.

To prove the Sylvester criterion for a negative definite matrix A just note that a matrix A is negative definite if and only if the matrix $B = -A$ is positive definite. On the other side, the upper left minors of B are connected with the minors of A as

$$\det B_{(k)} = (-1)^k \Delta_k.$$

By the (just proved) Sylvester criterion for positive definite matrices, we conclude that the condition that A is negative definite is equivalent to the conditions $\det B_{(k)} = (-1)^k \Delta_k > 0$ for all $k = 1, \ldots, n$, as stated.

Example 9.13. Find all values of α such that the quadratic form $q(x_1, x_2, x_3) = 2x_1 x_2 - x_1^2 - 2x_2^2 + 2\alpha x_1 x_3 - x_3^2$ is positive or negative definite.

First, we construct the matrix of the quadratic form by the same way as in Example 9.12. We obtain

$$A = \begin{bmatrix} -1 & 1 & \alpha \\ 1 & -2 & 0 \\ \alpha & 0 & -1 \end{bmatrix}.$$

Next, let us calculate the principal minors: $\Delta_1 = -1$, $\Delta_2 = \begin{vmatrix} -1 & 1 \\ 1 & -2 \end{vmatrix} = 1$ and $\Delta_3 = \det A = -2\alpha^2 - 1$. To check if A is positive definite we have to check the conditions $\Delta_1 > 0, \Delta_2 > 0, \Delta_3 > 0$ which are never hold simultaneously. Now, the matrix is negative definite if and only if three inequalities $\Delta_1 < 0, \Delta_2 > 0, \Delta_3 < 0$ hold. These equalities are equivalent to the condition $\alpha^2 < 1/2$. So, A is never positive definite and is negative definite for $\alpha^2 < 1/2$.

Theorem 9.17. *The sum of all eigenvalues (each one is taken so many times as its multiplicity) of any symmetric matrix A is equal to the trace of A.*

Proof. Using the fact that the modal matrix Q of A diagonalizes A, i.e., $Q^T A Q = \Lambda$, we get

$$\text{Tr } \Lambda = \text{Tr}(Q^T A Q) = \text{Tr}(A Q^T Q) = \text{Tr } A,$$

since for any two matrices A and B conformable for multiplication, we have $\text{Tr}(AB) = \text{Tr}(BA)$ (Exercise 2.13). □

9.5 Problems

1. Prove that the eigenvector of the matrix

$$\begin{bmatrix} 1 & a \\ 0 & 1 \end{bmatrix}$$

where $a \neq 0$, generates a one-dimensional space. Find a basis for this space.

2. Prove that the eigenvectors of the matrix
$$\begin{bmatrix} 2 & 2 & 0 \\ 0 & 2 & 0 \\ 0 & 0 & 2 \end{bmatrix}$$
generate a two-dimensional space and find a basis for this space.

3. Find the eigenvalues and eigenvectors of the matrices

(a) $\begin{bmatrix} 1 & 1 & 1 \\ 0 & 1 & 1 \\ 0 & 0 & 1 \end{bmatrix}$ and (b) $\begin{bmatrix} 1 & 1 & 0 \\ 0 & 1 & 1 \\ 0 & 0 & 1 \end{bmatrix}$.

4. Find all complex eigenvalues and eigenvectors of a linear operator in $\mathbb{C}^2$ given by the matrix
$$\begin{bmatrix} 0 & 2 \\ -2 & 0 \end{bmatrix}.$$

5. Show that for any matrix A, eigenvalues of A^T and A are the same.

6. Let $\mathcal{L}$ be an n-dimensional space, A be a linear operator in $\mathcal{L}$. Let the characteristic polynomial associated with the transformation matrix of A have n distinct roots. Show that $\mathcal{L}$ has a basis consisting of the eigenvectors of A.

7. Given
$$A = \begin{bmatrix} a_1 & b_1 \\ c_1 & d_1 \end{bmatrix} \quad \text{and} \quad B = \begin{bmatrix} a_2 & b_2 \\ c_2 & d_2 \end{bmatrix},$$
show that the eigenvalues of AB are the same as the eigenvalues of BA.

8. Find the eigenvalues of the matrices

(a) $\begin{bmatrix} 2 & -1 \\ -1 & 2 \end{bmatrix}$ and (b) $\begin{bmatrix} 1 & 1 \\ 1 & 0 \end{bmatrix}$.

9. Let
$$A = \begin{bmatrix} \lambda_1 & 0 & \cdots & 0 \\ 0 & \lambda_2 & \cdots & 0 \\ \vdots & \vdots & \ddots & \vdots \\ 0 & 0 & \cdots & \lambda_n \end{bmatrix}$$
with $\lambda_i \geq 0$ for all $i = 1, \ldots, n$. Show that there exists an $n \times n$ matrix B such that $B^2 = A$.

10. Diagonalize the matrices

(a) $\begin{bmatrix} -1 & 3 & -1 \\ -3 & 5 & -1 \\ -3 & 3 & 1 \end{bmatrix}$ and (b) $\begin{bmatrix} 1 & 1 & 1 & 1 \\ 1 & 1 & -1 & -1 \\ 1 & -1 & 1 & -1 \\ 1 & -1 & -1 & 1 \end{bmatrix}$.

9.5 Problems

11. Find the eigenvalues and eigenvector of the linear operator d/dt in the space of polynomials of degree at most n with real coefficients.
12. Let A be a linear operator in a linear space $\mathcal{L}$. Let $\mathcal{L}_1$ be a subspace of $\mathcal{L}$ consisting of all linear combinations of the eigenvectors of A. Prove that $\mathcal{L}_1$ is invariant with respect to A.
13. Show that the angle between any two vectors does not change under orthogonal operator.
14. (a) Construct the matrix of the bilinear form $\phi(\mathbf{x}, \mathbf{y}) = (\mathbf{x}, 2\mathbf{y}) - 4x_1 y_2$, where $\mathbf{x} = (x_1, x_2)$ and $\mathbf{y} = (y_1, y_2)$ are two vectors in $\mathbb{R}^2$, in the basis $\{\mathbf{u} = (0, -1), \mathbf{v} = (1, 2)\}$.
    (b) Find the correspondent quadratic form and construct its matrix in the same basis $\mathbf{u}, \mathbf{v}$.
15. Find the values of a, b and c such that the quadratic form $q(x_1, x_2, x_3) = ax_1^2 + 2x_1 x_2 + bx_2^2 + cx_3^2$ is positive or negative definite.
16. Let E and E' be two bases in $\mathbb{R}^n$ and let C be the basis transformation matrix from E to E'. Suppose that A and A' be the matrices of the same quadratic form in these two bases. Show that

$$A' = C^T A C.$$

Linear Model of Production in a Classical Setting 10

10.1 Introduction

In classic economics the interrelations in production are vital to understand the laws of production and distribution, and therefore to understand how an economic system works. Wassily Leontief[1], a Russian born American economist made the greatest contribution in this line of thought by developing the input/output analysis[2].

In order to explore this idea in a simple way consider an economy consisting three activities: coal mining, electricity generation and truck production. Outputs of these activities are used in the production of others and are also demanded for final use, i.e. consumption and/or investment. For example, extracting coal requires energy. Part of its energy requirement can be satisfied in house, by using coal to generate electricity. The rest is obtained from the power plant. The coal mine also employs several trucks to deliver its output. In this setting, this plant is using coal, energy and trucks to produce coal.

We can express this relation as follows

$$(Coal, Electricity, Truck) \rightarrow Coal \qquad (10.1)$$

Coal output is used by other plants. Part of it, as was indicated above, used by the coal mine to produce energy for its own use. Power plant's thermal units may be using coal for producing energy. However, truck production may not require coal as an input. Therefore the remaining output is used to satisfy final consumption by

---

[1] Wassily Leontief (1906–1999) was born and educated in St. Petersburg. He received his Ph. D. from Berlin University in 1929. He joined the famous Kiel Institute of World Economics in 1927 and worked there until 1930. He moved to the USA in 1931; worked at the Harvard University (1932–1975) and New York University (1975–1991). Leontief was awarded with Nobel Prize in Economics in 1973 for his contribution to input-output analysis.

[2] For a historical survey and review of the contributions of economists to input-output analysis see [17, 18].

households for heating. For these activities the above relation can be written as

$$(Coal, Electricity, Truck) \to Electricity \tag{10.2}$$

$$(Electricity, Truck) \to Truck \tag{10.3}$$

On the other hand, assuming supply of coal is equal to its demand we can then write

$$\begin{aligned}\text{Coal Output} = {} & \text{Coal used in coal mine} \\ & + \text{Coal used in energy plant} \\ & + \text{Coal used by households for heating}\end{aligned} \tag{10.4}$$

The remaining production activities can be defined in a similar way. Electric power is used to operate the power plant as well as in other production activities. Households also demand electricity for lighting and/or heating. Trucks can be used for all commercial activity. However, we may think that they may not be suitable as family cars, and therefore they are not demanded for final consumption. Therefore we can write

$$\begin{aligned}\text{Electricity Output} = {} & \text{Electricity used in coal mine} \\ & + \text{Electricity used in power plant} \\ & + \text{Electricity used in truck plant} \\ & + \text{Household's consumption of electricity}\end{aligned} \tag{10.5}$$

$$\begin{aligned}\text{Truck Output} = {} & \text{Trucks used in coal mine} \\ & + \text{Trucks used in power plant} \\ & + \text{Trucks used in truck plant}\end{aligned} \tag{10.6}$$

Let X_{ij} denote the amount of good i demanded by activity j and D_i denote households' demand. Then (10.4)–(10.6) can be written as

$$\begin{cases} X_1 = X_{11} + X_{12} + D_1, \\ X_2 = X_{21} + X_{22} + X_{23} + D_2, \\ X_3 = X_{31} + X_{32} + X_{33}, \end{cases} \tag{10.7}$$

where $i = 1, 2, 3$ denotes coal, electricity and truck production activities, respectively. Notice that in (10.7), based on the explanation given above, X_{13} and D_3 are taken as equal to zero.

The description of production activity through such input-output relations delineates following points:
(a) Production of a commodity requires the use of other commodities as inputs. But as was the case above, a production activity need not use all the commodities available in the system. (Coal was not an input in truck production)
(b) A commodity may also be demanded as a final product. In the example above coal is demanded for heating. Again this may not be true for all commodities. Trucks may be used only for commercial purposes, and not as family cars.

10.1 Introduction

Table 10.1 The use of output (in terms of physical quantities)

	Sector 1	Sector 2	...	Sector n	Final Demand	Total Output
Sector 1	X_{11}	X_{12}	...	X_{1n}	d_1	X_1
Sector 2	X_{21}	X_{22}	...	X_{2n}	d_2	X_2
	:	:	:	:	:	:
Sector n	X_{n1}	X_{n2}	...	X_{nn}	d_n	X_n

Now let us generalize the example given above to an n-sector economy where each sector produces *only one commodity* (X_i) by one production technique. In other words there is one-to-one correspondence between commodities produced and the techniques used in their production. Therefore in this simple framework, singling out a sector is equivalent to identifying a specific *production technique* and a particular *commodity*. In the remainder of this chapter these terms will be used interchangeably.

In such economy each sectors output is used either by other sectors as an input or by households for final demand, i.e. for consumption, investment or exports. In Table 10.1, each row represents the use of the output of the corresponding sector. The first n entities give the inter-industry use of the output, i.e. the amounts allocated to other sectors as inputs, (X_{ij}). The entries in the $n + 1$'th column represents the final use (final demand) of the output (such as consumption, investment and exports) of the corresponding sector.

If supply is equal to demand for all commodities, one can write the following set of equations from Table 10.1.

$$X_{i1} + X_{i2} + \cdots + X_{in} + d_i = X_i, \quad i = 1, 2, \ldots, n. \quad (10.8)$$

In each equation the sum of first n entities gives the *interindustry demand* for the corresponding commodity and d_i is the final demand.

Let us define the proportion of commodity i used in the production of commodity j by

$$a_{ij} = \frac{X_{ij}}{X_j}. \quad (10.9)$$

Then for a given set of output values $\{X_1, X_2, \ldots, X_n\}$, (10.8) can be written as[3]

$$\sum_{j=1}^{n} a_{ij} X_j + d_i = X_i, \quad i = 1, 2, \ldots, n. \quad (10.10)$$

---

[3] Notice that, in contrast to its row sums, no meaning can be attributed to column sums of Table 10.1, since columns consists of elements with different units of measurement (for example, coal is measured by tons, electricity by Kwh and trucks by numbers).

Let us assume that input per output ratios, i.e. a_{ij}'s, are constant for all commodities. Then (10.10) can be written in a more compact form using matrices and vectors, as

$$A\mathbf{x} + \mathbf{d} = \mathbf{x}, \qquad (10.11)$$

where $A = (a_{ij})$ is an $n \times n$ matrix of input coefficients, $\mathbf{x}$ is an $n \times 1$ vector of industrial output levels and $\mathbf{d}$ is an $n \times 1$ vector of final demand.

Let us denote the j-th column of the matrix A by $\mathbf{a}_j$,

$$\mathbf{a}_j = \begin{bmatrix} a_{1j} \\ a_{2j} \\ \vdots \\ a_{nj} \end{bmatrix} \qquad (10.12)$$

This vector represents the list of inputs used to produce one unit of commodity j and will give the commodity input requirements of producing one unit of output j. It is conveniently referred to as the *production technique* for j.

Example 10.1. Consider the following two sector economy. Let the input coefficient matrix A be given as

$$A = \begin{bmatrix} 0.3 & 0.7 \\ 0.4 & 0.1 \end{bmatrix}.$$

Find the total output of each sector when the final demand is given by

$$\mathbf{d} = \begin{bmatrix} 12 \\ 6 \end{bmatrix}.$$

By rearranging (10.11) we get

$$(I - A)\mathbf{x} = \mathbf{d}$$

Applying it to the problem above the following two linear equation system is obtained

$$\begin{cases} 0.7X_1 - 0.7X_2 = 12 \\ -0.4X_1 + 0.9X_2 = 6 \end{cases}$$

Using Cramer's rule

$$X_1 = \frac{\begin{vmatrix} 12 & -0.7 \\ 6 & 0.9 \end{vmatrix}}{\begin{vmatrix} 0.7 & -0.7 \\ -0.4 & 0.9 \end{vmatrix}} = \frac{6.6}{0.35} = 18.86$$

and

$$X_2 = \frac{\begin{vmatrix} 0.7 & 12 \\ -0.4 & 6 \end{vmatrix}}{\begin{vmatrix} 0.7 & -0.7 \\ -0.4 & 0.9 \end{vmatrix}} = \frac{10.4}{0.35} = 28.71$$

Exercise 10.1. Show that when the input coefficients matrix for a two sector economy is given by

$$A = \begin{bmatrix} 0.9 & 0.7 \\ 0.8 & 0.8 \end{bmatrix}$$

no positive output levels satisfy (10.11). Why? (See below)

10.2 The Leontief Model

The inter-industry transactions table simply is a snapshot of the accounting relations among sectors. In order to be able to derive results concerning the structure of production of such economy stronger assumptions are needed to connect output levels with the amount of inputs used. First the existence of relation between inputs and output has to be assumed. This can be done by ruling out the unrealistic case of production without any commodity input.

Assumption 1. Production of each commodity requires the use of at least one other commodity as input.

Although Assumption 1 is sufficient for establishing a relation between the commodity inputs and the output, the nature of this relation is not specified. However, in the analysis of the production structure one needs to specify the properties of such a relation. A historically important and widely used assumption is the following.

Assumption 2 (Linearity). The amount of an input required is proportional to the level of the output, i.e., for any input i ($i = 1, \ldots, n$)

$$\text{Input}_i = a_{ij} \times \text{Output}_j, \qquad i, j = 1, \ldots, n, \tag{10.13}$$

where a_{ij} is assumed to be constant.

Linearity assumption, despite its simple and innocuous nature, has far more reaching implications. First, as is stated above this assumption implies that the inputs required per unit of output remains invariant as the scale of the production changes. Therefore linearity assumption implies *constant returns* to scale in production. In other words, by making this assumption production technologies that exhibit decreasing (or *increasing*) *returns to scale* are excluded. Second, the fact that a_i is constant implies that substitution is not allowed among inputs. The production technology allows only one technique to be operated. Although such assumption may be assumed to hold in the (very) short run, it is too restrictive to represent the actual choices that a production unit faces.

Assumption 3 (No Joint Production). Each production technique produces only one output.

One other assumption implicit in the above framework is the absence of *joint production*. This means each production technique produces only one commodity. In reality this may not be the case. Consider a petroleum refinery. Refining crude oil is a technique that leads to multiple oil products, various types of fuel and liquid gas. In this case, the production technique can not be labelled by referring to a specific product. In this chapter we shall not deal with the problem of joint production[4].

In terms of the symbols used in Table 10.1, (10.13) can be rewritten as

$$a_{ij} = X_{ij}/X_i.$$

Obviously, $a_{ij} \geq 0$.

Under the assumptions (1)–(3), the production system characterized by (10.11) satisfies the following conditions:

(a) All elements of the A matrix are non-negative, i.e., $a_{ij} \geq 0$ and for all j there exists some i, such that $a_{ij} > 0$.
(b) Each commodity is produced only with one production technique, i.e. A is a square matrix.
(c) There is no joint production.
(d) There are constant returns to scale for all production techniques, i.e. A remains unchanged when the output vector x changes.

Note that the conditions (a) and (b) simply mean that A is a non-negative square matrix.

It is rather easy to show that such system has a unique solution. Due to (b) each commodity can be characterized by its production technique, i.e. the column vector of A that corresponds to the commodity in question. Since commodities are distinct, so are their corresponding production techniques. In addition, one can also assume in generic case that the columns of A are linearly independent. Therefore A has full rank, so $(I - A)^{-1}$ exists, and by definition is unique. Then the solution to (10.11) is given by

$$\mathbf{x} = (I - A)^{-1}\mathbf{d},$$

i.e., when the final demand vector is given, it is possible to calculate the output levels that enables the system to satisfy it.

It is clear that output level has non-negative value. A production unit can decide to produce the commodity in question (i.e., $x_i > 0$) or may quit production (i.e., $x_i = 0$). Negative output level has no economic meaning in this framework. Does (10.8) and the assumptions that characterize input coefficients matrix A guarantee such economically meaningful solution? The answer is no. In order to see why, consider the following

---

[4] See [16,22] for the discussion of joint production.

10.2 The Leontief Model

Example 10.2. Consider a two-sector economy that operates under the assumptions (1)–(3). Let the input coefficients matrix and the final demand vector be given as

$$A = \begin{bmatrix} 1.1 & 1.3 \\ 0.3 & 0.2 \end{bmatrix}, \qquad \mathbf{d} = \begin{bmatrix} 2 \\ 3 \end{bmatrix}$$

Notice that A satisfies the first condition (a) and the final demand is positive for both commodities.

However since

$$(I - A)^{-1} \approx \begin{bmatrix} -2.58 & -4.19 \\ -0.97 & -0.32 \end{bmatrix},$$

the output vector that satisfies the given final demand is obtained as

$$\mathbf{x} \approx \begin{bmatrix} -17.73 \\ -2.9 \end{bmatrix},$$

i.e. negative output levels for both commodities is obtained, which is clearly meaningless from economic point of view.

It is easy to show that this result is independent of the final demand vector. The structure of the A matrix is such that for any reasonable (i.e. non-negative) final demand vector, it is not possible to get non-negative total output levels for both commodities.

Using the relation $(I - A)\mathbf{x} = \mathbf{d}$ we can write

$$\begin{cases} 0.1x_1 - 1.3x_2 = d_1, \\ -0.3x_1 + 0.4x_2 = d_2. \end{cases}$$

By summing these two linear equations one can get

$$-0.2x_1 - 1.1x_2 = d_1 + d_2 \geq 0.$$

It is clear that, since right hand side is non-negative, x_1 and x_2 can not simultaneously be non-negative, which is inconsistent with their economic meaning.

This result demonstrates that assumptions made so far are not sufficient to give an economically meaningful characterization of the conditions of production. In order to find the missing characteristic, let us look at the example given above. In this example $a_{11} = 1.1$, i.e. it requires 1.1 units of commodity 1 to produce one unit of itself. It is clear that such a production technique will not be operated, since it is wasteful. In order to rule out such *wasteful production techniques* the following assumption is introduced.

Assumption 4 (Non-wasteful Production Technique). Each production technique is capable of producing more of its output than it consumes as an input. In terms of the coefficients of the A matrix, this assumption can be expressed as

$$a_{ii} < 1$$

for all i.

Although this assumption makes a logical point clear, it does not rule out the possibility that a sector may not be capable of supplying the amount of output required in producing other commodities and/or satisfying the final demand.

It is clear from (10.8) that the surplus output (i.e. the amount of output which is not consumed as input in its own production) should be sufficient to satisfy input requirements of the other sector and the final demand. In other words, the production system should be capable of produce at least as much as it uses as inputs. Non-wastefulness, although necessary, is not sufficient for guaranteeing such outcome. A stronger condition on the production structure is required. This is achieved by introducing a new assumption based on the following definition.

Definition 10.1. A non-negative matrix A is called *productive* if there exist $\mathbf{x} > \mathbf{0}$ such that
$$\mathbf{x} > A\mathbf{x}.$$

If the input coefficients matrix is productive, then the corresponding production system $A\mathbf{x} + \mathbf{d} = \mathbf{x}$ is also called *productive*.

Assumption 5. Input coefficients matrix A is productive.

The five assumptions made thus far are sufficient to characterize a production system that is capable of giving economically meaningful results.

Definition 10.2 (Leontief model). A production system
$$A\mathbf{x} + \mathbf{d} = \mathbf{x} \tag{10.14}$$
that satisfies the following assumptions:
(a) All the elements of the A matrix are non-negative, i.e. $a_{ij} \geq 0$ and for all j there exists some i such that $a_{ij} > 0$.
(b) Each commodity is produced only with one production technique, i.e. A is a square matrix.
(c) There is no joint production.
(d) There are constant returns to scale for all production techniques, i.e. A remains unchanged when the output vector $\mathbf{x}$ changes.
(e) A is productive.
is called the *Leontief model*.

10.3 Existence of a Unique Non-Negative Solution to the Leontief System

The purpose of this section is to demonstrate that for any $\mathbf{d} \geq \mathbf{0}$ it is possible to get a unique non-negative solution to the Leontief system.

In order to prove the existence and uniqueness of a non-negative solution to a Leontief model when the input coefficients matrix is productive, we shall prove some mathematical results that are needed to prove the main proposition. These results are given as lemmas below.

10.3 Existence of a Unique Non-Negative Solution to the Leontief System

Lemma 10.1. *Let $\mathbf{x}^1$ and $\mathbf{x}^2$ be two vectors such that $\mathbf{x}^1 \geq \mathbf{x}^2$ and let A be a non-negative matrix. Then $A\mathbf{x}^1 \geq A\mathbf{x}^2$.*

Proof. Let x_i^1 and x_i^2 be the i'th components of the vectors $\mathbf{x}^1$ and $\mathbf{x}^2$, respectively. Let a_{ij} denote the characteristic element of the matrix A. Then the i'th element of the vector $A\mathbf{x}^1$ and $A\mathbf{x}^2$ can be written as

$$\mathbf{a}_i \mathbf{x}^1 = \sum_{k=1}^{n} a_{ik} x_k^1 \quad \text{and} \quad \mathbf{a}_i \mathbf{x}^2 = \sum_{k=1}^{n} a_{ik} x_k^2 \quad \text{for} \quad i = 1, \ldots, n,$$

where $\mathbf{a}_i$ is the ith row of the matrix A considered as a row vector.

Since A is a non-negative matrix then $a_{ij} \geq 0$ for all i and k. Therefore,

$$\mathbf{a}_i \mathbf{x}^1 - \mathbf{a}_i \mathbf{x}^2 = \sum_{k=1}^{n} a_{ik}(x_k^1 - x_k^2) \quad \text{for all} \quad i,$$

or shortly,

$$A\mathbf{x}^1 \geq A\mathbf{x}^2.$$

$\square$

Lemma 10.2. *If A is a productive matrix then all elements of the matrix A^s converges to 0 as $s \to \infty$.*

Proof. It follows from Definition 10.1 that

$$\mathbf{x} > A\mathbf{x} \geq \mathbf{0}$$

since A is a non-negative matrix. Then there exist λ, $0 < \lambda < 1$, such that

$$\lambda \mathbf{x} > A\mathbf{x} \geq \mathbf{0}. \tag{10.15}$$

Premultiplying (10.15) by A and using Lemma 10.1 we obtain

$$\lambda(A\mathbf{x}) > A(A\mathbf{x}) = A^2 \mathbf{x} \geq \mathbf{0}. \tag{10.16}$$

On the other hand, premultiplying (10.15) by λ we get,

$$\lambda^2 \mathbf{x} > \lambda A\mathbf{x} \geq \mathbf{0}. \tag{10.17}$$

Inequalities (10.16) and (10.17) together yield,

$$\lambda^2 \mathbf{x} > A^2 \mathbf{x} \geq \mathbf{0}.$$

Analogously from

$$\lambda^{s-1} \mathbf{x} > A^{s-1} \mathbf{x} \geq \mathbf{0}$$

we obtain
$$\lambda^s \mathbf{x} > A^s \mathbf{x} \geq \mathbf{0}. \tag{10.18}$$

Hence for any productive matrix A, (10.18) holds. If $s \to \infty$, then $\lambda^s \to 0$, since $\lambda \in (0, 1)$. Therefore, from (10.18) we get
$$\lim_{s \to \infty} A^s \mathbf{x} = \mathbf{0},$$

or
$$\lim_{s \to \infty} \sum_{j=1}^{n} a_{ij}^s x_j = 0, \quad i = 1, \ldots, n, \tag{10.19}$$

with a_{ij}^s being the elements of A^s. Since $x_j > 0$, (10.19) holds only if
$$\lim_{s \to \infty} a_{ij}^s = 0, \quad i, j = 1, \ldots, n.$$

□

Lemma 10.3. *If A is productive matrix and if*
$$\mathbf{x} \geq A\mathbf{x} \tag{10.20}$$

for some $\mathbf{x}$ then
$$\mathbf{x} \geq \mathbf{0}.$$

Proof. Multiplying (10.20) sequentially $s - 1$ times by the matrix A and using Lemma 10.1 we get
$$\mathbf{x} \geq A\mathbf{x} \geq A^2 \mathbf{x} \ldots \geq A^s \mathbf{x}$$

or
$$\mathbf{x} \geq A^s \mathbf{x} \tag{10.21}$$

When $s \to \infty$, $A^s \mathbf{x} \to \mathbf{0}$, so from (10.21) we get $\mathbf{x} \geq \mathbf{0}$. □

Lemma 10.4. *If A is a productive matrix then $(I - A)$ is a non-singular matrix.*

Proof. Assume on the contrary that $(I - A)$ is a singular matrix, i.e. $\det(I - A) = 0$. This means that some columns of $(I - A)$ are linearly dependent. Thus there exists some $\mathbf{x} \neq \mathbf{0}$ such that
$$(I - A)\mathbf{x} = 0 \tag{10.22}$$

or equivalently,
$$\mathbf{x} = A\mathbf{x}.$$

From Lemma 10.3, we know that $\mathbf{x} \geq \mathbf{0}$. Now consider the vector $-\mathbf{x}$. It is clear that this vector also satisfies (10.22), i.e.,
$$(I - A)(-\mathbf{x}) = 0$$

10.3 Existence of a Unique Non-Negative Solution to the Leontief System

Again from Lemma 10.3, it must be true that $-\mathbf{x} \geq \mathbf{0}$. The inequalities $-\mathbf{x} \geq \mathbf{0}$ and $-\mathbf{x} \leq \mathbf{0}$ can jointly be satisfied only if $\mathbf{x} = \mathbf{0}$, which is in contradiction with $\mathbf{x} \neq \mathbf{0}$. □

The preceding four lemmas allows us to prove the following theorem showing that a Leontief model has a unique non-negative solution.

Theorem 10.5. *Given any non-negative* $\mathbf{d}$, *the system*

$$(I - A)\mathbf{x} = \mathbf{d} \tag{10.23}$$

has a unique non-negative solution if the matrix A is productive.

Proof. Let A be a productive matrix. Take any $\mathbf{d} \geq \mathbf{0}$ and consider the equation (10.23).

By Lemma 10.4, $(I - A)$ is not singular, since A is productive. So, (10.23) has a unique solution $\tilde{\mathbf{x}}$. It is also true that

$$(I - A)\tilde{\mathbf{x}} \geq \mathbf{0}.$$

since $\mathbf{d} \geq \mathbf{0}$. Then by Lemma 10.3 we get $\tilde{\mathbf{x}} \geq \mathbf{0}$.

Now, assume that the system (10.23) has a non-negative solution $\mathbf{x}$ for some $\mathbf{d} > \mathbf{0}$ and some A. Then $\mathbf{x} - A\mathbf{x} = (I - A)\mathbf{x} = \mathbf{d} > \mathbf{0}$, that is, A is productive. □

Remark 10.1. Notice that the productivity of A is not necessary to get a non-negative solution to (10.23). For example, for $\mathbf{d} = \mathbf{0}$ this system has a solution $\mathbf{x} = \mathbf{0}$ for any matrix A. However, if $\mathbf{d} > \mathbf{0}$, then a non-negative solution $\mathbf{x}$ exists if and only if the matrix A is productive (because we have $\mathbf{x} = A\mathbf{x} + \mathbf{d} > A\mathbf{x}$, that is, the productivity condition holds).

Another criterion (the Hawkins-Simon condition) can be derived from the following theorem.

Theorem 10.6. *Let A be a non-negative matrix. Then the following conditions are equivalent:*
 i. *A is productive;*
 ii. *the matrix $(I - A)^{-1}$ exists and is non-negative;*
 iii. *all successive principal minors of $B = I - A$ are positive.*

In the last condition, these principal minors are

$$B_{(1)} = b_{11} > 0,$$

$$B_{(2)} = \begin{vmatrix} b_{11} & b_{12} \\ b_{21} & b_{22} \end{vmatrix} > 0,$$

$$\cdots$$

$$B_{(n)} = \begin{vmatrix} b_{11} & \cdots & b_{1n} \\ \vdots & \ddots & \vdots \\ b_{n1} & \cdots & b_{nn} \end{vmatrix} > 0.$$

The condition *(iii)* is called *Hawkins–Simon condition*. The equivalence of this condition *(iii)* and the condition *(ii)* of productivity of A is called *Hawkins–Simon theorem* [12].

Proof. Let us first prove that the condition *(i)* implies *(ii)*. Suppose that A is productive. Then the matrix $(I - A)^{-1}$ exists by Lemma 10.4. Let us define

$$\Psi_s = I + A + A^2 + A^3 + \ldots + A^s,$$

then

$$A\Psi_s = A + A^2 + A^3 + \ldots + A^{s+1}.$$

Therefore

$$(I - A)\Psi_s = I - A^{s+1}.$$

Taking the limit of both sides as $s \to \infty$ one gets

$$\lim_{s \to \infty} [(I - A)\Psi_s] = I$$

since $A^{s+1} \to 0$ as $s \to \infty$, by Lemma 10.2. Therefore,

$$\Psi_s = I + A + A^2 + A^3 + \ldots + A^s \longrightarrow (I - A)^{-1},$$

and since $A \geq 0$, we have $\Psi_s > 0$ and

$$(I - A)^{-1} \geq 0.$$

Now let us show that *(ii)* implies *(i)*. Assume that $(I - A)^{-1} \geq 0$. Take any $\mathbf{d} > 0$ (e. g., $\mathbf{d} = (1, \ldots, 1)$). Then the system $(I - A)\mathbf{x} = \mathbf{d}$ has a unique solution $\mathbf{x} = (I - A)^{-1}\mathbf{d} \geq \mathbf{0}$. By the Remark after Theorem 10.5, it follows that A is productive.

For the equivalence of Hawkins–Simon condition *(iii)* and the productivity condition *(i)*, see, for example[5], [30, pp. 384–385]. Another proof based on economic intuition is given in [6]. □

10.4 Conditions for Getting a Positive (Economically Meaningful) Solution to the Leontief Model

For practical purposes if a commodity is not demanded and supplied it can be omitted. In other words, an economically meaningful solution to a Leontief model should be the one that assigns positive output levels for all commodities, even when

---

[5]In [21], O'Neill & Wood dropped the continuity assumption and proved theorem by assuming column sums of B are not greater than one. The latter assumption can be fulfilled in practice by appropriately choosing the measurement units for each commodity.

10.4 Conditions for Getting a Positive Solution to the Leontief Model

some commodities are not demanded for final use. This corresponds to the case when some commodities are solely used in production (intermediary goods). The question, then, is to find the conditions that the matrix A should satisfy in order to get positive output levels for all commodities. Obviously such conditions should also have an economic interpretation.

Before presenting a result that guarantees a positive solution to the Leontief model, let us focus on the $I - A$ matrix to delineate some of its properties. Notice that

$$B = I - A = \begin{bmatrix} 1 - a_{11} & -a_{12} & \cdots & -a_{1n} \\ -a_{21} & 1 - a_{22} & \cdots & -a_{2n} \\ \vdots & \vdots & \ddots & \vdots \\ -a_{n1} & -a_{n2} & \cdots & 1 - a_{nn} \end{bmatrix}$$

and let $b_{ij} = 1 - a_{ij}$ be its characteristic element. Then:

- (i)
$$b_{ii} > 0 \quad [a_{ii} < 1], \tag{10.24}$$

in economic terms, each sector produces more of its output than it consumes as input (non-wastefulness).

- (ii) Any commodity within the system can be used as an input, i.e.,

$$b_{ij} \leq 0 \text{ if } i \neq j.$$

- (iii) For any column j of B we have either

$$b_{jj} < 1 \quad [a_{jj} > 0]$$

or

$$b_{ij} < 0 \quad [a_{ij} > 0] \text{ for some } i,$$

that is no output can be produced without using at least one produced input.

- (iv) Row sums of B are non-negative with at least one positive row sum, that is, the system is capable of producing surplus to satisfy final demand.

Definition 10.3. An $n \times n$ matrix A is called *reducible* if it is possible to permute some of its rows and some of its columns (the permutation of rows and the one of columns must be the same) in such a way to obtain a matrix in the following form

$$A = \begin{bmatrix} A_{11} & A_{12} \\ \mathbf{0} & A_{22} \end{bmatrix},$$

where $\mathbf{0}$ is a zero submatrix.

If the rows and columns of a matrix can not be ordered in the above form, then it is called *irreducible*[6].

The following lemma gives combinatorial conditions for reducible and irreducible matrices.

Lemma 10.7. *Let $A = (a_{ij})$ be an $n \times n$ matrix. Then*
(a) *A is reducible if and only if there is a subset $S = \{i_1, \ldots, i_n\}$ of the set $1, 2, \ldots, n$ such that $a_{ij} = 0$ for all $i \in S$, $j \notin S$;*
(b) *A is irreducible if and only if for each $j \neq k$ there exists a finite sequence ("chain")*

$$j = i_0, i_1, \ldots, i_{m-1}, i_m = k$$

such that all entries a_{i_{j-1}, i_j} of A for $j = 1, \ldots, n$ are nonzero.[7]

Proof. Left as an exercise. □

The next lemma gives an algebraic condition for irreducible matrix.

Lemma 10.8. *A non-negative $n \times n$ matrix A is irreducible if and only if for each $j \neq k$ there exist some $m > 0$ such that the (j, k)-th entry of the matrix A^m is positive.*

Proof. Let $A^m = (a^m_{j,k})_{n \times n}$. We have $a^2_{j,k} = \sum_{i=1}^{n} a_{j,i} a_{i,k}$, and (by the induction on m)

$$a^m_{j,k} = \sum a_{j,i_1} a_{i_1, i_2} \cdots a_{i_{m-2}, i_{m-1}} a_{i_{m-1}, k},$$

where the sum is taken over all $i_1, \ldots, i_{m-1}$ from 1 to n. Obviously, the inequality $a^m_{j,k} > 0$ holds if and only if there is a positive summand in this sum, that is, there are some $i_1, \ldots, i_{m-1}$ such that all entries $a_{j,i_1}, a_{i_1,i_2}, \ldots, a_{i_{m-1},k}$ are positive. By Lemma 10.7b, this is equivalent to say that A is irreducible. □

Problem 10.1. Prove that for an irreducible $n \times n$ matrix A and for each $i \neq j$ one can choose the number in Lemma 10.8 to be not greater than n.

Hint. How large the number m can be chosen in Lemma 10.7b?

Theorem 10.9. *For a non-negative matrix A, the following three conditions are equivalent:*
i. *for each nonzero $\mathbf{d} \geq 0$ the system $\mathbf{x} = A\mathbf{x} + \mathbf{d}$ has a positive solution, $\mathbf{x} > \mathbf{0}$;*
ii. *the matrix $(I - A)^{-1}$ is positive;*
iii. *the matrix A is productive and irreducible.*

---

[6] Another term for reducible and irreducible matrices are *decomposable* and *indecomposable* matrices.

[7] In terms of graph theory, the part (b) means that a directed graph $G(A)$ with vertices $\{1, \ldots, n\}$ and edges $\{j \to k \mid a_{jk} \neq 0\}$ is strongly connected, that is, for each two vertices $j \neq k$ there is a path from j to k.

Proof. Let us first prove the equivalence of *(i)* and *(ii)*. If *(i)* holds, then for each $\mathbf{d} > \mathbf{0}$ there is $\mathbf{x}$ such that $\mathbf{x} - A\mathbf{x} = \mathbf{d} > \mathbf{0}$, hence A is productive. Then $(I - A)^{-1}$ exists by Theorem 10.6. For each $i = 1, \ldots, n$, let $\mathbf{d} = \mathbf{e}_i = (0, \ldots, 1, \ldots, 0)^T$ (1 in i-th place). Then the solution $\mathbf{x}_i$ of the system $\mathbf{x} - A\mathbf{x} = \mathbf{d}$ must be positive. Since $\mathbf{x}_i = (I - A)^{-1}\mathbf{e}_i$ is the i-th column of the matrix $(I - A)^{-1}$, it follows that all entries of each i-th column of this matrix are positive, that is, $(I - A)^{-1} > \mathbf{0}$.

Now, suppose that *(ii)* holds, that is, $(c_{ij}) = (I - A)^{-1} > \mathbf{0}$. Then for each nonzero $\mathbf{d} = (d_1, \ldots, d_n) \geq \mathbf{0}$ the above system has a unique solution $\mathbf{x} = (x_1, \ldots, x_n)$. Here $x_i = c_{i1}d_1 + \cdots + c_{in}d_n$, where at least one summand is positive. Thus, each $x_i > 0$, hence $\mathbf{x} > \mathbf{0}$.

Let us now prove the equivalence of the conditions *(ii)* and *(iii)*. Following the proof of Theorem 10.6, let

$$\Psi_s = I + A + A^2 + A^3 + \ldots + A^s;$$

then

$$\lim_{s \to \infty} \Psi_s = (I - A)^{-1}.$$

Since the sequence of matrices $\{\Psi_s\}$ is non-decreasing in each entry, we have also

$$\Psi_s \leq (I - A)^{-1} \text{ for all } s.$$

It follows that the diagonal elements c_{ii} of the matrix $C = (I - A)^{-1}$ are always nonzero. Moreover, each other element c_{ij} is nonzero if and only if the (ij)-th element a_{ij}^k of the matrix A^m is nonzero for some $m \geq 1$. It now follows from Lemma 10.8 that the matrix $(I - A)^{-1}$ has no zero entry if and only if A is irreducible. □

Note that a reducible productive matrix may also have a positive solution of the above system for *some* particular $\mathbf{d}$, but not for *all* $\mathbf{d}$ simultaneously.

10.5 Prices of Production in the Linear Production Model

In practice, data for physical quantities of outputs are rarely available. Inter-industry flows tables, therefore, are based on accounting information. In other words, inter-industry flows are expressed in their monetary values. Then each cell of the inter-industry transactions table (Table 10.2 below) gives the value of monetary input i used in the production of output j, i.e. $v_{ij} = p_i X_{ij}$. Therefore, in practice, the actual input coefficients are unobservable. The observed coefficients are

$$\tilde{a}_{ij} = \frac{p_i X_{ij}}{p_j X_j} = \left(\frac{p_i}{p_j}\right) a_{ij},$$

Table 10.2 The use of output (in terms of market value)

	Sector 1	...	Sector n	Final Demand	Total Output
Sector 1	$p_1 X_{11}$	...	$p_1 X_{1n}$	$p_1 d_1$	$p_1 X_1$
Sector 2	$p_2 X_{21}$	...	$p_2 X_{2n}$	$p_2 d_2$	$p_2 X_2$
...	...	...	...	...	...
Sector n	$p_n X_{n1}$	...	$p_n X_{nn}$	$p_n d_n$	$p_n X_n$
Input Cost	$c_1 = \sum_{i=1}^{n} p_i X_{i1}$	...	$c_n = \sum_{i=1}^{n} p_i X_{in}$		
Value Added	V_1	...	V_n		$V = \sum_{j=1}^{n} V_j$
Total Output	$p_1 X_1 = c_1 + V_1$	...	$p_n X_n = c_n + V_n$	$D = \sum_{i=1}^{n} p_i d_i$	

in other words, they are affected as relative prices change.[8]

In this framework row sums gives the money value of the outputs of each sector, whereas the column sums along sectors gives the total value of physical inputs used in producing the corresponding commodity. The value added row refers to payments to the primary inputs such as labor, capital and land, i.e. wages, profits and rent. By definition

$$\sum_{j=1}^{n}(c_j + V_j) + D = \sum_{i=1}^{n} p_i X_i + V.$$

Consider the j-th column of Table 10.2. It can be written as

$$\sum_{i=1}^{n} p_i X_{ij} + V_j = p_j X_j.$$

Dividing both sides by X_j, one gets

$$\sum_{i=1}^{n} p_i a_{ij} + v_j = p_j, \tag{10.25}$$

where

$$v_j = \frac{V_j}{X_j}$$

is the value added per unit of output. Repeating the same procedure for all commodities one gets the following set of linear equations

---

[8] For the sake of simplicity, in the remainder of the book, this distinction will be omitted and the input coefficients will be denoted by a_{ij}.

10.5 Prices of Production in the Linear Production Model

$$\begin{cases} p_1 a_{11} + p_2 a_{21} + \ldots + p_n a_{n1} + v_1 = p_1, \\ p_1 a_{12} + p_2 a_{22} + \ldots + p_n a_{n2} + v_2 = p_2, \\ \quad \ldots \\ p_1 a_{1n} + p_2 a_{2n} + \ldots + p_n a_{nn} + v_n = p_n, \end{cases}$$

which can be represented in matrix notation more compactly as

$$\mathbf{p}^T A + \mathbf{v}^T = \mathbf{p}^T. \tag{10.26}$$

In (10.26) $\mathbf{p}^T$ and $\mathbf{v}^T$ are row vectors of commodity prices and value added per unit coefficients. A, on the other hand, is the input coefficients matrix.

Referring to the discussion concerning Leontief model, if A is productive matrix then there is a unique non-negative solution to (10.26) given by

$$\mathbf{p} = (I - A)^{-1} \mathbf{v},$$

where $\mathbf{v}$ is the vector with components $(v_1, \ldots, v_n)$.

Let us assume that the value added in each sector is distributed between profits, Π_i, and wages, W_i, and introduce the following assumptions:
1. The system is productive.
2. Labor is homogenous and therefore the wage rate is uniform across industries, i.e.,

$$W_i = \omega L_i \tag{10.27}$$

where L_i is the total amount of labor used in industry i.

Now let us assume that the amount of labor used in industry i is a linear function of the output produced, i.e.

$$L_i = a_{0i} x_i. \tag{10.28}$$

3. Competition prevails both in labor and product markets
4. There is only working capital (i.e. fixed capital is omitted)

Capital can freely move from one sector to another, therefore rate of profit on capital is equal in all sectors

Let

$$K_i = \sum_{j=1}^{n} p_j a_{ij} x_j, \quad i = 1, \ldots, n \tag{10.29}$$

denote the working capital. The rate of profit in sector j is defined as

$$r_j = \frac{\Pi_j}{K_j} \tag{10.30}$$

or

$$\Pi_j = r_j \left(\sum_{i=1}^{n} p_j a_{ij} x_j \right) \quad j = 1, \ldots, n. \tag{10.31}$$

Equal rate of profit assumption implies that

$$r_j = r \quad \text{for} \quad j = 1, \ldots, n. \tag{10.32}$$

5. Wage and profit earners share the net income after the completion of the production[9].

Using (10.26) and (10.27)–(10.32) the following system of equations can be derived

$$p_i = (1 + r) \left(\sum_{j=1}^{n} p_j a_{ij} \right) + \omega a_{0i} \quad i = 1, \ldots, n$$

which can be expressed in matrix terms

$$\mathbf{p}^T = (1 + r)\mathbf{p}^T A + \omega \mathbf{a}_0.$$

This model is akin to the one developed by Sraffa[10] in his famous book [28] to analyze the value and distribution along classical lines[11].

Notice that in this model there are $n + 2$ unknowns. Namely n prices, wage rate and profit rate. But there are n equations. In other words, the system is underdetermined and it has two degrees of freedom (number of unknowns minus number of equations). The number of degrees of freedom can easily be decreased by taking a measure for prices. This is called numéraire (standard of value) of the system. We can take it by defining a vector $\mathbf{d} \in \mathbb{R}^n_+$ such that

$$\mathbf{p}^T \mathbf{d} = 1$$

or by taking

$$\omega = 1,$$

i.e., measuring prices in terms of wage unit. Whichever way is chosen the system is still underdetermined. There are more $(n + 1)$ unknowns than the number of equations[12] (n).

---

[9]The alternative assumption is that wages are paid in advance.

[10]Pierro Sraffa (1898–1983) was an Italian economist, who spent most of his life in Cambridge, England. In his famous book [28] he launched a strong critique of marginalist theory and laid the foundations of Neo-Ricardian school in economics.

[11]Sraffa analyzed the determination of prices at a given moment of time, given the prevailing technology. Therefore, he did not make any assumptions concerning the returns to scale and did not use input coefficients.

[12]Within the framework of [28] this is not a deficiency of the system. On the contrary it makes clear that the distribution of income between wage and profit earners can not be solved within the

10.5 Prices of Production in the Linear Production Model

For the sake of simplicity assume that the price of the first commodity is taken as equal to 1. Then the full system can be written as follows

$$\mathbf{p}^T = (1+r)\mathbf{p}^T A + \omega \mathbf{a}_0, \qquad (10.33)$$

$$\mathbf{p}^T \mathbf{d} = 1, \qquad (10.34)$$

$$\mathbf{d}^T = (1,\ldots,1) \in \mathbb{R}^n_+. \qquad (10.35)$$

Since one of the prices can be taken as numéraire, the system can be solved if one of the distributive variables, r or ω, is determined outside the model. From now on (10.33)–(10.35) will be referred to as the *Sraffa Model*.

Then the question is to find a solution to (10.33)–(10.35) satisfying $r > 0$ and $\mathbf{p}^T \geq \mathbf{0}^T$ under the following assumptions:
i. $\omega > 0$, i.e. the given wage rate measured in terms of commodity 1 is positive.
ii. $A \geq \mathbf{0}$ (no output without some commodity input).
iii. $\mathbf{a}_0 > \mathbf{0}$ (labor is directly used in the production of all commodities).

In the light of these assumptions it is clear from (10.33) that

$$\mathbf{p}^T [I - (1+r)A] > \mathbf{0}$$

or

$$\mathbf{p}^T \left[\left(\frac{1}{1+r} \right) I - A \right] > \mathbf{0}.$$

Example 10.3. Consider a two sector Sraffa model

$$\mathbf{p}^T = (1+r)\mathbf{p}^T A + \omega \mathbf{a}_0, \qquad (10.36)$$

where the input coefficients matrix is given as

$$A = \begin{bmatrix} 0.2 & 0.4 \\ 0.3 & 0.1 \end{bmatrix}$$

and the labor coefficient vector is

$$\mathbf{a}_0 = [0.1, 0.2].$$

Suppose that for convenience the sum of prices is taken as equal to unity, that is, for $\mathbf{d} = (1,1)$ we have

$$(\mathbf{d}, \mathbf{p}) = 1. \qquad (10.37)$$

---

price system. It requires a much broader framework that may include politico-economic as well as financial variables.

i. Find the relation between wage rate and the rate of profit (wage-profit trade-off) for this economy.
ii. Calculate the prices in terms of the numéraire.

Solution. i. Notice for any $0 < r < 1$ the matrix $(1+r)A^T$ is productive (by Hawkins–Simon conditions, see Theorem 10.6) and positive, hence by Theorem 10.9 the matrix $(I - (1+r)A)^{-1}$ exists and is positive.[13]

By (10.36), we have

$$\mathbf{p}^T = \omega \mathbf{a}_0 (I - (1+r)A)^{-1}. \qquad (10.38)$$

Using (10.37), we get

$$1 = \mathbf{p}^T \mathbf{d}^T = \omega \mathbf{a}_0 (I - (1+r)A)^{-1} \mathbf{d}^T,$$

thus,

$$\omega = \frac{1}{\mathbf{a}_0 (I - (1+r)A)^{-1} \mathbf{d}^T}. \qquad (10.39)$$

Notice that (10.39) indicates a trade-off between r and w. An increase in the rate of profit leads to a decline in the wage rate. Substituting the given numerical values, we have

$$\omega = \frac{1}{[0.1, 0.2] \begin{bmatrix} 1 - 0.2(1+r) & -0.4(1+r) \\ -0.3(1+r) & 1 - 0.1(1+r) \end{bmatrix}^{-1} \begin{bmatrix} 1 \\ 1 \end{bmatrix}} = 2\frac{6 - 5r - r^2}{7 + r}.$$

ii. Substituting (10.39) in (10.38) one gets

$$\mathbf{p}^T = \frac{1}{\mathbf{a}_0 (I - (1+r)A)^{-1} \mathbf{d}^T} \mathbf{a}_0 (I - (1+r)A)^{-1} = \left[\frac{3+r}{7+r}, \frac{4}{7+r} \right]. \qquad (10.40)$$

Notice that (10.39) and (10.40) are in general non-linear.

10.6 Perron–Frobenius Theorem

In this section another mathematical tool that can be used to find a economically meaningful solution to (10.33) is introduced. It is based on better understanding of eigenvectors and eigenvalues of non-negative matrices (recall that we have discussed eigenvectors and eigenvalues in general in Chap. 9). We are interested in non-

---

[13] For $r \geq 1$ the matrix $(1+r)A$ is not productive (again by Hawkins–Simon conditions), so, the problem has no economical sense.

10.6 Perron–Frobenius Theorem

negative real eigenvalues of A such that the associated eigenvectors are non-negative as well, $\mathbf{x} \geq \mathbf{0}$. The existence of such a pair can be assured under specific conditions.

The following theorem tells us that if A is a non-negative matrix such an eigenvalue and corresponding eigenvector can be found. This is a version of a more complicated theorem due to Perron[14] and Frobenius (Theorem 10.12 below) for all non-negative matrices.[15]

Theorem 10.10. *Let A be a non-negative matrix. Then*
  i. *A has a non-negative real eigenvalue $\hat{\lambda}_A$ (called* Perron–Frobenius eigenvalue*). If λ is another eigenvalue of A, then $\hat{\lambda}_A \geq |\lambda|$;*
 ii. *There exists a non-negative eigenvector $\hat{\mathbf{x}}_A$ associated with $\hat{\lambda}_A$ (called a* Perron–Frobenius eigenvector*).*
iii. *for $\mu \in \mathbb{R}$ the matrix $(\mu I - A)^{-1}$ exists and is non-negative if and only if $\mu > \hat{\lambda}_A$.*

Proof. [7, p. 600]. □

This theorem can be strengthened for the Leontief system, by taking into account Theorem 10.6.

Theorem 10.11. *Let A be a non-negative matrix and $B = I - A$. Then the following statements are equivalent*
  i. *the matrix A is productive, that is, there exists $\mathbf{x} \geq \mathbf{0}$ such that $B\mathbf{x} > \mathbf{0}$;*
 ii. *$B^{-1} = (I - A)^{-1}$ exists and is non-negative;*
iii. *the Hawkins–Simon conditions hold;*
 iv. *each eigenvalue λ of A satisfies the inequality $|\lambda| < 1$;*
  v. *$\hat{\lambda}_A < 1$.*

In this case, the Perron–Frobenius eigenvector of A satisfies the inequality $B\hat{\mathbf{x}}_A \geq \mathbf{0}$.

Proof. The equivalence of the conditions *(i)*, *(ii)* and *(iii)* is given in Theorem 10.6.
  The equivalence of *(ii)* and *(iv)*, and *(v)* follows from Theorem 10.10(iii).
  The last statement of the theorem follows from *(iv)* and the equality $A\hat{\mathbf{x}}_A = \hat{\lambda}_A \hat{\mathbf{x}}_A$, because

$$B\hat{\mathbf{x}}_A = (I - A)\hat{\mathbf{x}}_A = (\hat{\lambda}_A I - A)\hat{\mathbf{x}}_A + (1 - \hat{\lambda}_A)\hat{\mathbf{x}}_A = (1 - \hat{\lambda}_A)\hat{\mathbf{x}}_A \geq \mathbf{0}.$$

□

---

[14]Oskar Perron (1880–1975) was a German mathematician who made a significant contribution in algebra, geometry, analysis, differential equations, and number theory.

[15]In 1907, Perron proved Theorem 10.12 under the additional condition that the matrix A is positive (or at least some its power is positive). In 1912, Frobenius extended this result to its complete form. For the survey of these articles of Perron and Frobenius and for the history of the theorem, we refer the reader to [13]. Theorem 10.10 (which is also sometimes referred as Perron–Frobenius theorem) can be obtained from either Perron or Frobenius results by limit argument.

Notice that Theorem 10.11 does not guarantee the existence of a positive solution to the inequality $B\mathbf{x} > \mathbf{0}$, since it does not rule out the possibility of getting some entries of $\hat{\mathbf{x}}_A$ to be zero. A stronger result can be obtained if A is irreducible (see Definition 10.3).

Theorem 10.12 (Perron–Frobenius theorem). *Let A be a non-negative irreducible matrix. Then*

i. *There exists a positive real eigenvalue $\hat{\lambda}_A$ [called* Perron–Frobenius Eigenvalue *of A] such that $\hat{\lambda}_A$ is a simple (non-repeated) root of the characteristic polynomial of A and for any other eigenvalue λ of A we have $\hat{\lambda}_A \geq |\lambda|$.*
ii. *There exists a positive eigenvector $\hat{\mathbf{x}}_A > \mathbf{0}$ associated with $\hat{\lambda}_A$. This eigenvector $\hat{\mathbf{x}}_A$ is unique up to a scalar multiple. In other words, there are no eigenvectors associated with $\hat{\lambda}_A$ but the vectors of the form $\hat{\mathbf{y}}_A = \mu\hat{\mathbf{x}}_A$ with $\mu \in \mathbb{R}$.*

Proof. See [3, p. 17]. □

Combining Theorems 10.12 and 10.9, we get the following corollary. It gives a direct way to check if a given irreducible matrix is productive by calculating the Perron–Frobenius eigenvalue. It also gives a positive solution of the inequality $\mathbf{x} \geq A\mathbf{x}$ for an irreducible productive matrix A – this is a Perron–Frobenius eigenvector, $\mathbf{x} = \hat{\mathbf{x}}_A$.

Corollary 10.13. *Let A be a non-negative irreducible matrix. Then the following statements are equivalent*

i. *for each nonzero $\mathbf{d} \geq 0$ the system $\mathbf{x} = A\mathbf{x} + \mathbf{d}$ has a positive solution, $\mathbf{x} > \mathbf{0}$;*
ii. *the matrix A is productive;*
iii. *$\hat{\lambda}_A < 1$;*
iv. *$(I - A)\hat{\mathbf{x}}_A > \mathbf{0}$.*

The next corollary gives a simple way to evaluate the Perron–Frobenius eigenvalue.

Corollary 10.14. *Let A be a non-negative irreducible matrix and*

$$C_{min} = \min_j \sum_{i=1}^n a_{ij}$$

be its smallest column sum and

$$C_{max} = \max_j \sum_{i=1}^n a_{ij}$$

be its largest column sum. Then

$$C_{min} \leq \hat{\lambda}_A \leq C_{max},$$

and the both equalities hold only if and only if $C_{min} = C_{max}$.

Proof. Let $\hat{\mathbf{x}}_A$ be the eigenvector associated with $\hat{\lambda}_A$ Then

$$\hat{\lambda}_A \hat{\mathbf{x}}_A = A\hat{\mathbf{x}}_A,$$

which means

$$\hat{\lambda}_A \hat{x}_i^A = \sum_{j=1}^{n} a_{ij} \hat{x}_j^A \text{ for } i = 1, \ldots, n,$$

where $\hat{x}_i^A$ is the i'th element of the vector $\hat{\mathbf{x}}_A$. Summing this expression over i, we have

$$\hat{\lambda}_A \sum_{i=1}^{n} \hat{x}_i^A = \sum_{i=1}^{n} \sum_{j=1}^{n} a_{ij} \hat{x}_j^A = \sum_{j=1}^{n} \hat{x}_j^A \sum_{i=1}^{n} a_{ij},$$

which can be written as

$$\hat{\lambda}_A = \sum_{j=1}^{n} \hat{x}_j^A \sum_{i=1}^{n} a_{ij} / \sum_{i=1}^{n} \hat{x}_i^A,$$

i.e. $\hat{\lambda}_A$ can be expressed as a non-negative weighted average of the column sums of A. □

Remark 10.2. Same result holds for the row sums of A.

10.7 Linear Production Model (continued)

In this section the economic meaning of conditions required to guarantee a positive solution to (10.25) will be discussed. The discussion is based on a classification introduced by Sraffa in [28].

Sraffa makes a distinction between those commodities required *directly or indirectly* in the production of every commodity (Basic Commodities) and those that do not posses such a property (Non-Basic commodities). He demonstrates that while the former type of commodities play a significant role in the determination of the rate of profit and prices, the latter has no such influence.

It is clear that a commodity i is directly required in the production of commodity j if and only if

$$a_{ij} > 0.$$

Now suppose that for a commodity pair (i, j)

$$a_{ij} = 0,$$

but there exist a commodity k such that

$$a_{kj} > 0 \text{ and } a_{ik} > 0.$$

In other words, although commodity i does not directly used as an input in the production of the commodity j, it is required for the production of k, which is a direct input in producing j. The commodity i is said to be indirectly required in producing j.

Example 10.4. Let the input coefficient matrix be defined as follows

$$A = \begin{bmatrix} a_{11} & a_{12} & a_{13} \\ 0 & a_{22} & a_{23} \\ a_{31} & a_{32} & a_{33} \end{bmatrix}$$

where $a_{ij} > 0$ for all i, j, except a_{21}.

Here second commodity is not directly required for the production of the commodity 1. However in order to produce one unit of commodity 3, $a_{23} > 0$ amount of commodity 3 is required. On the other hand $a_{31} > 0$ amount of commodity 3 is required in order to produce one unit of commodity 1. Therefore, in order to produce one unit of commodity 1, $a_{31}a_{23}$ amount of commodity 2 is indirectly required.

Now consider the following matrix multiplication

$$A^2 = A \cdot A.$$

Let the (i, j)'th element of A^2 be denoted as a_{ij}^2. Then it is easy to see that in the above example that $a_{21} = a_{32}a_{31}$, hence for all i, j

$$\text{either } a_{ij} > 0 \text{ or } a_{ij}^2 > 0.$$

This following definition is the generalization of the result obtained in the example given above.

Definition 10.4. Let A be a non negative input coefficients matrix, whose j-th column is associated with the commodity j. Then j is a *basic commodity* if for some $k \leq n$

$$(A + A^2 + \ldots + A^k)\mathbf{e}_j > \mathbf{0}, \tag{10.41}$$

where e_j is the unit vector whose j-th component is equal to one.

Theorem 10.15. *Consider an economy with n commodities, $i = 1, \ldots, n$. Let A be the $n \times n$ input coefficients matrix of this economy. Then all commodities in this economy are basic commodities if and only if A is irreducible.*

Proof. Note that the inequality (10.41) in Definition 10.4 means that j-th column of the matrix

$$A + A^2 + \ldots + A^n$$

is positive.

10.7 Linear Production Model

If all commodities are basic, then from Definition 10.4 we get

$$A + A^2 + \ldots + A^n = (A + A^2 + \ldots + A^n)I > \mathbf{0},$$

hence for all i and j there is $k \leq n$ such that $a_{ij}^k > 0$. Then by Lemma 10.8 A is irreducible.

Now suppose that A is irreducible. Then, by Lemma 10.8 for each $i \neq j$ there exist $m = m(i,j)$ such that $a_{ij}^m > 0$. By Problem 10.1 (the problem after Lemma 10.8), we may choose $m \leq n$.

For each i and each $j \neq i$, we have

$$a_{ii}^{m(i,j)+m(j,i)} \geq a_{ij}^{m(i,j)} a_{ji}^{m(j,i)} > 0,$$

(since $A^{m(i,j)+m(j,i)} = A^{m(i,j)} A^{m(j,i)}$), that is, there exists $m > 0$ such that $a_{ii}^m > 0$. Following the proof of Lemma 10.8, this means that there exists a sequence

$$i = i_0, i_1, \ldots, i_{m-1}, i_m = i$$

of indexes such that all matrix entries $a_{i_j, i_{j+1}}$ are nonzero. Consider such a sequence of minimal length m. There is no repeated elements in it but $i_0 = i_m = i$, therefore, there are at most n pairwise different elements. Hence $m \leq n$.

So, for each i, j the (i, j)-th entry of some matrix A^m for $1 \leq m \leq n$ is positive. Hence,

$$A + A^2 + \ldots + A^n > \mathbf{0}.$$

Thus, each column of the matrix in the left hand side is positive, that is, each commodity is basic. □

10.7.1 Sraffa System: The Case of Basic Commodities

Now let us turn to (10.33) and assume that A is irreducible. Then by Theorem 10.12 we know that A has a positive Perron–Frobenius eigenvalue and an associated eigenvector which is also positive. Of course, the same is true for the transpose matrix A^T.

Let $\hat{\mathbf{p}}$ be the *left* Perron–Frobenius eigenvector of A, that is, the Perron–Frobenius eigenvector of A^T. Therefore for $\hat{\lambda}_A > 0$ and for the associated price vector $\hat{\mathbf{p}}^T > \mathbf{0}$ we have

$$\hat{\mathbf{p}}^T \hat{\lambda}_A = \hat{\mathbf{p}}^T A$$

or

$$\hat{\mathbf{p}}^T (\hat{\lambda}_A I - A) = 0.$$

It follows that if

$$\frac{1}{1+r} = \hat{\lambda}_A$$

or

$$r = (1-\hat{\lambda}_A)/\hat{\lambda}_A, \tag{10.42}$$

then the associated price vector is positive. Notice that, in the light of (10.27) this solution implies that the wage rate measured in terms of each commodity i is equal to zero (i.e., $\omega = 0$). Therefore, this value $R = (1-\hat{\lambda}_A)/\hat{\lambda}_A$ of r is the maximum rate of profit.

On the other hand, from Corollary 10.14 to Perron–Frobenius Theorem it is known that the Perron–Frobenius eigenvector of A lies within a closed interval

$$\hat{\lambda}_A \in [C_{min}, C_{max}] \in \mathbb{R}_+,$$

therefore the profit rate also lies in the closed interval

$$R \in \left[\frac{1-C_{max}}{C_{max}}, \frac{1-C_{min}}{C_{min}}\right]. \tag{10.43}$$

Notice that (10.42) implies that the rate of profit derived from $\hat{\lambda}_A$ is positive provided that C_{min} is less than 1. Now, by Theorem 10.11(v) $\hat{\lambda}_A < 1$, that is, $R > 0$. If this condition is satisfied then (10.43) becomes

$$R \in \left(0, \frac{1-C_{min}}{C_{min}}\right]$$

and the non-negativity of the maximum profit rate can be assured.

Finally, by Theorem 10.10iii, if $r < R$ then

$$\mu = \frac{1}{1+r} > \frac{1}{1+R} = \hat{\lambda}_A$$

and

$$(\mu I - A)^{-1} = \left(\frac{1}{1+r}I - A\right)^{-1} > 0$$

will exist. This means for $r < R$ the Sraffa system will have a positive solution $\mathbf{p}^T > 0$ given by

$$\mathbf{p}^T = \omega \mathbf{p}^T \left(\frac{1}{1+r}I - A\right)^{-1} \mathbf{a}_0.$$

10.7.2 Sraffa System: Non-Basic Commodities Added

From the definition of basic commodities, it is clear that once non-basic commodities are allowed, the input coefficients matrix becomes reducible. Under these conditions Perron–Frobenius Eigenvalue of A may be equal to zero, and the associated eigenvector may not be unique (Theorem 10.11).

Suppose now a production system has both basic and non-basic commodities. Then the subsystem consisting solely of basic commodities will satisfy the conditions of Theorem 10.12. Suppose that basic commodities are labelled as commodity group-1, whereas the non-basics as commodity group-2. Then we can write the system as

$$[\mathbf{p}_1^T \ \mathbf{p}_2^T] = (1+r)[\mathbf{p}_1^T \ \mathbf{p}_2^T]\begin{bmatrix} A_{11} & A_{12} \\ 0 & A_{22} \end{bmatrix} + \omega[\mathbf{a}_{01} \ \mathbf{a}_{02}].$$

This system can be written in the following form

$$\begin{cases} \mathbf{p}_1^T = (1+r)\mathbf{p}_1^T A_{11} + \omega\mathbf{a}_{01}, \\ \mathbf{p}_2^T = (1+r)\left[\mathbf{p}_1^T A_{12} + \mathbf{p}_2^T A_{22}\right] + \omega\mathbf{a}_{02}. \end{cases}$$

It is clear that the prices of the basic commodities can be obtained only by using information about their production technology (A_{11} and $\mathbf{a}_{01}$), i.e. independently from the non-basics.

10.8 Problems

1. Consider a three sector economy where the interindustry sales (sectors selling to each other) and total outputs given in the following table:

	Interindustry sales ($ billion)			Total output
	Agriculture	Manufacturing	Services	
Agriculture	200	50	150	1,000
Manufacturing	50	200	70	550
Services	150	100	100	600

   i. Express the meaning of the figures that lie on the main diagonal (north-west to south-east) of the interindustry sales part of the table.
   ii. Calculate the final demand for each sector.
   iii. Find the input coefficients matrix A and the Leontief inverse $(I - A)^{-1}$ for this economy.
   iv. Does the A matrix satisfy the necessary conditions for guaranteeing positive sectoral output levels?

2. *Wage increase and relative prices.* Consider the following input-output table of interindustry flows of goods for a three sector economy (in $ billion):

	Agriculture	Manufacturing	Services	Final demand	Total output
Agriculture	20	40	10	80	150
Manufacturing	30	200	100	150	480
Services	20	60	50	170	300
Value Added	80	180	140		400
Wages	30	80	50		160
Profit etc.	50	100	90		240
Total outlay	150	480	300	400	1,330

i. Calculate the input coefficients matrix A and the Leontief inverse $(I - A)^{-1}$ for this economy.

ii. Calculate the Leontief prices for this economy.

iii. Suppose that the wage costs in manufacturing sector increased 20%. Assume that in the new state the amount of profit generated in each sector remains unchanged. How such a change affects the prices? Do relative prices (i.e., the price of one good in terms of the other) change?

3. *Technical change and relative prices.* Consider the following input coefficients matrix for a three sector economy:

	Agriculture	Manufacturing	Services
Agriculture	0.25	0.15	0.3
Manufacturing	0.3	0.2	0.1
Services	0.35	0.25	0.2

Suppose the value added is given by

$$\mathbf{V} = [45, 50, 40].$$

i. Suppose a technological change took place in the manufacturing sector and all of its input coefficients declined by 10%. What will be the effect of such a technological improvement on the relative prices and on the maximal rate of profit?

ii. Express your finding as a general result for an n sector Leontief economy.

4. Find the Perron–Frobenius eigenvalue and the Perron–Frobenius eigenvector of length 1 for the following matrices

(a) $\begin{bmatrix} 2 & 4 \\ 4 & 8 \end{bmatrix}$; (b) $\begin{bmatrix} 0.7 & 0.6 \\ 0.2 & 0.5 \end{bmatrix}$; (c) $\begin{bmatrix} 0.1 & 0.0 & 0.0 \\ 0.1 & 0.3 & 0.3 \\ 0.1 & 0.3 & 0.0 \end{bmatrix}$;

(d) $\begin{bmatrix} 0 & 0 & 0.3 & 0.1 \\ 0.1 & 0.5 & 0.3 & 0.2 \\ 0 & 0 & 0.2 & 0.4 \\ 0.5 & 0.3 & 0.1 & 0.2 \end{bmatrix}$.

5. Which matrix from Problem 4 is:
   i. Productive?
   ii. Irreducible?

10.8 Problems

6. Consider a three sector economy with the input coefficients matrix

$$A = \begin{bmatrix} 0.46 & 0.18 & 0.25 \\ 0.24 & 0.08 & 0.46 \\ 0.21 & 0.44 & 0.19 \end{bmatrix}$$

and the labor coefficients vector $\mathbf{a}_0 = (0.3, 0.1, 0.2)$.

Suppose that this economy is analysed in a Sraffian framework.
   i. What is the maximum rate of profit for this economy? (Answer: 18.9%)
   ii. What is the corresponding price vector?
   iii. Normalize the price vector by taking sum of prices as equal to one. Using this normalization rule find a relation between the rate of profit and the wage rate for the Sraffa model. Calculate the wage rate when the rate of profit is 10%.

7. Suppose that the input coefficient matrix for a n-sector economy is as follows

$$A = \begin{bmatrix} A_{11} & A_{12} \\ A_{21} & A_{22} \end{bmatrix},$$

where $A_{11}, A_{12}, A_{21}, A_{22}$ are matrices of size $m \times m$, $n \times (n-m)$, $(n-m) \times n$, $(n-m) \times (n-m)$, respectively. Suppose that $A_{21} = \mathbf{0}$, i.e., the first m sectors do not use the outputs of the others as inputs. Show that in this economy the maximum growth rate depends exclusively on the production technology of these sectors and is independent from the production of other sectors.

Linear Programming 11

The *linear programming problem* is a general problem of finding the maximal value of the function $f(\mathbf{x}) = (\mathbf{a}, \mathbf{x})$, where $\mathbf{x} \in \mathbb{R}^n$ is a vector of n unknowns and $\mathbf{a} \in \mathbb{R}^n$ is a constant vector, under the restrictions

$$\mathbf{x} \geq \mathbf{0} \text{ and } A\mathbf{x} \geq \mathbf{d},$$

where A is a matrix and $\mathbf{d}$ is a constant vector. This problem has a lot of applications to economic models and practice. Some simple and rather artificial applications are discussed below.

In mathematical terms the problem is maximizing a linear objective function under linear constraints where the relevant variables are restricted to be non-negative. It was first solved by Kantorovich[1]. In 1947 Dantzig[2] made an important contribution by introducing a method to solve this kind of problems.

11.1 Diet Problem

Consider a person who has the n-tuple of foods $G_1, \ldots, G_n$ in her diet. Each food in the diet has m different components (attributes), e.g., fat, protein, calory, sodium, etc.

Let $A = \|a_{ij}\|_{m \times n}$ be a matrix such that a_{ij} is the quantity of component i in one unit of food G_j, for all $i = 1, \ldots, m$ and all $j = 1, \ldots, n$.

---

[1] Leonid Kantorovich (1912–1986) was a Russian mathematician and economist. He developed the linear programming method in 1939 for a particular industrial problem. In 1975, he shared Nobel Prize in Economics for his contributions to the theory of optimal allocation of resources.

[2] George Dantzig (1914–2005) was an American mathematician who made important contributions in many fields. He developed *simplex method* which was the first and the most useful effective method to solve linear programming problems.

Fig. 11.1 The diet problem with two foods and two components

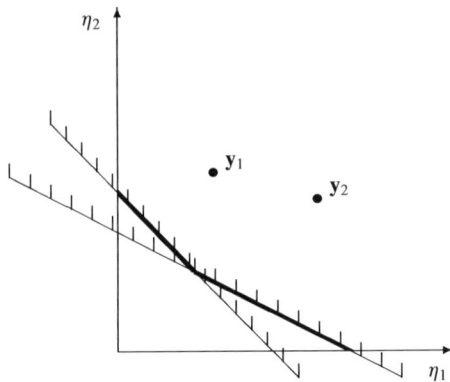

Let the vector $\mathbf{c} = (\gamma_1, \ldots, \gamma_m)$ be such that γ_i denotes the minimal amount of component i that should be in the diet. Then the problem is to find a diet (i.e., the quantity of each good in the diet) such that no nutrition component in the diet is less than the desirable level.

Let $\mathbf{y} = (\eta_1, \ldots, \eta_n)$ be the unknown vector denoting the quantities of the goods in the diet, i.e., η_i is the quantity of food G_i.

The expression

$$\sum_{j=1}^{n} \alpha_{ij} \eta_j = (\mathbf{a}_i, \mathbf{y}),$$

where $\mathbf{a}_i$ is i-th row of matrix A, denotes the amount of component i in the diet. Then, we have m constraints

$$(\mathbf{a}_1, \mathbf{y}) \geq \gamma_1,$$
$$\ldots\ldots$$
$$(\mathbf{a}_m, \mathbf{y}) \geq \gamma_m,$$

or the inequality in matrix form

$$A\mathbf{y} \geq \mathbf{c} \qquad (11.1)$$

to be satisfied by $\mathbf{y}$. On the other hand, we must have

$$\mathbf{y} \geq \mathbf{0}. \qquad (11.2)$$

If any row i in A has at least one component greater than 0, then for $n = m = 2$ the above inequalities can altogether be drawn as in Fig. 11.1, pointing to the fact that there always exists sufficiently large numbers $\eta_1, \ldots, \eta_n$ which satisfy inequalities (11.1).

Let $\mathbf{p} = (p_1, \ldots, p_n)$ denote the price vector for the n-tuple of goods, with p_i denoting the unit price of food G_i. Then the cost of the diet $\mathbf{y}$ is given by

$$(\mathbf{p},\mathbf{y}) = \sum_{j=1}^{n} p_j \eta_j. \tag{11.3}$$

Now, in terms of the matrix A and the vectors $\mathbf{c}$ and $\mathbf{p}$, the diet problem is to find a vector $\mathbf{y}$ which satisfies (11.1) and (11.2) minimizing at the same time the cost function (11.3), i.e.

$$\begin{cases} A\mathbf{y} \geq \mathbf{c}, \\ \mathbf{y} \geq \mathbf{0}, \\ (\mathbf{p},\mathbf{y}) \to \min. \end{cases}$$

11.2 Linear Production Model

The linear production model described in Example 1.5 assumed that each economic unit produces only one type of good. Let us consider a slightly different version of the same problem.

There are m economic units indexed by $i = 1,\ldots,m$ which produce and consume n goods indexed by $j = 1,\ldots,n$.

Let intensities of the units are given by the m-dimensional vector

$$\mathbf{x} = (\xi_1,\ldots,\xi_m) \geq \mathbf{0}$$

where ξ_i denotes the intensity of unit i. Intensities may be measured in several ways, e.g. the total work-hours and the number of employees or the quality of the goods produced by an economic unit may be the indicators of its intensity.

We assume that if the intensity of a unit increases by any constant s, then the production and the consumption of that unit will also rise by s.

Now, let

$$\mathbf{a}^j = (\alpha_{1j}, \alpha_{2j}, \ldots, \alpha_{mj}),$$

be a vector such that its i-th component is equal to the quantity of good j consumed or produced by the unit i when it is working with the intensity 1. Define α_{ij} to be positive if good j is produced by unit i, and negative it if is consumed by i. Then the matrix

$$A = \|\alpha_{ij}\|_{n \times m}$$

contains the production-consumption data for the economic system.

Note that given the intensity vector $\mathbf{x}$, we can uniquely find the quantities of goods produced and consumed by each unit. Indeed, the sum

$$\xi_1 \alpha_{1j} + \xi_2 \alpha_{2j} + \cdots + \xi_m \alpha_{mj} = (\mathbf{x}, \mathbf{a}^j)$$

is equal to the total production (consumption) of good j by the system if it is positive (negative).

Define the vector
$$\mathbf{d} = (\delta_1, \ldots, \delta_n)$$
such that δ_j be the upper-bound for production (consumption) of good j if it is positive (negative).

Then, the vector $\mathbf{x}$ is considered to be feasible if it satisfies the inequalities

$$(\mathbf{x}, \mathbf{a}^1) \leq \delta_1,$$
$$(\mathbf{x}, \mathbf{a}^2) \leq \delta_2,$$
$$\ldots\ldots$$
$$(\mathbf{x}, \mathbf{a}^n) \leq \delta_n,$$

or
$$\mathbf{x}A \leq \mathbf{d}.$$

Finally, let the vector $\mathbf{c} = (\gamma_1, \ldots, \gamma_m)$ be such that γ_i is the profit obtained by unit i when it is working with intensity 1. The profit of unit i with intensity ξ_i is just $\xi_i \gamma_i$, and hence the total profit of the system is

$$\sum_{i=1}^{n} \xi_i \gamma_i = (\mathbf{x}, \mathbf{c}).$$

Then the optimal program for the economic system at hand is a solution to

$$\max_{\{\mathbf{x}\}} (\mathbf{x}, \mathbf{c}).$$

subject to $\mathbf{x} \geq 0$ and $\mathbf{x}A \leq \mathbf{d}$.

Consider now the set of feasible $\mathbf{x}$'s. There are three possible cases: the feasible set of intensity vectors is either empty or a bounded polytope or an unbounded polytope (for the exact definition of polytope, see Definition 11.4 below).

In the second and the third case, let us first restrict ourselves to the two-dimensional problem assuming that $\mathbf{x} \in \mathbb{R}^2$ and $\mathbf{c} \in \mathbb{R}^2$ are planar vectors and the feasible set is a bounded (second case) or unbounded (third case) planar polygon.

Consider the second case. Through any point of a bounded polygon we can draw a line which is orthogonal to $\mathbf{c}$. Let $(\mathbf{c}, \mathbf{x}) = \mu$ denote such a line. Changing μ we can obtain a set of lines which are orthogonal to $\mathbf{c}$. Then the maximal value of μ, ensuring that the line still intersects with the polygon is a solution to our problem (Fig. 11.2).

It is possible to obtain a situation in which the line $(\mathbf{c}, \mathbf{x}) = \mu_{\max}$ passes through a side of the polygon (Fig. 11.3). Then each point on this side is a solution.

Consider now the case in which the feasible set is an unbounded polygon (Fig. 11.4). In this case we have a unique solution.

In the case from Fig. 11.5 there is no solution, since for each μ satisfying $(\mathbf{c}, \mathbf{x}) = \mu$ we can find some real number $\mu' > \mu$ such that $(\mathbf{c}, \mathbf{x}) = \mu'$.

11.2 Linear Production Model

Fig. 11.2 The case of bounded feasible set

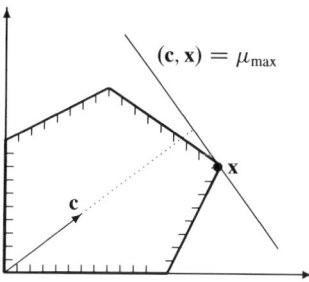

Fig. 11.3 An infinite number of solutions

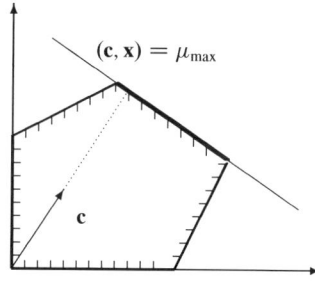

Fig. 11.4 The unique solution

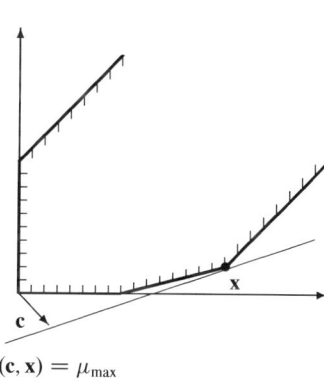

Fig. 11.5 No solutions with unbounded feasible set

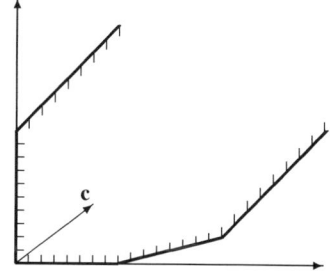

One can conclude that whether the feasible set is bounded or not, the solution, if exists, must always lie either at a corner point or on a side of the polygon. We will show that this is true for spaces of any dimension. In order to do this, we first study the multi-dimensional versions of polygons, that is, polytopes and convex sets in $\mathbb{R}^n$.

11.3 Convexity

Consider two vectors $\mathbf{x}$ and $\mathbf{y}$ in the n-dimensional vector space $\mathbb{R}^n$. We can define the line segment between the points $\mathbf{x}$ and $\mathbf{x} + \mathbf{y}$ to be the set of all points

$$\mathbf{z} = \mathbf{x} + t\mathbf{y}$$

where $t \in [0, 1]$. For $t = 1$, we obtain $\mathbf{z} = \mathbf{x} + \mathbf{y}$ while for $t = 0$, we get $\mathbf{z} = \mathbf{x}$, which are two boundary points of the line (Fig. 11.6).

Consider now the vector $\mathbf{w} = \mathbf{y} - \mathbf{x}$. The line segment between $\mathbf{x}$ and $\mathbf{y}$ is the set of all points defined as $\mathbf{v} = \mathbf{x} + t(\mathbf{y} - \mathbf{x})$, where $t \in [0, 1]$.

We can rewrite this expression as

$$\mathbf{v} = (1 - t)\mathbf{x} + t\mathbf{y}.$$

Inserting $\lambda \equiv 1 - t$ into $\mathbf{v}$ yields

$$\mathbf{v} = \lambda \mathbf{x} + (1 - \lambda)\mathbf{y}.$$

where $\lambda \in [0, 1]$ (Fig. 11.7).

Definition 11.1. Let S be a subset of the vector space $\mathbb{R}^n$. S is called to be *convex* if for any two elements $\mathbf{x}, \mathbf{y} \in S$, the line segment between $\mathbf{x}$ and $\mathbf{y}$ is contained in S, i.e., $\lambda \mathbf{x} + (1 - \lambda)\mathbf{y} \in S$ for all $\lambda \in [0, 1]$.

The examples of a non-convex set and a convex set are given on Fig. 11.8.

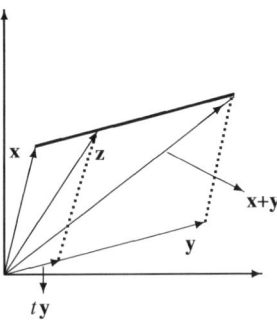

Fig. 11.6 $\mathbf{z} = \mathbf{x} + t\mathbf{y}$

11.3 Convexity

Fig. 11.7 $v = x + tw$

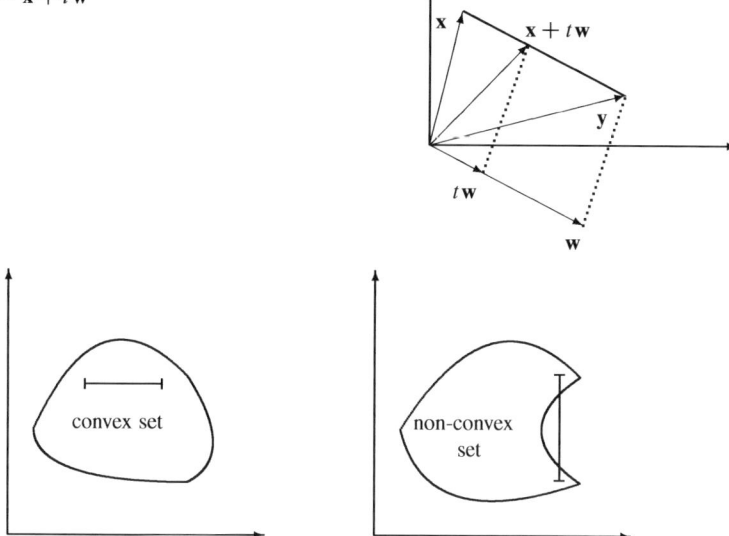

Fig. 11.8 Convex and non-convex sets

Theorem 11.1. *Let* $x_1, \ldots, x_m$ *be any vectors in the vector space* $\mathbb{R}^n$. *Then the set* S *of all linear combinations given by*

$$t_1 x_1 + \cdots + t_m x_m$$

where $t_i \geq 0$ *for all* $i = 1, \ldots, m$ *and* $\sum_{i=1}^{m} t_i = 1$, *is convex.*

Proof. For any vectors $x_1, \ldots, x_m$ in $\mathbb{R}^n$, let us define

$$\mathbf{a} = t_1 x_1 + \cdots + t_m x_m$$
$$\mathbf{b} = s_1 x_1 + \cdots + s_m x_m$$

where $t_i \geq 0$, $s_i \geq 0$, $\sum_{i=1}^{n} t_i = 1$ and $\sum_{i=1}^{n} s_i = 1$.

Consider $\theta = (1-\alpha)\mathbf{a} + \alpha \mathbf{b}$ for some $\alpha \in [0,1]$. We have

$$\theta = (1-\alpha)t_1 x_1 + \cdots + (1-\alpha)t_m x_m + \alpha s_1 x_1 + \cdots + \alpha s_m x_m$$
$$= [(1-\alpha)t_1 + \alpha s_1] x_1 + \cdots + [(1-\alpha)t_m + \alpha s_m] x_m$$

We note that $(1-\alpha)t_i + \alpha s_i \geq 0$ for all $i = 1, \ldots, n$ since $(1-\alpha) \geq 0$, $t_i \geq 0$, $\alpha \geq 0$ and, $s_i \geq 0$. Moreover,

$$\sum_{i=1}^{m}(1-\alpha)t_i + \alpha s_i = (1-\alpha)\sum_{i=1}^{m} t_i + \alpha \sum_{i=1}^{m} s_i = (1-\alpha) + \alpha = 1.$$

Hence, S is convex. □

Theorem 11.2. Let $x_1, \ldots, x_m$ be any m vectors in R. Any convex set S which contains $x_1, \ldots, x_m$ also contains all linear combinations

$$t_1 x_1 + \cdots + t_m x_m,$$

where $t_i \geq 0$ for all i and $\sum_{i=1}^{n} t_i = 1$.

Proof (by induction). For $m = 1$, we have $t_1 = 1$ and the statement of the theorem is obviously true. Assume the statement is true for some $m - 1 \geq 1$. Take any $x_1, \ldots, x_m \in S$, and any $t_1, \ldots, t_m$ satisfying $t_i \geq 0$ for all $i = 1, \ldots, m$ and $\sum_{i=1}^{m} t_i = 1$. If $t_m = 1$, then the statement is trivially true, because in this case $t_1 = \cdots = t_{m-1} = 0$. Assume now $t_m \neq 1$. Then we get

$$\theta_m = t_1 x_1 + \cdots + t_m x_m = (1 - t_m) \left(\frac{t_1}{1 - t_m} x_1 + \cdots + \frac{t_{m-1}}{1 - t_m} x_{m-1} \right) + t_m x_m.$$

Define

$$s_i = \frac{t_i}{1 - t_m}$$

for all $i = 1, \ldots, m - 1$. Then the vector

$$z = s_1 x_1 + \cdots + s_{m-1} x_{m-1},$$

by induction assumption, lies in S, since $s_i \geq 0$ and $\sum_{i=1}^{m-1} s_i = \sum_{i=1}^{m-1} t_i / (1 - t_m) = 1$. Thus

$$(1 - t_m) z + t_m x_m = \theta$$

is an element of S by the definition of a convex set.

The minimal convex set which contains $x_1, \ldots, x_m$ is called the *convex hull* of the set $\{x_1, \ldots, x_m\}$, and is denoted by $co\{x_1, \ldots, x_m\}$. By Theorems 11.1 and 11.2, we get

Corollary 11.3. *For any (finite) set $x_1, \ldots, x_m$ of vectors, its convex hull $co\{x_1, \ldots, x_m\}$ consists of all linear combinations given by*

$$t_1 x_1 + \cdots + t_m x_m,$$

where $t_i \geq 0$ for all $i = 1, \ldots, m$ and $\sum_{i=1}^{m} t_i = 1$.

Definition 11.2. For any $a \in \mathbb{R}^n$ and $\beta \in \mathbb{R}$, the set of points $x \in \mathbb{R}^n$ satisfying

$$(a, x) = \beta,$$

is called a *hyperplane*. We denote this hyperplane as $H(a, \beta) = \{x \in \mathbb{R}^n | (a, x) = \beta\}$.

11.3 Convexity

Fig. 11.9 The vector **a** is orthogonal to the hyperplane

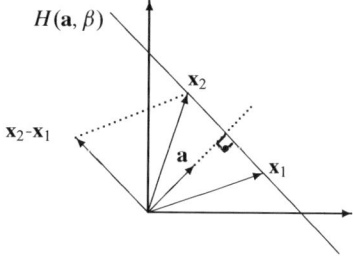

Fig. 11.10 The line $x + 2y = -4$ as a hyperplane $H((1, 2), -4)$

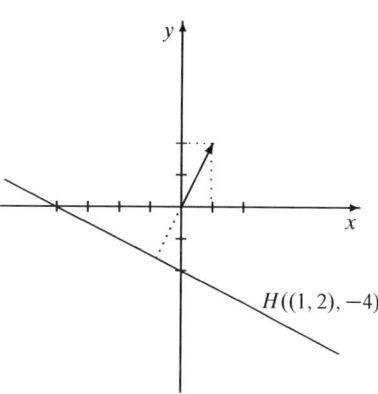

Lemma 11.4. *For all $\beta \in \mathbb{R}$, any vector $\mathbf{a} \in \mathbb{R}^n$ is orthogonal to the hyperplane $H(a, \beta)$ in $\mathbb{R}^n$.*

Proof. Take any $\beta \in \mathbb{R}$. Then any vector $\mathbf{a} \in \mathbb{R}^n$ is orthogonal to the hyperplane $H(\mathbf{a}, \beta)$ if and only if it is orthogonal to the vector $\mathbf{x}_2 - \mathbf{x}_1$ which connects two arbitrary points $\mathbf{x}_1, \mathbf{x}_2 \in H(\mathbf{a}, \beta)$ (Fig. 11.9).

Indeed, $(\mathbf{a}, \mathbf{x}_2 - \mathbf{x}_1) = (\mathbf{a}_1, \mathbf{x}_2) - (\mathbf{a}_1, \mathbf{x}_1) = \beta - \beta = 0$, which completes the proof. □

Example 11.1. The line $x + 2y = -4$ is a hyperplane in $\mathbb{R}^2$, and can be rewritten as $H((1, 2), -4)$ (Fig. 11.10).

Lemma 11.5. *For all $\mathbf{a} \in \mathbb{R}^n$ and $\beta \in \mathbb{R}$, the hyperplane $H(\mathbf{a}, \beta)$ is a convex set.*

Proof. Indeed, for any $\mathbf{a} \in \mathbb{R}^n$ and $\beta \in \mathbb{R}$, consider the hyperplane $H(\mathbf{a}, \beta)$. The hyperplane is always non-empty. If $\mathbb{R}^n = \mathbb{R}$, then any hyperplane in $\mathbb{R}$ is a singleton, and the Lemma is obviously true. For $n > 1$, take $\mathbf{x}_1, \mathbf{x}_2 \in H(\mathbf{a}, \beta)$. Consider $\mathbf{x}^\alpha \equiv (1 - \alpha)\mathbf{x} + \alpha\mathbf{x}_2$ for any $\alpha \in [0, 1]$. Note that

$$(\mathbf{a}, \mathbf{x}^\alpha) = (1 - \alpha)(\mathbf{a}, \mathbf{x}_1) + \alpha(\mathbf{a}, \mathbf{x}_2) = (1 - \alpha)\beta + \alpha\beta = \beta.$$

Thus, $\mathbf{x}^\alpha \in H(\mathbf{a}, \beta)$, and therefore $H(\mathbf{a}, \beta)$ is convex. □

Let $\beta_1, \ldots, \beta_m \in \mathbb{R}$, and $\mathbf{a}_1, \ldots, \mathbf{a}_m \in \mathbb{R}^n$. Consider the set of points $\mathbf{x}_1, \ldots, \mathbf{x}_m \in \mathbb{R}^n$ which satisfy the system of equations

$$(\mathbf{a}_1, \mathbf{x}) = \beta_1$$
$$\vdots$$
$$(\mathbf{a}_m, \mathbf{x}) = \beta_m.$$

The solution of this system can be considered as the set of points defined by the intersection of the hyperplanes $H(\mathbf{a}_1, \beta_1), \ldots, H(\mathbf{a}_m, \beta_m)$. (Show that this set is convex.)

If $m = n$ and $\mathbf{a}_1, \ldots, \mathbf{a}_m$ are linearly independent, then the above system has a unique solution - it is the unique point which lies in the intersection of n hyperplanes.

Definition 11.3. Let $H(\mathbf{a}, \beta)$ be a hyperplane in $\mathbb{R}^n$. The sets

$$H^{[+]}(\mathbf{a}, \beta) = \{\mathbf{x} \in \mathbb{R}^n | (\mathbf{a}, \mathbf{x}) \geq \beta\}$$

and

$$H^{[-]}(\mathbf{a}, \beta) = \{\mathbf{x} \in \mathbb{R}^n | (\mathbf{a}, \mathbf{x}) \leq \beta\}$$

are called *closed half-spaces* that $H(\mathbf{a}, \beta)$ yields. Analogously,

$$H^{(+)}(\mathbf{a}, \beta) = \{\mathbf{x} \in \mathbb{R}^n | (\mathbf{a}, \mathbf{x}) > \beta\}$$

and

$$H^{(-)}(\mathbf{a}, \beta) = \{\mathbf{x} \in \mathbb{R}^n | (\mathbf{a}, \mathbf{x}) < \beta\}$$

are called *open half-spaces* that $H(\mathbf{a}, \beta)$ defines.

Example 11.2. The four halfspaces in $\mathbb{R}^2$ defined by the hyperplane $H((1, 1), 1)$ are illustrated in Fig. 11.11.

Definition 11.4. The intersection of a finite number of closed half-spaces in $\mathbb{R}^n$ is called a *polytope*.

Example 11.3. The polytope defined by the three halfspaces $H^{[-]}((1, 1), 1)$, $H^{[+]}((1, 0), 0.3)$, $H^{[+]}((0, 1), 0.2)$ is the triangular region at the center of Fig. 11.12.

Theorem 11.6. *Every bounded polytope P is a convex hull of some finite set of vectors X.*

11.3 Convexity

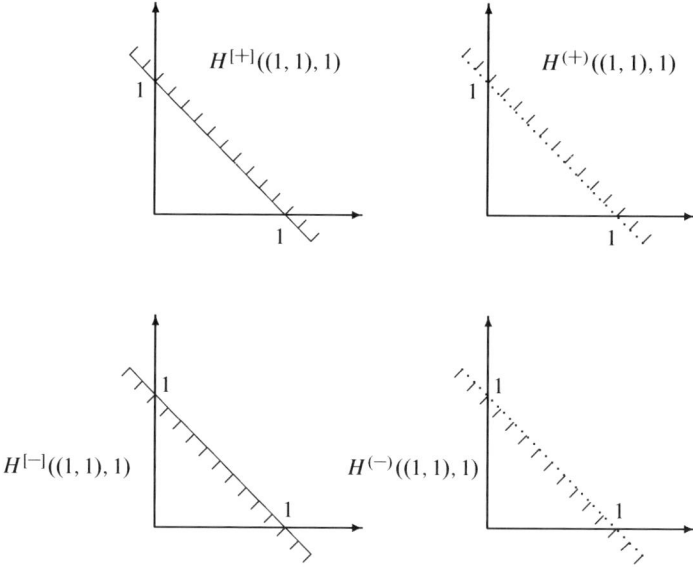

Fig. 11.11 Halfspaces defined by a given hyperplane

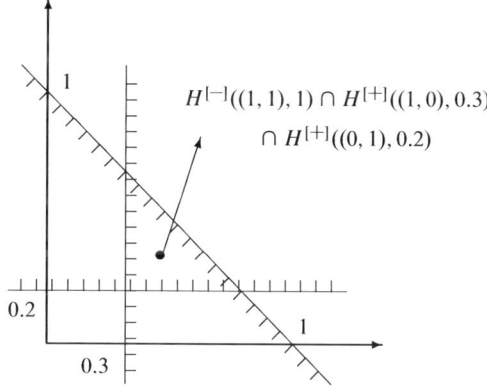

Fig. 11.12 A polytope defined by three halfspaces

If X is the minimal set of such vectors, then the elements of X are called *vertices* of the polytope.[3]

---

[3] For a general (bounded or non-bounded) polytope P, one can also define a vertex. Namely, a vector X is a vertex of P if it is the only element in the intersection of some of the hyperplanes which define the half-spaces whose intersection is P. It is proved, e.g. in [33, Theorems 7.2.3 and 7.2.6], that for bounded polytopes this definition describes the same vertices as the given one.

Example 11.4. The polytope defined by $2n$ hyperplanes $H^{[+]}(\mathbf{e}_i, 0)$ and $H^{[+]}(\mathbf{e}_i, 1)$ where $i = 1, \ldots, n$ and $\mathbf{e}_i = (0, \ldots, 1, \ldots, 0)$ is the i-th vector of the canonical basis, is called a *hypercube*. It consists of all vectors $X = (x_1, \ldots, x_n)$ such that $0 \le x_i \le 1$ for all $i = 1, \ldots, n$. Its vertices are 2^n vectors of the form $(a_1, \ldots, a_n)$, where each a_i is either 0 or 1.

The next corollary assures the existence of the solution of any linear programming problem defined on a bounded polytope.

Corollary 11.7. *Let $f(\mathbf{x}) = (\mathbf{x}, \mathbf{a})$ be a linear function of $\mathbf{x} \in \mathbb{R}^n$ and let $P = \mathrm{co}(\mathbf{v}_1, \ldots, \mathbf{v}_k)$ be a bounded polytope in $\mathbb{R}^n$. Then the maximal value of f on P is equal to its value on one of the vertices $\mathbf{v}_1, \ldots, \mathbf{v}_k$.*

Proof. Let $\mathbf{x} \in P$. Then $\mathbf{x} = t_1 \mathbf{v}_1 + \cdots + t_k \mathbf{v}_k$ for some $t_i \ge 0$ such that $t_1 + \cdots + t_k = 1$, so that

$$f(\mathbf{x}) = (\mathbf{x}, \mathbf{a}) = (t_1 \mathbf{v}_1 + \cdots + t_k \mathbf{v}_k, \mathbf{a}) = t_1(\mathbf{v}_1, \mathbf{a}) + \cdots + t_k(\mathbf{v}_k, \mathbf{a}).$$

Let $(\mathbf{v}_j, \mathbf{a}) = \max_{i=1..k}(\mathbf{v}_i, \mathbf{a})$. Then

$$f(\mathbf{x}) \le (t_1 + \cdots + t_k)(\mathbf{v}_j, \mathbf{a}) = (\mathbf{v}_j, \mathbf{a}).$$

Thus

$$\max_{x \in P} f(x) = (\mathbf{v}_j, \mathbf{a}) = f(\mathbf{v}_j).$$

□

11.4 Transportation Problem

Assume that there are m producer-cities and n consumer-cities of some good, say, potato. Consider the matrix $X = \|\xi_{ij}\|_{m \times n}$, where ξ_{ij} is the quantity of good sent from producer-city i to consumer-city j (Fig. 11.13).

Note that we must have

$$\xi_{ij} \ge 0, \quad i = 1, \ldots, m; \quad j = 1, \ldots, n.$$

Assume that the production of city i cannot exceed σ_i (due to capacity constraints), and the consumption of city j must be at least δ_i (to survive). In other words,

$$\sum_{j=1}^{n} \xi_{ij} \le \sigma_i, \quad i = 1, \ldots, m \qquad (11.4)$$

11.5 Dual Problem

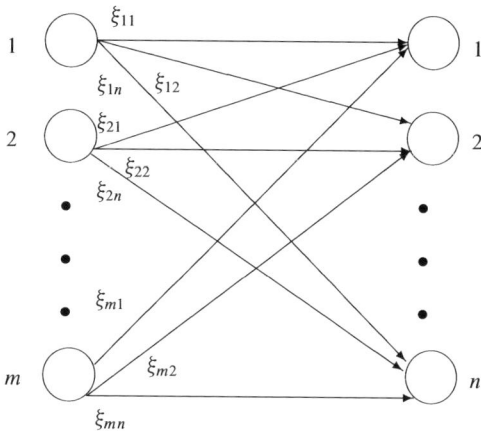

Fig. 11.13 The transportation problem

and

$$\sum_{i=1}^{m} \xi_{ij} \geq \delta_j, \; j = 1, \ldots, n. \tag{11.5}$$

Let the cost of transporting one unit of good from producer-city i to consumer-city j is equal to γ_{ij}. Then, the optimal transportation program is a solution to

$$\min_{\{\xi_{ij}\}} \sum_{i=1}^{n} \sum_{j=1}^{n} \gamma_{ij} \xi_{ij}.$$

subject to (11.4) and (11.5).

11.5 Dual Problem

Let us return to the two-dimensional case of the linear programming problem. Consider a case in which linear programming problem has the unique solution (Fig. 11.14).

Clearly, the point $\mathbf{x}^*$ in the above figure is a (unique) solution of the problem. Let us rotate the line $(\mathbf{c}, \mathbf{x}) = \mu_{\max}$ around $\mathbf{x}^*$ by an arbitrarily small angle, and obtain a new line $(\mathbf{c}', \mathbf{x}) = \mu'$. The new line is not orthogonal to $\mathbf{c}$. So, μ' cannot be the maximum of $(\mathbf{c}, \mathbf{x})$. By the fact that $\mathbf{x}^*$ uniquely maximizes $(\mathbf{c}, \mathbf{x})$, it follows that there exists $\mathbf{c}'$ such that $\mathbf{x}^*$ maximizes $(\mathbf{c}', \mathbf{x})$ and $(\mathbf{c}', \mathbf{x}^*) = \mu'$.

Fig. 11.14 The dual problem

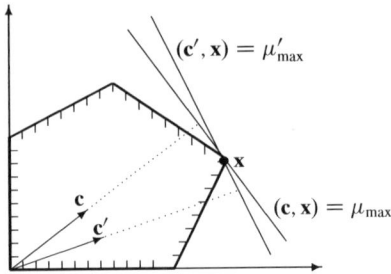

This observation can be generalized to the n-dimensional case as follows. Consider the following linear programming model:

$$\max_{\{x\}} \; (\mathbf{c}, \mathbf{x}) \tag{11.6}$$

$$\text{s.t.} \; \mathbf{x}A \leq \mathbf{d}, \tag{11.7}$$

$$\mathbf{x} \geq \mathbf{0}, \tag{11.8}$$

where

$\mathbf{x} = (\xi_1, \ldots, \xi_m)$ - vector of intensities,
$\mathbf{c} = (\gamma_1, \ldots, \gamma_m)$ - profits under unit intensities,
$\mathbf{d} = (\delta_1, \ldots, \delta_n)$ - restrictions on total production and consumption of the goods,
$A = \|\alpha_{ij}\|_{m \times n}$ - production-consumption matrix under unit intensities,
m - number of firms,
n - number of goods.

The following problem is called the *dual* of the problem (11.6)–(11.8):

$$\min_{\{y\}} \; (\mathbf{d}, \mathbf{y}) \tag{11.9}$$

$$\text{s.t.} \; \mathbf{y}A^T \geq \mathbf{c}, \tag{11.10}$$

$$\mathbf{y} \geq \mathbf{0}, \tag{11.11}$$

where $\mathbf{y} = (\eta_1, \ldots, \eta_n)$.

We can rewrite the dual problem as follows:

$$\max_{\{y\}} \; (-\mathbf{d}, \mathbf{y})$$

$$\text{s.t.} \; -\mathbf{y}A^T \leq -\mathbf{c},$$

$$\mathbf{y} \geq \mathbf{0},$$

which looks like a direct problem. Then the dual of the last problem is given by

$$\min_{\{x\}} (-c, x) \tag{11.12}$$

$$\text{s.t.} \quad -xA \le -d, \tag{11.13}$$

$$x \ge 0, \tag{11.14}$$

which is equivalent to (11.6)–(11.8). So, the dual of the dual problem is the original problem.

Lemma 11.8. *Let* x *be a solution of the direct problem (11.6)–(11.8) and* y *be a solution of the dual problem (11.9)–(11.11). Then*

$$(c, x) \le (d, y).$$

Theorem 11.9. *Let* x *be a feasible vector in (11.6)–(11.8) and* y *be a feasible vector in (11.9)–(11.11). If*

$$(c, x) = (d, y),$$

then x *and* y *are the solutions of the corresponding problems.*

11.6 Economic Interpretation of Dual Variables

In this section we follow the analysis given in [4].
  Consider the function
$$L(d) = (c, x) = (d, y).$$
Note that by changing d and then solving, say, the direct problem (11.6)–(11.8), we can get new values of $L(d)$. Assume that the solution of the direct problem, and hence of the dual problem, is unique.
  Then, for an increment $\triangle d$ in d, let us evaluate $L(\triangle d)$. The total differentiation of $L(d) = (d, y)$ gives us

$$\triangle L(d) = \triangle(d, y) = (\triangle d, y) + (d, \triangle y) + (\triangle d, \triangle y).$$

When the solution to a given problem is unique, it follows that y is one of the vertices of the polytope given by the equation $yA^T \ge c$. Note that the set of vertices is finite. By a standard analytical argument, y does not change under sufficiently small perturbations of the objective function. So, we can assume $\triangle y = 0$, which yields

$$\triangle L(d) = (\triangle d, y).$$

Let all components of $\triangle d$, except for j-th one, are zero, i.e.,

$$\triangle d = (0, \ldots, 0, \triangle \delta_j, 0, \ldots, 0).$$

Then we can write
$$\Delta L(\mathbf{d}) = \Delta \delta_j \eta_j.$$

For the profit maximization problem, $\Delta L(\mathbf{d})$ is measured in monetary terms, say, in dollars, while $\Delta \delta_j$ is measured in units of production-consumption goods. So, η_j must be in dollars, and can be interpreted as the unit price of the goods.

Inequality (11.10) can be explicitly written as

$$\sum_{j=1}^{n} \alpha_{ij} \eta_j \geq \gamma_i, (i = 1, \ldots, m) \tag{11.15}$$

for the optimal solution $\mathbf{y}$. Since for all j, η_j being a price is always nonnegative, the left hand side of (11.15) may have both positive and negative components. Recall that positive (negative) α_{ij}'s correspond to production (consumption). So, the sum of positive components in the left hand side of (11.15) denote the revenues that unit i obtains by selling goods when it works with intensity one. Analogously, the sum of negative components in the left hand side of (11.15) denote the (minus) expenditures for purchasing goods incurred by the unit i when working with intensity one. Thus, for each unit (11.15) can be written as

$$\text{revenues} - \text{expenditures} \geq \text{total profits}, \tag{11.16}$$

when the economic units work with the intensity 1. If, for some unit, (11.16) holds with a strict inequality, we say that this unit is not fully utilizing its resources. In that case, we claim that the intensity of this unit in the optimal solution must be 0.

Theorem 11.10 (Equilibrium Theorem). *Let $\mathbf{x}$ and $\mathbf{y}$ are feasible vectors for the direct and dual problems, respectively, which are described above. Then these vectors are solutions to the corresponding problems if and only if for all $i = 1, \ldots, m$ and $j = 1, \ldots, n$*

$$\eta_j = 0 \text{ if } \sum_{i=1}^{m} \alpha_{ij} \xi_i \leq \delta_j \tag{11.17}$$

and

$$\xi_i = 0 \text{ if } \sum_{j=1}^{n} \alpha_{ij} \eta_j \leq \gamma_i. \tag{11.18}$$

Remark 11.1. Note that (11.18) validates our previous claim that if a unit is not efficiently operating, then it does not take part in the optimal program. On the other hand, from (11.17) we observe that if some resource is not fully utilized in the optimal program, then its price equals zero. A natural question that follows is whether it has been ever the case that a resource had a zero price. As it is emphasized in [4], the answer is yes! In the 'old days' demand for water was much below of its (natural) supply, and thus factories did not pay for it.

11.7 A Generalization of the Leontief Model: Multiple Production Techniques and Linear Programming

In Leontief model (Sect. 10.2) each production technique is assumed to be producing only one commodity (no joint production assumption) and each commodity is produced only by one production technique (no alternative production techniques). In this section, the second assumption will be relaxed to allow the model to deal with the problem of choice of techniques. The remaining assumptions of the model will be retained.

Suppose there are again m commodities, each can be produced by $n(i)$ different techniques $(i = 1, \ldots, m)$. Let

$$\sum_{i \in I} n(i) = n, \quad n(i) \geq 1, \quad I = \{1, \ldots, m\}$$

Let us define

$$\tilde{A} = (\tilde{a}_{ij})_{m \times n} \tag{11.19}$$

as the technology matrix that represent the available production techniques, where $m < n$. The problem then is to identify a submatrix of $\tilde{A}$ that serves best for the policy maker's purpose. The solution of this problem requires a clear definition of the purpose of the decision maker and the appropriate mathematical apparatus to solve it. Consider the following example.

Example 11.5. Consider an economy that can be characterized by the following conditions:
1. There are m commodities and $n(> m)$ techniques of production.
   Let x_i denote the activity level of production technique i. If it is found meaningful to operate this activity then $x_j > 0$, if not $x_j = 0$.
2. The technology matrix is given by (11.19).
3. All commodities are treated as resources. At the beginning of the production period t, the initial endowment of resources is given by the m-dimensional column vector

$$\omega = \begin{bmatrix} \omega_1 \\ \omega_2 \\ \ldots \\ \omega_m \end{bmatrix}$$

each component of this vector gives the amount of commodity i $(I = 1, \ldots, m)$ available to be used for production.
4. Let

$$\mathbf{v}^T = (v_1, \ldots, v_n)$$

be the row vector of value added coefficients (value added created by the process when one unit of output is produced).

5. Suppose the policymaker's problem is to maximize the total value added at time t, under the conditions stated above.
Then this problem can be expressed as

$$\max \mathbf{v}^T \mathbf{x}$$

subject to

$$\tilde{A}\mathbf{x} \leq \omega,$$

and since activity level is non-negative, the solution should also satisfy the following set of conditions

$$\mathbf{x} \geq \mathbf{0}.$$

We see that this is a linear programming problem.

11.8 Problems

1. Show that $\mathbb{R}^n$ is a convex set.
2. Show that the intersection of any two convex sets in $\mathbb{R}^n$ is convex.
3. Let $A : \mathbb{R}^n \to \mathbb{R}^n$ is a linear operator, and let X be a set of all vectors $\mathbf{x} \in \mathbb{R}^n$ such that

$$A(\mathbf{x}) \geq \mathbf{0}.$$

 Show that X is convex.
4. Let $\mathbf{a} \in \mathbb{R}^n$ and $\alpha \in \mathbb{R}$. Show that the sets

$$\{\mathbf{x} \in \mathbb{R}^n | (\mathbf{a}, \mathbf{x}) \geq \alpha\} \text{ and } \{\mathbf{x} \in \mathbb{R}^n | (\mathbf{a}, \mathbf{x}) \leq \alpha\}$$

 are convex.
5. Let $X = \{\mathbf{x}\}$ where $\mathbf{x} \in \mathbb{R}^n$. Show that X is convex.
6. Draw the convex hull of the following sets of points:
   (a) $(-2, 0), (0, -2), (0, 0), (0, 2), (1, 0.5), (2, 0)$.
   (b) $(2, 2), (-2, 2), (0, 0), (0, 1), (0, 2), (-2, -2), (2, -2), (1, -1)$.
7. Determine whether the following sets are convex:
   (a) $X = \{(x_1, x_2) \mid x_1^2 + x_2^2 \leq 1\}$
   (b) $X = \{(x_1, x_2) \mid x_1, x_2 \leq 1, \ x_1 \geq 0, \ x_2 \geq 0\}$
   (c) $X = \{(x_1, x_2) \mid x_1^2 + 2x_2^2 \leq 4\}$
8. Consider the hyperplane

$$H = \{(x_1, x_2, x_3, x_4, x_5) \in \mathbb{R}^5 \mid x_1 - x_2 + 4x_3 - 2x_4 + 6x_5 = 1\}.$$

11.8 Problems

In which half space does each of the following points lie?
   a. $(1, 1, 1, 1, 1)$
   b. $(2, 7, -1, 0, -3)$
   c. $(0, 1, 1, 4, 1)$
9. Let $(\mathbf{a}, \mathbf{x}) = 0$ be a hyperplane in $\mathbb{R}^n$. Prove that it is a subspace of $\mathbb{R}^n$.
10. Determine whether the union of any m convex sets $X_1, \ldots, X_m$ in $\mathbb{R}^n$ is convex.
11. Mr Holst has 100 acres (404,686 square meters) of land where he can plant wheat and/or barley. His problem is to maximize his expected net revenue. His capital is $1,000. It costs $7 to plant a acre of wheat and $5 to plant a acre of barley. Mr Holst is able to secure 150 of days of labor per season. Wheat and barley require, respectively, 2 and 3 days of labor per acre.

    According to the existing information, revenue from an acre of wheat and barley are $75 and $55, respectively.

    How many acres of each cereal should be planted to maximize Mr Holst's net revenue?
12. Suppose that a fund manager is entrusted with $1 million to be invested in securities. The feasible alternatives are government bonds, automotive and textile companies. The list of alternatives and the expected returns are given below

	Expected return (%)
Government bonds	3.5
Auto producer A	5.5
Auto producer B	6.5
Textile company C	6
Textile company D	9

    The following investment guidelines were imposed:
    (a) Neither industry (auto or textile) should receive more than 50% of the amount to be invested.
    (b) Government bonds should be at least 35% of the auto industry investments.
    (c) The investment in high yield, high risk Textile Company D can not be more than 65% of the total textile industry investments.
    Find the composition of the portfolio that maximizes the projected return, under the conditions given above.
13. A research institute has an endowment fund of $10 million. The research institute, naturally, concerned with the risk it is taking. On the other hand it also needs to generate $400,000 to cover its expenses. The institute has two options: Investing in low risk money market fund (x) or to high risk stock funds (y).

    Money market fund's risk scale is 4, whereas the stock funds' is 9. The return of these funds, on the other hand, is as follows: Money market fund 4% and stock fund 10%. The price of a money market fund certificate is $50 and the price of a stock fund certificate is $150.

The management of the institute decided that, for liquidity purposes, the amount invested in money markets should not be less than $90,000.

Using this information determine the optimal portfolio allocation for the research institute.

14. Consider a simple two sector economy, where labor is the sole production factor. The input coefficients matrix for this economy is given as

$$A = \begin{bmatrix} 0.1 & 0.2 \\ 0.3 & 0.15 \end{bmatrix},$$

which implies that both goods are used in production of itself and the other one. The labor coefficients vector is given by

$$\mathbf{a}_0 = (0.05, 0.07).$$

The total labor supply is 150 (million man-hours).

In the previous year the final consumption of first sector output C_1 was $1,000 (million), and the second sector output C_2 was $440 (million).

Suppose that the social welfare is a linear function of consumption and given by

$$W(C_1, C_2) = C_1 + (1, 1)C_2.$$

Suppose the government wants to maximize social welfare without reducing the consumption of each good below their previous levels. What will be the optimal solution?

15. *Motivation.* The following problem is a simple exercise in the two-gap growth theory. This theory dates back to early 1960s and approaches the development problem by observing the fact that developing countries need both savings to invest and foreign exchange to import the necessary goods. Therefore they face two problems: the first is the saving constraint and the second is the foreign exchange constraint. Therefore for such countries using simple Harrod-Domar type of framework, which takes into account only the savings constraint, may not be sufficient.

Question. Suppose the economy at hand is described by the following set of structural equations:

1. Aggregate Demand.
$$Y^d = C + I + X - M,$$

where
Y^d: Aggregate Demand (GDP measured from demand side).
C: Consumption.
I: Investment.
X: Exports.
M: Imports.

11.8 Problems

2. **Capacity Growth.**

    The increase in the production capacity of the economy depends on the amount of investment. It is assumed that marginal capital output ratio, i.e., $I/\Delta Y = 4$; this means that one unit of investment produces 0.25 units of GDP. This relation can be expressed as

    $$\Delta Y = Y^s - Y^s_{-1} = 0.25 I,$$

    where Y^s and Y^s_{-1} are capacity aggregate output levels of present and the previous periods. Suppose that

    $$Y^s_{-1} = 300.$$

    Therefore the above equation can be written as

    $$Y^s = 0.25 I + 300.$$

3. **Saving and Investment.**

    Investments are either financed by domestic savings (S) or external borrowing (F):

    $$I = S + F.$$

    Suppose that the domestic savings is a linear function of the GDP (Y),

    $$S = 0.2Y,$$

    and foreign borrowing is forecasted to be at most 20, i.e.,

    $$F \leq 20.$$

    Therefore,

    $$I \leq 0.2Y + 20.$$

4. **Equilibrium Condition.**

    The economy is in equilibrium when aggregate supply (Y^s) is equal to aggregate demand (Y^d). Equilibrium output level is denoted by Y:

    $$Y = Y^s = Y^d.$$

5. **Import Equation.**

    The 5% of the total consumption and 15% of the investment goods are imported. Therefore,

    $$M = 0.05 C + 0.15 I.$$

6. Foreign Exchange Constraint.

The country can get access to foreign exchange either by exports or through borrowing from abroad. In other words, imports can not exceed the sum total of them, i.e.,

$$M \leq X + F.$$

For the sake of simplicity lets assume that the maximum amount of exports for the current year is estimated to be 100, that is,

$$X \leq 100.$$

Therefore, the foreign exchange constraint can be written as

$$M \leq 100 + 20 = 120.$$

The government wants to maximize current GDP, under saving and foreign exchange constraints. However, the government also does not want the current consumption level to be lower than the one of the previous period.

How much investment is needed? What is the new GDP level? Will there be an increase in total consumption?

Natural Numbers and Induction   A

In this appendix, we discuss one powerful and general method to deduce theorems and formulas depending on a natural parameter. This method, called *induction*, could be illustrated by the domino effect: if dominoes are stood on end one slightly behind the other, a slight push on the first will topple the others one by one.

As a toy example, let us prove that the area of the rectangle of the size $5 \times n$ is equal to $5n$. For $n = 1$, the rectangle can obviously be cut up into 5 unit squares, so its area is equal to 5 (the first domino falls down). Now, for $n = 2$ we explore the domino effect and cut up the rectangle into two ones of size 5×1. This gives the area $5 + 5 = 5 \cdot 2$. For $n = 3$, we cut up the rectangle into the rectangles 5×2 and 5×1: this gives the area $5 \cdot 2 + 5 \cdot 1 = 5 \cdot 3$. By the same way, we cut up any large rectangle of the size $5 \times n$ into two ones, that is, the rectangle $5 \times (n-1)$ and the 'ribbon' 5×1 (see Fig. A.1). At some moment, we can assume that the area of the first one is known to be $S_{n-1} = 5 \cdot (n-1)$ (the $(n-1)$-th domino has fallen down), so, we calculate the area as $S_n = 5 \cdot (n-1) + 5 \cdot 1 = 5 \cdot n$.

Similar (and slightly more complicated) methods can be applied in many problems. In order to give their formal description, we first discuss a formal introduction to natural numbers.

A.1   Natural Numbers: Axiomatic Definition

Natural numbers are known as the main and the basic objects in mathematics. Many complicated things such as rational and real numbers, vectors, and matrices can be defined via the natural numbers. "God made the integers; all else is the work of man", said Leopold Kronecker, one of the most significant algebraists of nineteenth century.

What are the natural numbers? This question admits a lot of answers, all in different levels of abstraction. The naïve definition says that these are just the numbers used in counting, that is, $0, 1, 2, \ldots$ In the geometry of the real line, the

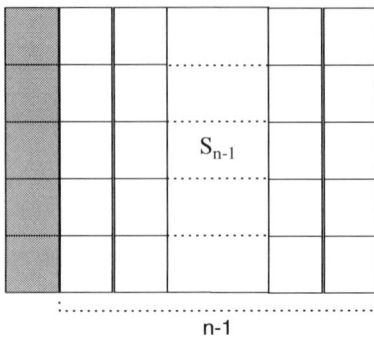

Fig. A.1 The area of a rectangle: proof by induction

set of natural numbers ℕ is defined as the set of the following points in the line: an initial point (say, O); the right-hand end of the unit segment whose left-hand end is O; the right-hand end of the unit segment whose left-hand end is the just defined point; and so on. In the set theory, every natural number n is considered as the simplest set of n elements: $0 = \emptyset, 1 = \{0\}, 2 = \{0, 1\}, 3 = \{0, 1, 2\}$ etc. These set-theoretical numbers play the role of etalons for counting all other finite sets.

Let us give the most general definition, that covers all previous ones. It is based on the axioms due to Peano[1] (1889). Assume the $\mathcal{N}$ is a set, 0 is an element of the set $\mathcal{N}$, and $s : \mathcal{N} \to \mathcal{N}$ is a function (called 'successor'). The set $\mathcal{N}$ is called *a set of natural numbers*, if the following three *Peano axioms* are satisfied.
1. If $s(m) = s(n)$, then $m = n$. (This means that the function s is injective).
2. There is no such n that $s(n) = 0$.
3. (Induction axiom). Suppose that there is a subset $A \subset \mathcal{N}$ such that (1) $0 \in A$ and (2) for every $n \in A$ we have $s(n) \in A$. Then $A = \mathcal{N}$.

For example, one can define an operation $n \mapsto n + 1$ as $n + 1 := s(n)$.

Exercise A.1. Define an operation s for the above versions of natural numbers, that is, for the collection N_1 of points in a real line and for the sequence of sets: $0 = \emptyset, 1 = \{0\}, 2 = \{0, 1\} \ldots$

Let us define also an operation $n \mapsto n + m$ via the two rules: $n + 0 := n$ and $n + (m + 1) := s(n + m)$. Let A be the set of the numbers $m \in \mathcal{N}$ for which we can calculate the sum $n + m$ using this rule. Then $0 \in A$ (by the first rule) and for every $m \in A$ we have $m + 1 \in A$ (by the second rule). It follows from the induction axiom that the sum $n + m$ is defined for all $m, n \in \mathcal{N}$.

---

[1]Giuseppe Peano (1858–1932) was a famous Italian mathematician. Being one of the founders of mathematical logic and set theory, he was also an author of many analytical discoveries including a continuous mapping of a line onto every point of a square. Another his discovery was *Latino sine Flexione* (or *Interlingua*), an artificial language based on Latin with simpler grammar.

Exercise A.2. Give definitions (in a similar way as the above definition of the addition) of the following operations with natural number m and n:
1. $m \cdot n$.
2. m^n.

This way to define an operation (from each natural number m to the next number $m + 1$ an so on) is called *recursion*. For example, a recursive formula gives a definition of the determinant of a matrix of order n, see the formula (3.6). The recursion definitions are appropriate for using in the induction reasonings, as described below.

A.2 Induction Principle

In order to deduce any significant property of natural numbers from the above axioms, one should use a special kind of reasoning, called *the induction principle*, or *mathematical induction*.

Let $P(n)$ be an arbitrary statement concerning a natural number n (like, for example, "n is equal to 5", or $n + 2 = 2 + n$, or "either $n \leq 2$ or $x^n + y^n \neq z^n$ for any natural x, y, z").

Theorem A.1. *Let $P(n)$ be a statement[2] depending on element n of a set of natural numbers $\mathcal{N}$. Suppose that the following two assumptions hold:*
1. *(The basis, or The initial step) $P(0)$ is true.*
2. *(The inductive step) $P(n + 1)$ is true provided that $P(n)$ is true.*
   Then $P(n)$ is true for every $n \in \mathcal{N}$.

Note that the element n here is called *induction variable*, and the assumption that $P(n)$ holds in the inductive step is called *induction assumption*.

Proof. Let A be a set consisting of all natural numbers $n \in \mathcal{N}$ such that $P(n)$ is true. According to the basis of induction, we have $0 \in A$. By the induction step, for every $n \in A$ we have also $n + 1 \in A$. Thus, we can apply the induction axiom and conclude that $A = \mathcal{N}$. □

Example A.1. Let us prove the formula

$$0 + 1 + 2 + \cdots + n = \frac{n(n+1)}{2}. \tag{A.1}$$

Let $P(n)$ be the above equality. For $n = 0$ it is obviously true: $0 = \frac{0 \cdot 1}{2}$. This gives the basis of the induction. To prove the induction step, let us assume that the statement $P(n)$ is true for some n, that is, the equality (A.1) holds. We have to

---

[2] Note that we do not give here a strong mathematical definition of a term 'statement'. At least, all statements consisting of arithmetical formulas with additions like "for every natural n" or "there exists natural x such that" are admissible.

deduce the statement $P(n+1)$, that is, the same formula with n replaced by $n+1$. Using the statement $P(n)$, we re-write the left hand side of the equality $P(n+1)$ as follows:

$$0 + 1 + 2 + \cdots + n + (n+1) = \frac{n(n+1)}{2} + (n+1) = \frac{(n+1)(n+2)}{2}.$$

This equality is equivalent to $P(n+1)$, so, the induction step is complete. By the induction principle, we conclude that $P(n)$ holds for all n.

Example A.2. In this example we deduce some standard properties of natural numbers from Peano axioms.

First, let us consider the following statement $P(n)$: $0 + n = n$. By the above definition of addition, we have $m + 0 = m$ for all $m \in \mathcal{N}$, hence $0 + 0 = 0$. This gives the basis of the induction: $P(0)$ is true. To prove the induction step, let us assume that the statement $P(n)$ is true for some n, that is, $0 + n = n$. Using this fact and the definition of addition, we obtain: $0 + (n+1) = s(0+n) = s(n) = n+1$, i.e., we have deduced the statement $P(n+1)$. Since both the basis of the induction and the inductive step are true, we conclude that $P(n)$ is true for all n.

Now, let us prove the associativity property

$$(l + m) + n = l + (m + n)$$

for all natural l, m, n (we denote this statement by $S_{l,m}(n)$). Again, let us apply the induction (on the number n). For $n = 0$, we get the trivial statement $S_{l,m}(0) : l + m = l + m$, which is obviously true, so, the basis of the induction $S_m(0)$ is proved. To show the induction step, it remains to show that $S_{l,m}(n+1)$ is true for all l, m provided that $S_{l,m}(n)$ is. Using the equalities $S_{l,m}(n)$, we have $(l+m)+(n+1) = s((l+m)+n) = s(l+(m+n)) = l+s(m+n) = l+(m+(n+1))$. This gives $S_{l,m}(n+1)$. Hence, the proof is complete.

Now, let us prove the following statement $Q_m(n)$

$$m + (n+1) = (m+1) + n.$$

We proceed by the induction on the variable n. For $n = 0$, we have $m + 1 = m + 1$: this is the basis $Q_m(0)$ of the induction. To show the induction step, let us assume that for some n the statement $Q_m(n)$ is true for all m. We have to show $Q_m(n+1)$. Using the assumption, we have

$$m + (n+2) = s(m + (n+1)) = s((m+1) + n) = (m+1) + (n+1).$$

So, we have deduced $Q_m(n+1)$. By the induction principle, the equality $Q_m(n)$ holds for all m and n.

A.2 Induction Principle

Finally, let us show the commutativity property

$$m + n = n + m$$

for all $m, n \in \mathcal{N}$. We proceed by the induction on n. For $n = 0$ we get the above statement $P(m)$, hence the initial step holds. To prove the induction step, we use the induction assumption (the equality $m + n = n + m$) and the statement $Q_n(m)$:

$$m + (n + 1) = s(m + n) = s(n + m) = n + (m + 1) = (n + 1) + m.$$

Thus, the induction step is proved, and the induction is complete. □

Exercise A.3. Using the definition of multiplication of natural numbers given in Exercise A.2, prove the following standard properties:
1. $a(bc) = (ab)c$.
2. $ab = ba$.

The following version of Theorem A.1 is called a *weak induction principle*.

Corollary A.2. *Let $P(n)$ be a statement depending on element n of a set of natural numbers $\mathcal{N}$. Suppose that the following two assumption hold:*
1. *$P(n_0)$ is true for some $n_0 \in \mathcal{N}$.*
2. *$P(n + 1)$ is true provided that $P(n)$ is true, where $n \geq n_0$.*
Then $P(n)$ is true for every $n \geq n_0$.

Proof. Let $P'(n)$ be the following statement: '$P(n_0 + n)$ is true'. Then the above conditions on $P(n)$ are equivalent to the conditions of Theorem A.1 for the statement $P'(n)$. Hence, we apply the induction principle and deduce that $P'(n)$ is true for all n. This means that $P(n)$ is true for all $n \geq n_0$. □

Example A.3. Let us solve the inequality

$$2^n > 3n, \qquad (A.2)$$

where $n > 0$ is an integer.

It is easy to check that the inequality fails for $0 < n \leq 3$, while for $n = 4$ it holds: $2^4 > 3 \cdot 4$. It is natural to assume that the inequality holds for all $n \geq 4$. How to prove the assumption?

Let us apply the weak induction principle with $n_0 = 4$. The statement $P(n)$ is then the inequality (A.2). The initial step $P(4)$ is done. To prove the induction step, we try to deduce $P(n + 1)$ from $P(n)$, where $n \geq 4$. Consider the left hand side 2^{n+1} of $P(n + 1)$. According to $P(n)$, we have

$$2^{n+1} = 2 \cdot 2^n > 2 \cdot 3n = 6n.$$

Since $n \geq 4$, we have $6n \geq 3n + 3n > 3n + 3 = 3(n + 1)$. Thus, we obtain an inequality $2^{n+1} > 3(n + 1)$, which is equivalent to $P(n + 1)$. By Corollary A.2, the inequality (A.2) holds for all $n \geq 4$.

A more general version of the induction principle is given by the following Strongest Induction Principle.

Corollary A.3. *Let $P(n)$ be a statement depending on element n of a set of natural numbers $\mathcal{N}$. Suppose that the following two assumption hold:*
1. *$P(n_0)$ is true for some $n_0 \in \mathcal{N}$.*
2. *$P(n + 1)$ is true provided that $P(k)$ is true for all $n \geq k \geq n_0$.*

Then $P(n)$ is true for every $n \geq n_0$.

Proof. Let $P'(n)$ be the following statement: '$P(k)$ is true for all $n \geq k \geq n_0$'. Then we can apply Corollary A.2 to the statement $P'(n)$. □

Example A.4. Problem. Evaluate the determinant of the following matrix of order $n \times n$

$$A_n = \begin{pmatrix} 3 & 2 & 0 & \ldots & 0 & 0 \\ 1 & 3 & 2 & \ldots & 0 & 0 \\ 0 & 1 & 3 & \ldots & 0 & 0 \\ & & & \ldots & & \\ 0 & 0 & 0 & \ldots & 1 & 3 \end{pmatrix}$$

Solution. Let us denote $D_n = \det A_n$. By direct calculations, we have $D_1 = 3$, $D_2 = 7$. It is natural to formulate a *conjecture*, called $P(n)$:

$$D_n = 2^{n+1} - 1, \text{ where } n \geq 1.$$

To prove the above conjecture, we apply the strongest induction principle (Corollary A.3). We put $n_0 = 1$: then the initial step $P(1)$ is given by the equality $D_1 = 3$. To show the induction step, let us evaluate D_{n+1}. If $n = 1$, then $D_{n+1} = D_2 = 7$, and the conjecture holds. For $n \geq 2$ we have

$$D_{n+1} = \det A_{n+1} = \det \begin{pmatrix} 3 & 2 & 0 & \ldots & 0 \\ 1 & 3 & 2 & \ldots & 0 \\ 0 & 1 & & & \\ \ldots & \ldots & & A_{n-1} & \\ 0 & 0 & & & \end{pmatrix}$$

$$= 3 \cdot \det A_n - 2 \cdot 1 \cdot \det A_{n-1} - 2 \cdot 2 \cdot 0 = 3D_n - 2D_{n-1}.$$

Using $P(n)$ and $P(n - 1)$, we get

$$D_{n+1} = 3(2^{n+1} - 1) - 2(2^n - 1) = 3 \cdot 2^{n+1} - 2^{n+1} - 1 = 2 \cdot 2^{n+2} - 1.$$

A.3 Problems

So, we have deduce $P(n+1)$. This complete the induction step. Thus, the conjecture is true for all $n \geq 1$. □

Other methods of evaluating such determinants will be discussed in Appendix B.

A.3 Problems

1. Show that
$$1 + 3 + \cdots + (2n - 1) = n^2.$$

2. Find the sum
$$\frac{1}{1 \cdot 2} + \frac{1}{2 \cdot 3} + \cdots + \frac{1}{(n-1) \cdot n}.$$

3. Prove the following *Bernoulli's inequality*
$$(1 + x)^n \geq 1 + nx \text{ for every natural } n \text{ and real } x \geq 0.$$

4. Show that the number $x^n + \frac{1}{x^n}$ is integer provided that $x + \frac{1}{x}$ is integer.
5. Describe all natural n such that $2^n > n^2$.
6. Suppose that an automatic machine sells two type of phone cards, for \$3 and \$5 (for 30 min and 70 min of phone calls, respectively). Show that any integer amount greater than \$7 can be exchanged for the cards without change.
7. Show that
$$\begin{bmatrix} 1 & 1 \\ 0 & 1 \end{bmatrix}^n = \begin{bmatrix} 1 & n \\ 0 & 1 \end{bmatrix}.$$

8. Compute
   (a)
   $$\begin{bmatrix} \lambda & 1 \\ 0 & \lambda \end{bmatrix}^n$$
   (b)
   $$\begin{bmatrix} \cos \alpha & -\sin \alpha \\ \sin \alpha & \cos \alpha \end{bmatrix}^n$$

9. Show that the matrix
$$\begin{bmatrix} a & -b \\ b & a \end{bmatrix}^n + \begin{bmatrix} a & b \\ -b & a \end{bmatrix}^n$$
has the form
$$\begin{bmatrix} c & 0 \\ 0 & c \end{bmatrix}$$
for some number c.

10. Show that the determinant of order n

$$\begin{vmatrix} 0 & 1 & 0 & \ldots & 0 & 0 \\ -1 & 0 & 1 & \ldots & 0 & 0 \\ 0 & -1 & 0 & \ldots & 0 & 0 \\ \vdots & \vdots & \vdots & \ddots & \vdots & \vdots \\ 0 & 0 & 0 & \ldots & -1 & 0 \end{vmatrix}$$

is equal to $(1+(-1)^n)/2$.

Methods of Evaluating Determinants   B

In addition to examples in Sect. 3.4, we discuss here some advanced methods of evaluating the determinants of various special matrices. This Appendix is mainly based on [25, Sect. 1.5]. For further methods of determinant evaluation, we refer the reader to [15].

B.1   Transformation of Determinants

Sometimes we can prove some equalities of determinants without directly evaluating them.
   Let us consider the following problems.

Example B.1. Problem. Prove that

$$\begin{vmatrix} 1 & a & bc \\ 1 & b & ca \\ 1 & c & ab \end{vmatrix} = \begin{vmatrix} 1 & a & a^2 \\ 1 & b & b^2 \\ 1 & c & c^2 \end{vmatrix}.$$

Solution. We can add one column of a determinant with any other column multiplied by some constant without changing the value of the determinant (see the property (a) in Sect. 3.4). Let us use this method here to get

$$\begin{vmatrix} 1 & a & bc \\ 1 & b & ca \\ 1 & c & ab \end{vmatrix} = \begin{vmatrix} 1 & a & bc + a(a+b+c) - 1(ab+ac+bc) \\ 1 & b & ac + b(a+b+c) - 1(ab+ac+bc) \\ 1 & c & ab + c(a+b+c) - 1(ab+ac+bc) \end{vmatrix}$$

$$= \begin{vmatrix} 1 & a & bc + a^2 + ab + ac - ab - ac - bc \\ 1 & b & ac + b^2 + ab + ac - ab - ac - bc \\ 1 & c & ab + c^2 + ab + ac - ab - ac - bc \end{vmatrix}$$

$$= \begin{vmatrix} 1 & a & a^2 \\ 1 & b & b^2 \\ 1 & c & c^2 \end{vmatrix}.$$

Example B.2. Problem. Prove that

$$\begin{vmatrix} 0 & x & y & z \\ x & 0 & z & y \\ y & z & 0 & x \\ z & y & x & 0 \end{vmatrix} = \begin{vmatrix} 0 & 1 & 1 & 1 \\ 1 & 0 & z^2 & y^2 \\ 1 & z^2 & 0 & x^2 \\ 1 & y^2 & x^2 & 0 \end{vmatrix}.$$

Solution. The answer follows from

$$\frac{1}{x^2 y^2 z^2} \begin{vmatrix} 0 & xyz & xyz & xyz \\ x & 0 & xz^2 & xy^2 \\ y & yz^2 & 0 & x^2y \\ z & y^2z & x^2z & 0 \end{vmatrix} = \frac{xyz \cdot xyz}{x^2 y^2 z^2} \begin{vmatrix} 0 & 1 & 1 & 1 \\ 1 & 0 & z^2 & y^2 \\ 1 & z^2 & 0 & x^2 \\ 1 & y^2 & x^2 & 0 \end{vmatrix}.$$

B.2 Methods of Evaluating Determinants of High Order

B.2.1 Reducing to Triangular Form

One of useful methods of calculation determinant is to reduce a matrix to a triangular form via elementary transformations and then calculate its determinant as a product of diagonal elements by Example 3.5.

Example B.3. Reducing the below matrix

$$\begin{vmatrix} a_1 & x & x & \dots & x \\ x & a_2 & x & \dots & x \\ x & x & a_3 & \dots & x \\ \vdots & \vdots & \vdots & \ddots & \vdots \\ x & x & x & \dots & a_n \end{vmatrix}$$

to the triangular form, subtract the first row from all other rows to get

$$\begin{vmatrix} a_1 & x & x & \dots & x \\ x-a_1 & a_2-x & 0 & \dots & 0 \\ x-a_1 & 0 & a_3-x & \dots & 0 \\ \vdots & \vdots & \vdots & \ddots & \vdots \\ x-a_1 & 0 & \dots & \dots & a_n-x \end{vmatrix}.$$

B.2 Methods of Evaluating Determinants of High Order

Take out $a_1 - x$ from the first column, $a_2 - x$ from the second one, and so on, to obtain

$$(a_1 - x)(a_2 - x)\ldots(a_n - x) \cdot \begin{vmatrix} \frac{a_1}{a_1-x} & \frac{x}{a_2-x} & \cdots & \cdots & \frac{x}{a_n-x} \\ -1 & 1 & 0 & \cdots & 0 \\ -1 & 0 & 1 & \cdots & 0 \\ \vdots & \vdots & \vdots & \ddots & \vdots \\ -1 & 0 & 0 & \cdots & 1 \end{vmatrix}.$$

Put $a_1/(a_1 - x) = 1 + x/(a_1 - x)$ and add all columns to the first one to get

$$(a_1 - x)(a_2 - x)\ldots(a_n - x) \cdot$$

$$\begin{vmatrix} 1 + \frac{a_1}{a_1-x} + \cdots + \frac{x}{a_n-x} & \frac{x}{a_2-x} & \frac{x}{a_3-x} & \cdots & \frac{x}{a_n-x} \\ 0 & 1 & 0 & \cdots & 0 \\ 0 & 0 & 1 & \cdots & 0 \\ \vdots & \vdots & \vdots & \ddots & \vdots \\ 0 & 0 & 0 & \cdots & 1 \end{vmatrix}.$$

The last matrix has an upper triangular form. It follows from Example 3.5 that its determinant is the product of the diagonal entries $\left(1 + \frac{x}{a_1-x} + \cdots + \frac{x}{a_n-x}\right) \cdot 1^{n-1}$. Thus,

$$\det A = (a_1 - x)(a_2 - x)\ldots(a_n - x)\left(1 + \frac{x}{a_1 - x} + \cdots + \frac{x}{a_n - x}\right).$$

B.2.2 Method of Multipliers

Let D be the determinant of a matrix $A = \|a_{ij}\|_{n\times n}$ of order n.

Proposition B.1. *(a) D is a polynomial on n^2 variables a_{ij}, and the degree of this polynomial is equal to n.*
(b) The polynomial D is linear as a polynomial on the elements $a_{i1},\ldots,a_{in}$ of the i-th row of the matrix A.

Proof. Using decomposition by the i-th row, we have $D = \sum_{j=1}^{n} a_{ij} A_{ij}$, where the cofactors A_{ij} do not depend on the elements of the i-th row. So, D is linear as a polynomial of $a_{i1},\ldots,a_{in}$. This proves (b). To prove (a), one can assume, by the induction arguments, that the minor determinants M_{ij} of order $n - 1$ are polynomials of degree $n - 1$. Since $A_{ij} = (-1)^{i+j} M_{ij}$, we conclude that D is a sum of polynomials of degree n. □

To calculate a determinant D, one can now consider it as a polynomial of some variables. Then as a polynomial it can be divided on linear multipliers. Thus, comparing elements of D with elements of multiplication of linear multipliers one can evaluate we can find a formula for D.

Example B.4. Let
$$D = \begin{vmatrix} 0 & x & y & z \\ x & 0 & z & y \\ y & z & 0 & x \\ z & y & x & 0 \end{vmatrix}$$

Consider the columns A_1, A_2, A_3, A_4 of D. Add all columns to the first one. Then we obtain linear multiplier $(x + y + z)$.

Consider

$$\begin{array}{ll} A_1 + A_2 - A_3 - A_4 & (y + z - x) \\ A_1 + A_3 - A_2 - A_4 - \text{multiplier} & (x - y + z) \\ A_1 + A_4 - A_2 - A_3 & (x + y - z) \end{array}$$

These multipliers are mutual. Hence, D is divisible by their product

$$\widetilde{D} = (x + y + z)(y + z - x)(x - y + z)(x + y - z)$$

According to Proposition B.1 a), the degree of the polynomial D is 4, so, it is equal to the degree of $\widetilde{D}$. It follows that $D = c\widetilde{D}$, where c is a scalar multiplier.

In the decomposition of $\widetilde{D}$, we obtain z^4 with coefficient -1, and in D we have z^4 with coefficient $+1$. Hence $c = -1$, that is,

$$D = -(x + y + z)(y + z - x)(x - y + z)(x + y - z).$$

B.2.3 Recursive Definition of Determinant

The method is to decompose determinant by row or column and reduce it to the determinant of the same form but lower order. An example of application of this idea has been given in Example A.4 in Appendix A. Here we give some general formulae for determinants of that kind.

One of the possible forms is

$$D_n = pD_{n-1} + qD_{n-2}, \quad n > 2$$

If $q = 0$ then
$$D_n = p^{n-1}D_1.$$

If $q = 0$ then consider quadratic equation

$$x^2 - px - q = 0.$$

If its roots are α and β, then

$$p = \alpha + \beta,$$
$$q = -\alpha\beta$$

B.2 Methods of Evaluating Determinants of High Order

and
$$D_n = (\alpha + \beta) D_{n-1} - \alpha\beta D_n.$$

Suppose that $\alpha \neq \beta$. Then one can prove (by induction, see Appendix A) a formula for D_n
$$D_n = c_1 \alpha^n + c_2 \beta^n,$$
where
$$c_1 = \frac{D_2 - \beta D_1}{\alpha(\alpha - \beta)},$$
$$c_2 = -\frac{D_2 - \alpha D_1}{\beta(\alpha - \beta)}.$$

If $\alpha = \beta$ then we can obtain the following formula
$$D_n = (c_1 n + c_2) \alpha^{n-2},$$
where
$$c_1 = D_2 - \alpha D_1,$$
$$c_2 = 2\alpha D_1 - D_2.$$

This formula is again can be proved by induction.

Example B.5. Evaluate
$$\begin{vmatrix} 5 & 3 & 0 & 0 & \ldots & 0 & 0 \\ 2 & 5 & 3 & 0 & \ldots & 0 & 0 \\ 0 & 2 & 5 & 3 & \ldots & 0 & 0 \\ \vdots & \vdots & \vdots & \vdots & \ddots & \vdots & \vdots \\ 0 & 0 & 0 & 0 & \ldots & 2 & 5 \end{vmatrix}_{n \times n}$$

Decompose by first row
$$D_n = 5 D_{n-1} - 6 D_{n-2}.$$

Quadratic equation gives the following solution:
$$x^2 - 5x + 6 = 0$$
$$\alpha = 2 \quad \beta = 3$$

Hence
$$D_n = c_1 \alpha^n + c_2 \beta^n = 3^{n+1} - 2^{n+1}.$$

B.2.4 Representation of a Determinant as a Sum of Two Determinants

By the linearity property of determinants (see Property *(a)* in p. 57), a complicated determinant can sometimes be presented as a sum of simpler ones.

Example B.6. Let

$$D_n = \begin{vmatrix} a_1+b_1 & a_1+b_2 & \ldots & a_1+b_n \\ a_2+b_1 & a_2+b_2 & \ldots & a_2+b_n \\ \vdots & \vdots & \ddots & \vdots \\ a_n+b_1 & a_n+b_2 & \ldots & a_n+b_n \end{vmatrix} =$$

$$= \begin{vmatrix} a_1 & a_1 & \ldots & a_1 \\ a_2+b_1 & a_2+b_2 & \ldots & a_2+b_n \\ \vdots & \vdots & \ddots & \vdots \\ a_n+b_1 & a_n+b_2 & \ldots & a_n+b_n \end{vmatrix} + \begin{vmatrix} b_1 & b_2 & \ldots & b_n \\ a_2+b_1 & a_2+b_2 & \ldots & a_2+b_n \\ \vdots & \vdots & \ddots & \vdots \\ a_n+b_1 & a_n+b_2 & \ldots & a_n+b_n \end{vmatrix}$$

So after n steps we obtain 2^n determinants as summands.

If in each decomposition we take as first components the numbers a_i, and for second component numbers b_j then the rows will be either of the form

$$a_i, \ldots, a_i$$

or of the form

$$b_1, b_2, \ldots, b_n.$$

In the first case two are proportional, and in the second case even equal. If $n > 2$ in each determinant we have at least two rows of one type, i.e., for $n > 2$ we have $D_n = 0$. For $n = 1$ and 2, we have

$$D_1 = a_1 + b_1$$

$$D_2 = \begin{vmatrix} a_1 & a_1 \\ b_1 & b_2 \end{vmatrix} + \begin{vmatrix} b_1 & b_2 \\ a_2 & a_2 \end{vmatrix} = (a_1 - a_2)(b_2 - b_1).$$

B.2.5 Changing the Elements of Determinant

Consider

$$D = \begin{bmatrix} a_{11} & \ldots & a_{1n} \\ \vdots & \ddots & \vdots \\ a_{n1} & \ldots & a_{nn} \end{bmatrix}$$

B.2 Methods of Evaluating Determinants of High Order

and
$$D' = \begin{bmatrix} a_{11} + x & \ldots & a_{1n} + x \\ \vdots & \ddots & \vdots \\ a_{n1} + x & \ldots & a_{nn} + x \end{bmatrix}.$$

Using the method of Sect. B.2.4, one can deduce that

$$D' = D + \begin{vmatrix} x & \ldots & x \\ a_{21} & \ldots & a_{2n} \\ \vdots & \ddots & \vdots \\ a_{n1} & \ldots & a_{nn} \end{vmatrix} + \begin{vmatrix} a_{11} & \ldots & a_{1n} \\ x & \ldots & x \\ \vdots & \ddots & \vdots \\ a_{n1} & \ldots & a_{nn} \end{vmatrix} + \cdots + \begin{vmatrix} a_{11} & \ldots & a_{1n} \\ a_{21} & \ldots & a_{2n} \\ \vdots & \ddots & \vdots \\ x & \ldots & x \end{vmatrix}.$$

It follows that
$$D' = D + x \sum_{i,j=1}^{n} A_{ij},$$

where A_{ij} are cofactors of a_{ij}.

Example B.7. Let us evaluate

$$D' = \begin{vmatrix} a_1 & x & \ldots & x \\ x & a_2 & \ldots & x \\ \vdots & \vdots & \ddots & \vdots \\ x & x & \ldots & a_n \end{vmatrix}.$$

Subtract x from all elements. Then

$$D = \begin{vmatrix} a_1 - x & 0 & \ldots & 0 \\ 0 & a_2 - x & \ldots & 0 \\ \vdots & \vdots & \ddots & \vdots \\ 0 & 0 & \ldots & a_n - x \end{vmatrix} = (a_1 - x) \ldots (a_n - x).$$

For any $i \neq j$ we have $A_{ij} = 0$, and for $i = j$

$$A_{ii} = (a_1 - x) \ldots (a_{i-1} - x)(a_{i+1} - x) \ldots (a_n - x)$$

Hence
$$D' = (a_1 - x) \ldots (a_n - x) + x \sum_{i=1}^{n} A_{ii} =$$

by simple algebraic transformations

$$= x(a_1 - x)(a_2 - x)\ldots(a_n - x)\left(\frac{1}{x} + \frac{1}{a_1 - x} + \cdots + \frac{1}{a_n - x}\right).$$

B.2.6 Two Classical Determinants

The determinant

$$V(x_1,\ldots,x_n) = \begin{vmatrix} 1 & x_1 & x_1^2 & \ldots & x_1^{n-1} \\ 1 & x_2 & x_2^2 & \ldots & x_2^{n-1} \\ 1 & x_3 & x_3^2 & \ldots & x_3^{n-1} \\ \vdots & \vdots & \vdots & \ddots & \vdots \\ 1 & x_n & x_n^2 & \ldots & x_n^{n-1} \end{vmatrix} \tag{B.1}$$

is called Vandermonde[1] determinant.

Theorem B.2.

$$V(x_1,\ldots,x_n) = \prod_{1 \le i < j \le n} (x_j - x_i).$$

Proof. By Proposition B.1 (b), the total degree of the polynomial $V(x_1,\ldots,x_n)$ is equal to the sum

$$0 + 1 + \cdots + (n-1) = n(n-1)/2.$$

If we subtract the i-th row from the j-th one, we get a new matrix with j-th row of the form $(0, x_j - x_i, x_j^2 - x_i^2, \ldots, x_j^{n-1} - x_i^{n-1})^T$. Here each term $x_j^k - x_i^k = (x_j - x_i)(x_j^{k-1} + x_j^{k-2}x_i + \cdots + x_i^{k-1})$ is divisible by $(x_j - x_i)$, hence $V(x_1,\ldots,x_n)$ is divisible by $(x_j - x_i)$ for all $1 \le i < j \le n$. It follows that D is divisible by a polynomial

$$\widetilde{V}(x_1,\ldots,x_n) \equiv \prod_{1 \le i < j \le n} (x_j - x_i).$$

Since $\deg \widetilde{V}(x_1,\ldots,x_n) = n(n-1)/2 = \deg V(x_1,\ldots,x_n)$, it follows that $V(x_1,\ldots,x_n) = c\widetilde{V}(x_1,\ldots,x_n)$, where c is a number.

Using the decomposition by the last row, we see that the coefficient of x_n^{n-1} in $V(x_1,\ldots,x_n)$ is $V(x_1,\ldots,x_{n-1})$. At the same time, it is easy to see that the coefficient of x_n^{n-1} in $\widetilde{V}(x_1,\ldots,x_n)$ is $\widetilde{V}(x_1,\ldots,x_{n-1})$. By the induction arguments we conclude that $c = 1$. □

---

[1] Alexandre Theophile Vandermonde (1735–1796), French mathematician and musician, one of the founders of the theory of determinants.

B.3 Problems

The matrix

$$C = \begin{bmatrix} a_0 & a_1 & a_2 & \cdots & a_{n-1} \\ a_{n-1} & a_0 & a_1 & \cdots & a_{n-2} \\ a_{n-2} & a_{n-1} & a_0 & \cdots & a_{n-3} \\ \vdots & \vdots & \vdots & \ddots & \vdots \\ a_1 & a_2 & a_3 & \cdots & a_0 \end{bmatrix}$$

is called *circulant matrix*.

Theorem B.3. *The determinant $C(a_0, \ldots, a_{n-1})$ of the above matrix (circulant determinant) is equal to*

$$f(\varepsilon_0) \ldots f(\varepsilon_{n-1}),$$

where $f(x) = a_{n-1}x^{n-1} + \cdots + a_1 x + a_0$ and $\varepsilon_0, \ldots, \varepsilon_{n-1}$ are different complex n-th roots of unity (for the definition of the complex roots, see Sect. C.2).

Proof. Consider the product $Q = CV^T$, where V is the Vandermonde matrix (B.1) on $\varepsilon_0, \ldots, \varepsilon_{n-1}$, that is,

$$V = \begin{bmatrix} 1 & \varepsilon_0 & \cdots & \varepsilon_0^{n-1} \\ \vdots & \vdots & \ddots & \vdots \\ 1 & \varepsilon_{n-1} & \cdots & \varepsilon_{n-1}^{n-1} \end{bmatrix}.$$

Then $Q = \|q_{ij}\|_{n \times n}$, where $q_{1j} = f(\varepsilon_j)$ and $q_{ij} = (C_i, V_j) = \sum_{k=0}^{i-2} a_{n+1+k-i}\varepsilon_{j-1}^k + \sum_{k=i-1}^{n-1} a_{1+k-i}\varepsilon_{j-1}^k = \varepsilon_{j-1}^{i-1} f(\varepsilon_{j-1})$ for $i \geq 2$. Therefore,

$$\det Q = f(\varepsilon_0) \ldots f(\varepsilon_{n-1}) |\varepsilon_{j-1}^{i-1}|_{n \times n} = f(\varepsilon_0) \ldots f(\varepsilon_{n-1}) V(\varepsilon_0, \ldots, \varepsilon_{n-1}).$$

On the other hand, we have

$$\det Q = \det(CV^T) = \det C \det V = (\det C) V(\varepsilon_0, \ldots, \varepsilon_{n-1}).$$

Since $V(\varepsilon_0, \ldots, \varepsilon_{n-1}) \neq 0$, we have $\det C = f(\varepsilon_0) \ldots f(\varepsilon_{n-1})$. □

B.3 Problems

1. Prove that if all elements of a 3×3 matrix are equal ± 1, then the determinant of this matrix is an even number.
2. Without evaluating the determinants, show that:
   (a)
   $$\begin{vmatrix} a_1 & b_1 & a_1 x + b_1 y + c_1 \\ a_2 & b_2 & a_2 x + b_2 y + c_2 \\ a_3 & b_3 & a_3 x + b_3 y + c_3 \end{vmatrix} = \begin{vmatrix} a_1 & b_1 & c_1 \\ a_2 & b_2 & c_2 \\ a_3 & b_3 & c_3 \end{vmatrix}$$

(b)
$$\begin{vmatrix} 1 & a & bc \\ 1 & b & ca \\ 1 & c & ab \end{vmatrix} = (b-a)(c-a)(c-b)$$

(c)
$$\begin{vmatrix} 1 & a & a^3 \\ 1 & b & b^3 \\ 1 & c & c^3 \end{vmatrix} = (a+b+c) \begin{vmatrix} 1 & a & a^2 \\ 1 & b & b^2 \\ 1 & c & c^2 \end{vmatrix}.$$

3. Evaluate the determinant

$$\begin{vmatrix} a_{11} & a_{12} & a_{13} & a_{14} & a_{15} \\ a_{21} & a_{22} & a_{23} & a_{24} & a_{25} \\ a_{31} & a_{32} & 0 & 0 & 0 \\ a_{41} & a_{42} & 0 & 0 & 0 \\ a_{51} & a_{52} & 0 & 0 & 0 \end{vmatrix}.$$

4. Solve the equation

$$\begin{vmatrix} 1 & 1 & 1 & \ldots & 1 \\ 1 & 1-x & 1 & \ldots & 1 \\ 1 & 1 & 2-x & \ldots & 1 \\ \ldots & \ldots & \ldots & \ddots & \ldots \\ 1 & 1 & 1 & \ldots & n-x \end{vmatrix} = 0.$$

5. Using the third row, evaluate the determinant

$$\begin{vmatrix} 2 & -3 & 4 & 1 \\ 4 & -2 & 3 & 2 \\ a & b & c & d \\ 3 & -1 & 4 & 3 \end{vmatrix}.$$

6. Evaluate the determinant

$$\begin{vmatrix} 1 & 0 & 2 & a \\ 2 & 0 & b & 0 \\ 3 & c & 4 & 5 \\ d & 0 & 0 & 0 \end{vmatrix}.$$

Evaluate the determinants in exercises 7–10 by reducing each of them to the triangular form. (Corresponding matrices in 7–9 are of order $n \times n$.)

B.3 Problems

7.
$$\begin{vmatrix} 1 & 1 & 1 & \cdots & 1 \\ 1 & 1 & 0 & \cdots & 0 \\ 1 & 0 & 1 & \cdots & 0 \\ \vdots & \vdots & \vdots & \ddots & \vdots \\ 1 & 0 & 0 & \cdots & 1 \end{vmatrix}.$$

8.
$$\begin{vmatrix} 1 & 1 & 1 & \cdots & 1 \\ 1 & 0 & 1 & \cdots & 1 \\ 1 & 1 & 0 & \cdots & 1 \\ \vdots & \vdots & \vdots & \ddots & \vdots \\ 1 & 1 & 1 & \cdots & 0 \end{vmatrix}.$$

9.
$$\begin{vmatrix} 3 & 2 & 2 & \cdots & 2 \\ 2 & 3 & 2 & \cdots & 2 \\ 2 & 2 & 3 & \cdots & 2 \\ \cdots & \cdots & \cdots & \ddots & \cdots \\ 2 & 2 & 2 & \cdots & 3 \end{vmatrix}.$$

10.
$$\begin{vmatrix} a_0 & a_1 & a_2 & \cdots & a_n \\ -x & x & 0 & \cdots & 0 \\ 0 & -x & x & \cdots & 0 \\ \cdots & \cdots & \cdots & \ddots & \cdots \\ 0 & 0 & 0 & \cdots & x \end{vmatrix}.$$

Evaluate the determinants in questions 11 and 12 (by using linear multipliers).

11.
$$\begin{vmatrix} a_0 & a_1 & a_2 & \cdots & a_n \\ a_0 & x & a_2 & \cdots & a_n \\ a_0 & a_1 & x & \cdots & a_n \\ \vdots & \vdots & \vdots & \ddots & \vdots \\ a_0 & a_1 & a_2 & \cdots & x \end{vmatrix}$$

12.
$$\begin{vmatrix} -x & a & b & c \\ a & -x & c & b \\ b & c & -x & a \\ c & b & a & -x \end{vmatrix}$$

Evaluate the determinants in questions 13–15 by using the recursive definition. (Corresponding matrices in Problems 14 and 15 are of order $n \times n$.)

13.
$$\begin{vmatrix} a_1 & 1 & 1 & \cdots & 1 \\ 1 & a_2 & 0 & \cdots & 0 \\ 1 & 0 & a_3 & \cdots & 0 \\ \vdots & \vdots & \vdots & \ddots & \vdots \\ 1 & 0 & 0 & \cdots & a_n \end{vmatrix}$$

14.
$$\begin{vmatrix} 2 & 1 & 0 & \cdots & 0 \\ 1 & 2 & 1 & \cdots & 0 \\ 0 & 1 & 2 & \cdots & 0 \\ \vdots & \vdots & \vdots & \ddots & \vdots \\ 0 & 0 & 0 & \cdots & 2 \end{vmatrix}$$

15.
$$\begin{vmatrix} 3 & 2 & 0 & \cdots & 0 \\ 1 & 3 & 2 & \cdots & 0 \\ 0 & 1 & 3 & \cdots & 0 \\ \vdots & \vdots & \vdots & \ddots & \vdots \\ 0 & 0 & 0 & \cdots & 3 \end{vmatrix}$$

16. Evaluate the following determinant representing it as a sum of determinants:

$$\begin{vmatrix} x_1 & a_2 & \cdots & a_n \\ a_1 & x_2 & \cdots & a_n \\ \vdots & \vdots & \ddots & \vdots \\ a_1 & a_2 & \cdots & x_n \end{vmatrix}$$

Hint: insert $x_i = (x_i - a_i) + a_i$.

17. Let $x_0, x_1, \ldots, x_n$ are variables and $p_0, p_1, \ldots, p_n$ are polynomials of the form $p_j = a_j x^j +$ (lower terms). Show that

$$\begin{vmatrix} p_0(x_0) & p_1(x_0) & \cdots & p_n(x_0) \\ \vdots & \vdots & \ddots & \vdots \\ p_0(x_n) & p_1(x_n) & \cdots & p_n(x_n) \end{vmatrix} = a_0 \ldots a_n V(x_0, \ldots, x_n).$$

Complex Numbers

Besides the dot product, in the plane $\mathbb{R}^2$ there is another product of vectors, called *complex multiplication*. In contrast to the dot product, the complex product of two vectors is again a vector in $\mathbb{R}^2$. It is given by the formula

$$(x, y)(x', y') = (xx' - yy', xy' + yx').$$

The properties of complex multiplication (given below) are close to the usual multiplication of real numbers, and the elements of $\mathbb{R}^2$ are referred as *complex numbers*. The set $\mathbb{R}^2$ of complex numbers is also denoted by $\mathbb{C}$. In particular, the vectors in the horizontal axis are identified with real numbers, that is, the real number x corresponds to the vector $(x, 0)$ (therefore, we have $\mathbb{R} \subset \mathbb{C}$). For example, the number 1 is identified with the vector $(1, 0)$, the first vector of the canonical basis. The horizontal axis is called a *real axis*. The sum (product) of two real numbers a and b is identified with the sum (product) of corresponding complex numbers $(a, 0)$ and $(b, 0)$, so that one can consider the complex numbers as an extension of the real number system.

The second vector of the canonical basis $(0, 1)$, being considered as a complex number, is denoted by i and called a *imaginary unit*, or *the square root of* -1 (because $i^2 = (-1, 0) = -1$). The vertical axis is called an *imaginary axis*, and every vector of the form $(0, x) = x \cdot i$ is called a *pure imaginary number*. It follows that any complex number $z = (x, y)$ is a sum of a real number and a pure imaginary number, that is,

$$z = (x, y) = x(1, 0) + y(0, 1) = x + yi.$$

The first entry x is called a *real part* of z and denoted as $\Re z$. The second entry y is called an *imaginary part* of z and denoted as $\Im z$.

Main Properties of Complex Multiplication

The next properties can easily be checked directly.

Let $z = (x, y), z' = (x', y')$ and $z'' = (x'', y'')$ be three complex numbers.
1. $zz' = z'z$, (commutativity)
2. $z(z'z'') = (zz')z''$, (associativity)
3. $z(z' + z'') = zz' + zz''$, (distributivity)
4. $1 \cdot z = z, 0 \cdot z = 0$, where $0 = \mathbf{0} = (0, 0)$.

Example C.1. If $z = (1, 2) = 1 + 2i$ and $z' = (4, 3) = 4 + 3i$, then $zz' = (1 + 2i)(4 + 3i) = 4 + 3i + 8i + 6i^2 = 4 + 3i + 8i + 6(-1) = -2 + 11i$.

C.1 Operations with Complex Numbers

Let us introduce some other operations with complex numbers.

C.1.1 Conjugation

A conjugation is a reflection of a vector in the real axis, that is, the conjugated complex number $z = (a, b) = a + bi$ is

$$\bar{z} = (a, -b) = a - bi$$

(see Fig. C.1).

In particular, $\bar{\bar{z}} = z$.

The conjugation has the following nice connections with the standard operations.
1. $\overline{z + z'} = \bar{z} + \bar{z}'$.
2. $\overline{zz'} = \bar{z}\bar{z}'$.
3. If $z = (a, 0) \in \mathbb{R}$, then $\bar{z} = z$.
4. $z\bar{z} = a^2 + b^2 \in \mathbb{R}_+$.

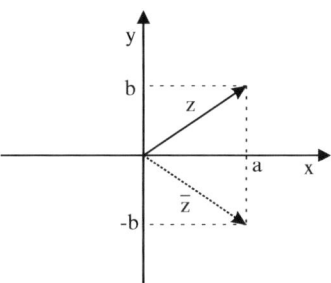

Fig. C.1 A conjugation

C.1 Operations with Complex Numbers

One can express the real and imaginary parts of a complex number via conjugation as
$$\Re z = (1/2)(z + \bar{z})$$
and
$$\Im z = (i/2)(\bar{z} - z).$$

C.1.2 Modulus

The *modulus* $|z|$ of a complex number $z = (a, b)$ is its length as a vector in $\mathbb{R}^2$, that is,
$$|z| = \sqrt{a^2 + b^2} = \sqrt{z\bar{z}}.$$
The modulus of the product of two complex numbers is the product of their moduli, that is,
$$|zz'| = |z||z'|$$
(because $|z||z'| = \sqrt{z\bar{z}}\sqrt{z'\bar{z}'} = \sqrt{zz'\overline{zz'}} = |zz'|$).

Example C.2. $|3 + 4i| = \sqrt{(3+4i)\overline{(3+4i)}} = \sqrt{(3+4i)(3-4i)} = \sqrt{3^2 + 4^2} = 5$.

C.1.3 Inverse and Division

The inverse of a nonzero complex number $z = a + bi$ is defined as $z^{-1} = |z|^{-2}\bar{z}$. Then $z^{-1}z = |z|^{-2}\bar{z}z = |z|^{-2}|z|^2 = 1$, so that one can define a *division* of complex numbers as $z'/z = z'z^{-1} = |z|^{-2}z'\bar{z}$, that is,
$$\frac{a' + b'i}{a + bi} = \frac{(a' + b'i)(a - bi)}{(a + bi)(a - bi)} = (a^2 + b^2)^{-1}(a' + b'i)(a - bi).$$

Note that the denominator $z = a + bi$ must be nonzero.

Example C.3.
$$\frac{18 + i}{3 - 4i} = \frac{(18 + i)(3 + 4i)}{(3 - 4i)(3 + 4i)} = \frac{50 + 75i}{3^2 + 4^2} = 2 + 3i.$$

C.1.4 Argument

An *argument* of a nonzero complex number $z = a + bi$ is an angle φ between the real axis and the vector z (see Fig. C.2). Obviously, $\sin\varphi = b/|z|$ and $\cos\varphi = a/|z|$, so, φ is defined uniquely up to the period 2π. The set of all such possible ϕ is denoted

Fig. C.2 The angle φ is the argument of z

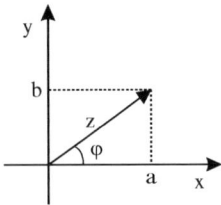

by Arg z, while the unique argument φ such that $0 \leq \phi < 2\pi$ is denoted by arg z. This means that

$$\text{Arg } z = \{\arg z + 2\pi k \mid k \in \mathbb{Z}\}.$$

If $\varphi \in \text{Arg } z$ is one the values of the argument of z, then it follows from the above sine and cosine values that

$$z = |z|(\cos \varphi + i \sin \varphi).$$

Let z and z' be two complex numbers with arguments α and β. One can check that

$$zz' = |z|(\cos \alpha + i \sin \alpha)|z'|(\cos \beta + i \sin \beta) = |z||z'|(\cos(\alpha + \beta) + i \sin(\alpha + \beta)), \tag{C.1}$$

so that the argument of the product of complex number is equal to the sum of their arguments. Analogously, we have

$$z/z' = |z|/|z'|(\cos(\alpha - \beta) + i \sin(\alpha - \beta)).$$

It follows that for each integer n we have

$$z^n = |z|^n(\cos(n\alpha) + i \sin(n\alpha)). \tag{C.2}$$

Example C.4. Let us calculate z^{100}, where $z = i + 1$. We have $|z| = \sqrt{1^2 + 1^2} = \sqrt{2}$ and $\cos(\arg z) = \sin(\arg z) = 1/\sqrt{2}$, so that $\arg z = \pi/4$. Therefore,

$$z^{100} = (\sqrt{2})^{100}\left(\cos \frac{100\pi}{4} + i \sin \frac{100\pi}{4}\right),$$

where $\frac{100\pi}{4} = 25\pi = 12 \cdot 2\pi + \pi$. Thus,

$$z^{100} = 2^{50}(\cos \pi + i \sin \pi) = -2^{50}.$$

C.1.5 Exponent

The exponent of a complex number $z = a + bi$ is defined as

$$e^{a+bi} = e^a(\cos b + i \sin b)$$

(Euler[1] formula). It follows that for each nonzero complex z we have

$$z = |z|e^{i \arg z}.$$

Example C.5. $e^{i\pi} = e^0(\cos \pi + i \sin \pi) = -1.$

Other properties of exponents of complex numbers follows from the equation (C.1):

$$e^{z+z'} = e^z e^{z'}, e^{-z} = 1/e^z, e^{z-z'} = e^z/e^{z'}.$$

C.2 Algebraic Equations

Many equations which have no real solutions have complex ones. The simplest example is the equation $z^2 + 1 = 0$, which has two complex solutions, $z = i$ and $z = -i$, and no real ones. More generally, each quadratic equation

$$ax^2 + bx + c = 0$$

with real coefficients a, b and c always have complex solutions $x = \frac{-b \pm \sqrt{b^2 - 4ac}}{2a}$, where in the case $D = b^2 - 4ac < 0$ we take $\sqrt{D} = i\sqrt{|D|}$.

Moreover, some simple algebraic equations have quite more complex solutions than real ones. Consider an equation

$$z^n = c,$$

where c is a nonzero complex number. If $\arg z = \alpha$ and $\arg c = \phi$, the formula (C.2) gives the equation

$$|z|^n(\cos(n\alpha) + i\sin(n\alpha)) = |c|(\cos\phi + i\sin\phi),$$

---

[1] Leonhard Euler (1707–1783) was a great Swiss mathematician who made enormous contributions to a wide range of mathematics and physics including analytic geometry, trigonometry, geometry, calculus and number theory. Most of his life he had been working in Russia (St. Petersburg) and Prussia (Berlin).

Fig. C.3 The fourth roots of a complex number z

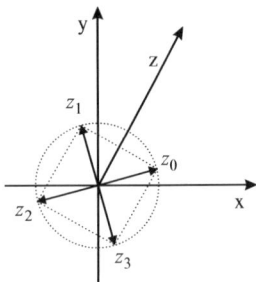

therefore,
$$|z| = \sqrt[n]{|c|} \text{ and } \alpha = \frac{\phi + 2\pi k}{n}$$

for some integer k. Since $0 \leq \alpha < 2\pi$, we get n different values for α, that is, $\alpha_k = \frac{\phi + 2\pi k}{n}$, where $k = 0, \ldots, n-1$. This gives exactly n pairwise different solutions of the above equation, that is,

$$z_k = \sqrt[n]{|c|}(\cos \alpha_k + i \sin \alpha_k) \text{ for } k = 0, \ldots, n-1.$$

All these n complex numbers could be referred as roots of degree n of c. All of them belong to a circle centered in the zero of radius $\sqrt[n]{|c|}$ and placed in the vertices of a regular polygon of n vertices inscribed to the circle (see Fig. C.3 for the case $n = 4$).

Example C.6. Solve the equation $z^3 = i$.

In the notation above, we have $n = 3$, $|c| = 1$ and $\phi = \pi/2$, so that $|z| = \sqrt[3]{1} = 1$, $\alpha_0 = \frac{\pi/2 + 2\pi 0}{3} = \pi/6$, $\alpha_1 = \frac{\pi/2 + 2\pi}{3} = 5\pi/6$ and $\alpha_2 = \frac{\pi/2 + 4\pi}{3} = 3\pi/2$. We obtain three solutions of the from $z_k = 1(\cos \alpha_k + i \sin \alpha_k)$ for $k = 0, 1, 2$, that is, $z_0 = \frac{1}{2} + i\frac{\sqrt{3}}{2}$, $z_1 = -\frac{1}{2} + i\frac{\sqrt{3}}{2}$ and $z_2 = -i$.

Theorem C.1 (Fundamental theorem of algebra). *Every non-constant polynomial with complex coefficients of a variable x has at least one complex root.*

Proof. See [33, Theorem 3.3.1]. □

We have discussed the finding of the roots for the polynomials of degree two and for polynomials of the form $x^n - z$. For general polynomials of any degree d greater than 4, there is no an universal formula for finding the root (Abel's theorem[2]).

If $x = x_1$ is a root of a polynomial $f(x)$ of degree d, then one can decompose

---

[2]Niels Henrik Abel (1802–1829) was a famous Norwegian mathematician. In spite of his short life, he made an extremely important contribution both to algebra and calculus. One of the most significant international prize for mathematician is called the Abel Prize. Abel had proved his theorem at age of 19.

C.2 Algebraic Equations

$$f(x) = (x - x_1)g(x),$$

where $g(x)$ is another polynomial of degree $d - 1$ (with complex coefficients). If the degree $d - 1$ is positive, the polynomial $g(x)$ must have another complex root, say, x_2. Proceeding such decomposition up to degree 0, we get the decomposition of the from

$$f(x) = c(x - x_1)(x - x_2)\ldots(x - x_d),$$

where c is a complex number and $x_1, \ldots, x_d$ are roots of the polynomial $f(x)$. Combining the identical terms, we get (after a possible re-numerating of x_i's) the decomposition

$$f(x) = c(x - x_1)^{k_1}(x - x_2)^{k_2}\ldots(x - x_s)^{k_s}$$

for some $s \leq d$, where the sum of the powers $k_1 + \cdots + k_s$ is d. Each number k_j is called a *multiplicity* of the corresponding root x_j. The roots of multiplicity one are called *simple*. One can check that each root x_j of the polynomial should appear in this decomposition.

We have

Corollary C.2. *The number of roots of any polynomial $p(x)$ of degree $d > 0$ is not greater than d. Moreover, the sum of all multiplicities of the roots is equal to d.*

Example C.7. *Problem.* Solve the equation $z^2 - 4iz - 7 - 4i = 0$.

Solution. By the well-known formula (which is easy to be checked), the number $z = \frac{-(-4i)+d}{2}$ satisfies the equation, where $d^2 = (-4i)^2 - 4(-7 - 4i) = 12 + 16i$. We need to find d. Let $d = x + iy$, where x and y are real numbers. Then $d^2 = (x^2 - y^2) + 2ixy$, so that the equality of two 2-dimensional vectors $d^2 = 12 + 16i$ gives a system of two equations

$$\begin{cases} x^2 - y^2 = 12, \\ 2xy = 16. \end{cases}$$

This system has two solutions for $d = x + iy$, that is, $d_1 = 4 + 2i$ and $d_2 = -4 - 2i$. This leads us to two roots of the equation above: $z_1 = (4i + d_1)/2 = 2 + 3i$ and $z_2 = (4i + d_2)/2 = -2 + i$. Since the equation has degree two, there are no other roots but these two.

To find a decomposition of a polynomial with real coefficients, the following statement is often useful.

Theorem C.3. *If a complex number z is a root of a polynomial $f(x)$ with real coefficients, then the conjugated number $\bar{z}$ is also a root of $f(x)$. Moreover, the multiplicities of these roots are the same.*

Proof. If $f(x) = a_d x^d + \cdots + a_1 x + a_0$ and $f(z) = 0$, then $f(\bar{z}) = a_d \bar{z}^d + \cdots + a_1 \bar{z} + a_0 = \bar{a}_d \bar{z}^d + \cdots + \bar{a}_1 \bar{z} + \bar{a}_0 = \overline{a_d z^d} + \cdots + \overline{a_1 z} + \overline{a_0} = \overline{f(z)} = \bar{0} = 0$.

The claim about the multiplicities follows from the induction argument applied to the polynomial $g(x)$ of degree $d-2$ such that $f(x) = (x-z)(x-\bar{z})g(x)$. □

Example C.8. Let us find the decomposition of a polynomial $f(z) = z^4 - 5z^3 + 7z^2 - 5z + 6$. One can check that $f(i) = 0$, so that $z = i$ is a root of the polynomial. If follows that the number $\bar{z} = -i$ is another root. From the decomposition

$$f(z) = (z-i)(z+i)g(z)$$

one can find $g(z) = z^2 - 5z + 6$. It follows that $g(z) = (z-2)(z-3)$, so that

$$f(z) = (z-i)(z+i)(z-2)(z-3).$$

C.3 Linear Spaces Over Complex Numbers

Recall from the definition in the beginning of Chap. 6 that a linear space is a set admitting two operations, that is, addition and multiplication by a real number (dot product) such that these operations satisfies the linearity properties. One can extend this definition by allowing the multiplication of by complex numbers, not only by real ones. Such a vector space is called a *vector space over complex numbers*, or simply a *complex vector space*. In contrast, the vector space in the sense of Chap. 6 is called a *vector space over real numbers*, or a *real vector space*. The definition of complex vector space repeats the definition of real one literally but the world 'real' is replaced by 'complex'.

The notions of linear dependence and independence of vectors, dimension and basis of a vector space, subspace, isomorphism etc. for complex vector spaces repeat the correspondent definitions given in Chap. 6 verbatim. The definition and the properties of linear transformation and its matrix from Chap. 8 are transferred to the case of complex vector spaces verbatim as well.

Example C.9. The set $\mathbb{C}$ itself is a one-dimensional complex vector space. Any its nonzero element $z = a + bi \neq 0$ form a basis of it, since for each $w = x + yi \in \mathbb{C}$ one has $w = \alpha z$, where $\alpha = w/z \in \mathbb{C}$.

Example C.10. The set $\mathbb{C}^n$ of n-tuples of complex numbers $\mathbf{x} = (x_1, \ldots, x_n)$ with $x_1, \ldots, x_n \in \mathbb{C}$ is an n-dimensional vector space with standard operations

$$(x_1, \ldots, x_n) + (y_1, \ldots, y_n) = (x_1 + y_1, \ldots, x_n + y_n)$$

and

$$\alpha(x_1, \ldots, x_n) = (\alpha x_1, \ldots, \alpha x_n).$$

Its dimension is equal to n by the same reason as for the real space $\mathbb{R}^n$, see Example 6.5.

Given a real vector space V, one can embed it in a complex vectors space by the following way. Let $V^{\mathbb{C}}$ be the set of pairs of vectors from V. Each such pair we denote by $(\mathbf{u}, \mathbf{v}) = \mathbf{u} + i\mathbf{v}$. The addition of pairs and their multiplication by complex numbers are defined as

$$(\mathbf{u} + i\mathbf{v}) + (\mathbf{u}' + i\mathbf{v}') = (\mathbf{u} + \mathbf{u}') + i(\mathbf{v} + \mathbf{v}')$$

and

$$(a + bi)(\mathbf{u} + i\mathbf{v}) = (a\mathbf{u} - b\mathbf{v}) + i(a\mathbf{v} + b\mathbf{u}).$$

Exercise C.1. Show that $V^{\mathbb{C}}$ is a linear space over complex numbers.

Such a complex vector space $V^{\mathbb{C}}$ is called a *complexification* of V.

Example C.11. $\mathbb{C} = \mathbb{R}^{\mathbb{C}}$ and $\mathbb{C}^n = (\mathbb{R}^n)^{\mathbb{C}}$.

Given a vector $\mathbf{w} = \mathbf{u} + i\mathbf{v}$ in a complexification $V^{\mathbb{C}}$, one can define its real and imaginary parts as $\Re \mathbf{w} = \mathbf{u}$ and $\Im \mathbf{w} = \mathbf{v}$, the both are vectors in the real vector space V. Any basis of V is also a basis of the complexification $V^{\mathbb{C}}$, but one can also construct other bases in $V^{\mathbb{C}}$ which do not belong to V. For any linear transformation $f : U \to V$ of two real vector spaces U and V, one can also define its complexification $f^{\mathbb{C}} : U^{\mathbb{C}} \to V^{\mathbb{C}}$ by the obvious formula $f^{\mathbb{C}}(\mathbf{u} + i\mathbf{v}) = f(\mathbf{u}) + if(\mathbf{v})$. Note that if we fix two bases in the real vector spaces U and V, then the matrices of the linear transformations f and $f^{\mathbb{C}}$ in these bases are the same.

In particular, we prove in Corollary 9.5 that any linear operator in a finite-dimensional complex vector space has an eigenvector. This means that if f is a linear operator in a real vector space V, then its complexification $f^{\mathbb{C}}$ has an eigenvector in $V^{\mathbb{C}}$. In particular, it follows that the matrix of f has an eigenvalue.

C.4 Problems

1. Calculate
$$\frac{(1+i)(2+i)}{3-i}.$$

2. Calculate
$$\frac{i-3}{2-3i} + \frac{i+3}{2+3i}.$$

3. Calculate
$$\frac{(1+i)^4}{(i-1)^5}.$$

4. Solve the system of linear equations with complex coefficients
$$\begin{cases} (1+2i)x - 2iy = 5 + 9i \\ (-1+3i)x + (1+i)y = -6 + 4i. \end{cases}$$

5. Find all real solutions of the equation
$$(2i - 2)x - (i + 1)y = 2i - 10.$$

6. Solve the equation
$$z^2 + 2z + 37 = 0.$$

7. Solve the equation
$$iz^2 + (3i - 2)z + 12 + 4i = 0.$$

8. Solve the equation
$$z^2 + 6z - 4iz + 5 - 12i = 0.$$

9. Solve the equation
$$z^2 - 5z + 4iz + 9 - i = 0.$$

10. Calculate i^{100}.
11. Calculate $(1 - i)^n$, where n is a positive integer.
12. Solve the equation $z^6 = i$.
13. Solve the equation $z^4 = -128 + 128\sqrt{3}i$.
14. Plot all the solutions of the equation $z^6 = 117 + 44i$ in the complex plane.
15. Find all solutions of the equation $z^5 = 5e^{5i}$.
16. Find the multiplicity of the root $z = 2$ of the polynomial $z^5 - 6z^4 + 13z^3 - 14z^2 + 12z - 8$.
17. Find the multiplicity of the root $z = i + 1$ of the polynomial $z^5 - 6z^4 + 16z^3 - 24z^2 + 20z - 8$.
18. Prove that $|z_1 + z_2| \leq |z_1| + |z_2|$.
19. Prove the equation (C.1).
20. Let z and z' be two nonzero vectors in $\mathbb{R}^2$ (that is, complex numbers) such that $|z| = |z'|$ and $z \perp z'$. Find all possible values of z'/z.
21. Let $\varepsilon_0, \ldots, \varepsilon_{n-1}$ be different complex n-th roots of unity, where $n \geq 2$. Show that $\varepsilon_0 + \cdots + \varepsilon_{n-1} = 0$ and $\varepsilon_0 \ldots \varepsilon_{n-1} = (-1)^{n-1}$.
22. Prove that $e^{z_1 + z_2} = e^{z_1} e^{z_2}$.
23. Solve the equation $e^z = e$ for a complex number z.
24. Let A be a linear operator in $\mathbb{C}^2$ given by the matrix
$$\begin{bmatrix} 1 & 2 \\ 3 & 4 \end{bmatrix}.$$

Find its matrix in the basis

$$\mathbf{f}_1 = (i, 1), \mathbf{f}_2 = (1, i).$$

25. Let V be an n-dimensional complex vector space. Show that V is also a real vector space, where the multiplication by the real numbers is defined by the same way as the multiplication by complex ones (as the real numbers form a subspace of complex numbers). Find the dimension of V as a real vector space.

Pseudoinverse

Consider a general linear system

$$Ax = b. \tag{D.1}$$

If the matrix A is square and non-singular, we have the solution

$$x = A^{-1}b.$$

But if A is singular or even non-square, problems of two types arise. First, if the system (D.1) is consistent and the rank of A is less than the number of variables, we have a multiple solution which can be found algorithmically by, say, Gaussian elimination (see Sect. 5.2). But in many practical problems, we need an explicit formula for a (particular) solution depending on the vector b in the right side. How to get such an explicit formula?

Second, a problem arises if the system (D.1) is inconsistent. In practical problems, we need in this case an approximate solution (see Sect. (7.3)). Is there a formula for expressing it?

An answer to the both questions is given by a construction due to Moore[1] and Penrose[2] called a *pseudoinverse*[3]. For many generalizations and applications of this construction, we refer the reader to [2,5].

---

[1] Eliakim Hastings Moore (1862–1932), an American mathematician who made a significant contribution to algebra and logics. In 1935, he gave a definition pseudoinverse under the name *general reciprocal*.

[2] Roger Penrose (1931) is a famous English physicists and mathematician. He was awarded the Wolf foundation prise (1988, shared with Stephen Hawking) for the work which has "greatly enlarged our understanding of the origin and possible fate of the Universe". In 1955, he rediscovered the Moore definition of the pseudoinverse and demonstrated its connection with the least square approximation.

[3] The terms *generalized inverse* and *Moore–Penrose inverse* are also used.

D.1 Definition and Basic Properties

Definition D.1. Let A be an $m \times n$ matrix. The matrix A^+ is called a *pseudoinverse*[4] of A if the following four conditions hold

1. $AA^+A = A$
2. $A^+AA^+ = A^+$
3. $\left(AA^+\right)^T = AA^+$
4. $\left(A^+A\right)^T = A^+A$.

Later (see Theorem D.5) we will show that every matrix A admits a pseudoinverse.

Note that since both A^+A and AA^+ exist, A^+ must be an $m \times n$ matrix.

The first property give us the right to call it *the* pseudoinverse.

Theorem D.1. *Let A be a matrix. If its pseudoinverse exists, then it is unique.*

Proof. Suppose B and C are two pseudoinverses of A, that is, both matrices B and C satisfy the conditions (1–4) of Definition D.1. Then we have:

$$AB = (ACA)B = (AC)(AB) = (AC)^T(AB)^T$$
$$= C^T A^T B^T A^T = C^T(ABA)^T = C^T A^T = (AC)^T = AC.$$

By the same way, we have

$$BA = B(ACA) = (BA)(CA) = (BA)^T(CA)^T$$
$$= A^T B^T A^T C^T = (ABA)^T C^T = A^T C^T = (CA)^T = CA.$$

Thus

$$B = BAB = (BA)B = (CA)B = C(AB) = CAC = C.$$

□

Example D.1. Suppose that A is a non-singular square matrix. Then the pseudoinverse A^+ exists and
$$A^+ = A^{-1}.$$

Proof. We have $AA^{-1}A = AI = A$ and $A^{-1}AA^{-1} = A^{-1}I = A^{-1}$. If A is of order n, then A^{-1} is of order n as well, so that

---

[4]There are other versions of definition of generalized inverse, or pseudoinverse, matrix. The most useful of the definitions require only *some* of the Penrose conditions (1–4) from Definition D.1, e.g., (1) and (2). On the definitions and properties of these more general versions of inverse matrix, see [5, Chap. 6] and [2]; see also [8, pp. 94–98], [26, pp. 203–205].

D.1 Definition and Basic Properties

$$AA^{-1} = I_n = A^{-1}A.$$

This means that the matrix A^{-1} satisfies the conditions (1–4) of Definition D.1. □

Example D.2. Let

$$A = \begin{bmatrix} a_1 \\ \vdots \\ a_n \end{bmatrix}$$

be a nonzero column vector. Then the row vector

$$C = \lambda A^T$$

is a pseudoinverse of A, where $\lambda = 1/(a_1^2 + \cdots + a_n^2)$.

Proof. We have $AC = \lambda(a_i a_j)_{n \times n}$ and $CA = \lambda(AA^T) = 1$, so that the conditions (3) and (4) hold for the matrix C. In addition, we have $ACA = A1 = A$ and $CAC = 1C = C$, so that the conditions (1) and (2) hold as well. □

Example D.3. Suppose that an $m \times n$ matrix A has the form

$$A = \begin{bmatrix} B & 0 \\ \hline 0 & 0 \end{bmatrix},$$

where B is an $r \times r$ non-singular submatrix. Then the pseudoinverse A^+ is an $n \times m$ matrix of the form

$$C = \begin{bmatrix} B^{-1} & 0 \\ \hline 0 & 0 \end{bmatrix}.$$

Proof. It is easy to see that AC and CA are square matrices of the form

$$\begin{bmatrix} I_r & 0 \\ \hline 0 & 0 \end{bmatrix}$$

of orders m and n, respectively. It follows that $(AC)^T = AC$ and $(CA)^T = CA$. According to Lemma 9.16, we have $ACA = (AC)A = A$ and $CAC = (CA)C = C$. By Definition D.1, this means that $A^+ = C$ is the pseudoinverse. □

D.1.1 The Basic Properties of Pseudoinverse

5. $(A^+)^+ = A$
6. $(A^T)^+ = (A^+)^T$

Proof. Exercise. □

7. rank $A^+ =$ rank A

Proof. Recall that $\text{rank}(AB) \leq \text{rank } A$ and $\text{rank}(AB) \leq \text{rank } B$ for any two matrices A and B conformable for multiplication (see Problem 24 in Chap. 2). Then

$$\text{rank } A = \text{rank}(AA^+A) \leq \text{rank}(AA^+) \leq \text{rank } A^+$$

and

$$\text{rank } A^+ = \text{rank}(A^+AA^+) \leq \text{rank}(AA^+) \leq \text{rank } A.$$

□

Exercise D.1. Show that if rank $A = 1$, then $A^+ = \frac{1}{\text{Tr}(AA^T)}A^T$.

Hint. If rank $A = 1$, then all columns of A are linear combinations of a single column, say, A^k, that is,

$$A = (\lambda_1 A^k | \ldots | \lambda_n A^k) = (\lambda_1, \ldots, \lambda_n)A^k.$$

D.2 Full Rank Factorization and a Formula for Pseudoinverse

Let us give the formula for a pseudoinverse of a matrix with linearly independent columns.

Theorem D.2. *Suppose that an $m \times n$ matrix A has full column rank, that is, rank $A = n$. Then A has a pseudoinverse*

$$A^+ = (A^T A)^{-1} A^T.$$

We begin with

Lemma D.3. *Let A be an $m \times n$ with full column rank. Then the $n \times n$ matrix $A^T A$ is non-singular.*

Proof. Suppose that $\text{rank}(A^T A) < n$. By Theorem 5.6, it follows that the system

$$A^T A \mathbf{x} = \mathbf{0}$$

has a nonzero solution $\mathbf{x}$. Then

$$\mathbf{x}^T A^T A \mathbf{x} = \mathbf{0},$$

$$(A\mathbf{x})^T A\mathbf{x} = \mathbf{0},$$

that is, $(A\mathbf{x}, A\mathbf{x}) = \mathbf{0}$. Hence $A\mathbf{x} = \mathbf{0}$. Since $\mathbf{x} \neq \mathbf{0}$, we have rank $A < n$, a contradiction. □

D.2 Full Rank Factorization and a Formula for Pseudoinverse

Proof of Theorem D.2. By Lemma D.3, the matrix $(A^T A)^{-1}$ exists. Let $C = (A^T A)^{-1} A^T$. We are going to show that C satisfies Definition D.1. Then $CA = I_n$, hence $(CA)^T = I_n^T = CA$. Since $C^T = A(A^T A)^{-1}$, we have $(AC)^T = C^T A^T = A(A^T A)^{-1} A^T = AC$. Moreover, $ACA = AI_n = A$ and $CAC = I_n C = C$. By Definition D.1, $C = A^+$. □

Example D.4. Let us calculate the pseudoinverse of the matrix

$$A = \begin{bmatrix} 1 & 0 \\ 1 & 2 \\ -1 & 3 \end{bmatrix}.$$

We have rank $A = 2$ is equal to the number of columns, so that we can use the above formula. We have

$$A^T A = \begin{bmatrix} 3 & -1 \\ -1 & 13 \end{bmatrix},$$

so that

$$A^+ = (A^T A)^{-1} A^T = \begin{bmatrix} 3 & -1 \\ -1 & 13 \end{bmatrix}^{-1} \begin{bmatrix} 1 & 1 & -1 \\ 0 & 2 & 3 \end{bmatrix} = \begin{bmatrix} 13/38 & 1/38 \\ 1/38 & 3/38 \end{bmatrix} \begin{bmatrix} 1 & 1 & -1 \\ 0 & 2 & 3 \end{bmatrix}$$

$$= \frac{1}{38} \begin{bmatrix} 13 & 15 & -10 \\ 1 & 7 & 8 \end{bmatrix}.$$

In order to obtain similar formula in the general case of arbitrary matrix A, we need a presentation of A as a product of two matrices of full ranks, one of full column rank and one of full row rank.

Theorem D.4 (Full rank factorization). *Let A be an $m \times n$ matrix of rank r. Then there exist an $m \times r$ matrix F and an $r \times n$ matrix G (both of rank r) such that*

$$A = FG.$$

Proof. Consider any r linearly independent columns of A. Let F be the submatrix of A formed by these columns. Then F has size $m \times r$ and rank r. Each column A^k of A is a linear combination of columns of F, that is, $A^k = F\mathbf{G}^k$, where $\mathbf{G}^k$ is a column vector of dimension r. Then all n vectors $\mathbf{G}^1, \ldots, \mathbf{G}^n$ form a matrix G of size $r \times n$ such that $A = FG$. According to Problem 24 in Chap. 2, we have rank $G \geq r$. Since G has r rows, we have rank $G \leq r$, thus rank $G = r$. □

Note that for practical purposes, we may choose as F any matrix columns of which form a basis of the linear span of the columns of A.

Example D.5. Let
$$A = \begin{bmatrix} 1 & 2 & 3 \\ 4 & 5 & 6 \\ 7 & 8 & 9 \end{bmatrix}$$

Since $\det A = 0$ and $\det A_{(2)} = \begin{vmatrix} 1 & 2 \\ 4 & 5 \end{vmatrix} = -3 \neq 0$, we conclude that $\operatorname{rank} A = 2$ and the first two columns (which form the submatrix $A_{(2)}$) are linearly independent. Then one can choose
$$F = \begin{bmatrix} 1 & 2 \\ 4 & 5 \\ 7 & 8 \end{bmatrix}.$$

Let us construct presentations of the columns of A as linear combinations of columns F^1 and F^2 of F. For the first two columns of A, we have $A^1 = F^1$ and $A^2 = F^2$, hence $A^1 = FG^1$ and $A^2 = FG^2$ with $G^1 = (1,0)^T$ and $G^2 = (0,1)^T$. To obtain a presentation of the third column A^3 via the columns of F, we have the system $FG^3 = A^3$ with unknown G^3, or

$$\begin{bmatrix} 1 & 2 \\ 4 & 5 \\ 7 & 8 \end{bmatrix} G^3 = \begin{bmatrix} 3 \\ 6 \\ 9 \end{bmatrix}.$$

We have the unique solution $G^3 = (-1, 2)^T$. Finally, we have $A = FG$, where F is as above and
$$G = \begin{bmatrix} 1 & 0 & -1 \\ 0 & 1 & 2 \end{bmatrix}.$$

For another useful method to construct a full rank decomposition, see Problem 6 below.

Now, we a ready to deduce a general formula for pseudoinverse.

Theorem D.5. *For an arbitrary $m \times n$ matrix A, its pseudoinverse A^+ exists. If $A = FG$ is a full rank decomposition of A, then*

$$A^+ = G^T(GG^T)^{-1}(F^TF)^{-1}F^T.$$

Proof. By Lemma D.3, both matrices GG^T and F^TF are non-singular, so the matrix $C = G^T(GG^T)^{-1}(F^TF)^{-1}F^T$ exists. We will show that it satisfies the conditions of Definition D.1. Let $X = GG^T$ and $Y = F^TF$. Note that $X = X^T$ and $Y = Y^T$ are symmetric matrices. Then $AC = FXX^{-1}Y^{-1}F^T = FY^{-1}F^T$, so, $(AC)^T = AC$. Moreover, $CA = G^TX^{-1}Y^{-1}YG = G^TX^{-1}G$, hence $(CA)^T = CA$. Finally, $CAC = G^TX^{-1}Y^{-1}F^TFGG^TX^{-1}Y^{-1}F^T = G^TX^{-1}Y^{-1}YXX^{-1}Y^{-1}F^T = G^TX^{-1}Y^{-1}F^T = C$ and $ACA = FGG^TX^{-1}Y^{-1}F^TFG = FXX^{-1}Y^{-1}YG = FG = A$. By Definition D.1, $C = A^+$. □

D.3 Pseudoinverse and Approximations

Example D.6. Let us calculate the pseudoinverse of the matrix A from Example D.5. We have

$$A^+ = G^T(GG^T)^{-1}(F^TF)^{-1}F^T,$$

where

$$GG^T = \begin{bmatrix} 1 & 0 & -1 \\ 0 & 1 & 2 \end{bmatrix} \cdot \begin{bmatrix} 1 & 0 \\ 0 & 1 \\ -1 & 2 \end{bmatrix} = \begin{bmatrix} 2 & -2 \\ -2 & 5 \end{bmatrix}$$

and

$$F^TF = \begin{bmatrix} 1 & 4 & 7 \\ 2 & 5 & 8 \end{bmatrix} \cdot \begin{bmatrix} 1 & 2 \\ 4 & 5 \\ 7 & 8 \end{bmatrix} = \begin{bmatrix} 66 & 78 \\ 78 & 93 \end{bmatrix},$$

so that

$$A^+ = \begin{bmatrix} 1 & 0 \\ 0 & 1 \\ -1 & 2 \end{bmatrix} \begin{bmatrix} 2 & -2 \\ -2 & 5 \end{bmatrix}^{-1} \begin{bmatrix} 66 & 78 \\ 78 & 93 \end{bmatrix}^{-1} \begin{bmatrix} 1 & 4 & 7 \\ 2 & 5 & 8 \end{bmatrix}$$

$$= \begin{bmatrix} 1 & 0 \\ 0 & 1 \\ -1 & 2 \end{bmatrix} \begin{bmatrix} 5/6 & 1/3 \\ 1/3 & 1/3 \end{bmatrix} \begin{bmatrix} 31/18 & -13/9 \\ -13/9 & 11/9 \end{bmatrix} \begin{bmatrix} 1 & 4 & 7 \\ 2 & 5 & 8 \end{bmatrix}$$

$$= \frac{1}{36} \begin{bmatrix} -23 & -6 & 11 \\ -2 & 0 & 2 \\ 19 & 6 & -7 \end{bmatrix}.$$

For another useful method of pseudoinverse calculation, see [5, Theorem 1.3.1].

D.3 Pseudoinverse and Approximations

Suppose that the system $A\mathbf{x} = \mathbf{b}$ is inconsistent, that is, it has no exact solution. If the values of the coefficients are considered as approximate ones, then it is reasonable to find an approximate solution of the system, say, by the least-square method.

Recall from Sect. 7.3 that a vector $\mathbf{u} \in \mathbb{R}^n$ is called a *least square solution* of the system (D.1) if for every $\mathbf{x} \in \mathbb{R}^n$ we have $|A\mathbf{u} - \mathbf{b}| \leq |A\mathbf{x} - \mathbf{b}|$. In Sect. 7.3, we have considered the case of full rank matrix A and shown that in this case the least square solution is unique. In the general case of arbitrary A, the solution is not necessarily unique. A least square solution $\mathbf{u}$ of the system (D.1) is called *minimal* if it has the smallest length among all others, that is, for any other least square solution **v** of the same system we have $|\mathbf{u}| \leq |\mathbf{v}|$.

Theorem D.6. *The minimal least-squares solution of smallest length of the linear system $A\mathbf{x} = \mathbf{b}$, where A is an $m \times n$-matrix, is unique and given by the formula*

$$\mathbf{u} = A^+ \mathbf{b},$$

where A^+ is the pseudoinverse of A.

Lemma D.7. *For each $m \times n$-matrix A, $\mathrm{Im}(AA^+ - I) \perp \mathrm{Im}\, A$, where I is the corresponding identity matrix.*

Proof of Lemma D.7. For each matrix B, its kernel is the orthogonal complement to the span of its rows, that is, the span of columns of its transpose, $(\ker B)^\perp = \mathrm{Im}\, B^T$. Therefore, we have $\mathrm{Im}(AA^+ - I) = \left(\mathrm{Ker}(AA^+ - I)^T\right)^\perp = \left(\mathrm{Ker}\left((AA^+)^T - I^T\right)\right)^\perp = \left(\mathrm{Ker}(AA^+ - I)\right)^\perp$. So, it is sufficient to prove that $\mathrm{Im}\, A \subset \mathrm{Ker}(AA^+ - I)$. Indeed, let $\mathbf{y} = A\mathbf{x} \in \mathrm{Im}\, A$. Then

$$(AA^+ - I)\mathbf{y} = (AA^+ - I)A\mathbf{x} = (AA^+ A - A)\mathbf{x} = \mathbf{0}\mathbf{x} = \mathbf{0},$$

hence $\mathbf{y} \in \mathrm{Ker}(AA^+ - I)$. □

Proof of Theorem D.6. First, let us proof that the vector $\mathbf{u} = A^+ \mathbf{b}$ is a least square solution. For each $\mathbf{x} \in \mathbb{R}^n$, we have $A\mathbf{x} - \mathbf{b} = (A\mathbf{x} - AA^+\mathbf{b}) + (AA^+\mathbf{b} - \mathbf{b}) = \mathbf{c}_\mathbf{x} + \mathbf{d}$, where $\mathbf{c}_\mathbf{x} = A\mathbf{x} - AA^+\mathbf{b} = A(\mathbf{x} - A^+\mathbf{b}) \in \mathrm{Im}\, A$ and $\mathbf{d} = (AA^+ - I)\mathbf{b} \in \mathrm{Im}(AA^+ - I)$. By Lemma D.7, we have $\mathbf{c}_\mathbf{x} \perp \mathbf{d}$. By Pythagorean theorem,

$$|A\mathbf{x} - \mathbf{b}|^2 = |\mathbf{c}_\mathbf{x} + \mathbf{d}|^2 = |\mathbf{c}_\mathbf{x}|^2 + |\mathbf{d}|^2 \geq |\mathbf{d}|^2.$$

Put $\mathbf{x} = \mathbf{u}$. Then $\mathbf{c}_\mathbf{x} = A\mathbf{u} - AA^+\mathbf{b} = \mathbf{0}$, so that the value of $|A\mathbf{x} - \mathbf{b}|$ has its minimal value $|\mathbf{d}| = |(AA^+ - I)\mathbf{b}|$. This means that $\mathbf{u}$ is a least square solution.

Now, let $\mathbf{x}$ be another least square solution of the same system $A\mathbf{x} = \mathbf{b}$. By the above, $\mathbf{c}_\mathbf{x} = \mathbf{0}$, that is, $\mathbf{x}$ satisfies the linear system $A\mathbf{x} - AA^+\mathbf{b} = \mathbf{0}$, or $A\mathbf{x} = AA^+\mathbf{b}$. Since $\mathbf{u}$ form a solution of the last system, any other solution has the form

$$\mathbf{x} = \mathbf{u} + \mathbf{w},$$

where $\mathbf{w}$ is a solution of the corresponding homogeneous system $A\mathbf{w} = \mathbf{0}$. Then we have

$$(\mathbf{u}, \mathbf{w}) = \mathbf{u}^T \mathbf{w} = \mathbf{b}^T \left(A^+\right)^T \mathbf{w},$$

where $\left(A^+\right)^T = \left(A^+ A A^+\right)^T = \left(A^+\right)^T \left(A^+ A\right)^T = \left(A^+\right)^T A^+ A$. Hence

$$(\mathbf{u}, \mathbf{w}) = \mathbf{b}^T \left(A^+\right)^T A^+ A \mathbf{w} = 0,$$

that is, $\mathbf{u} \perp \mathbf{w}$. By Pythagorean theorem, it follows that

D.3 Pseudoinverse and Approximations

$$|\mathbf{x}|^2 = |\mathbf{u} + \mathbf{w}|^2 = |\mathbf{u}|^2 + |\mathbf{w}|^2 \geq |\mathbf{u}|^2.$$

Thus, $\mathbf{u}$ is a least square solution of the smallest possible length, and any other least square solution $\mathbf{x} = \mathbf{u} + \mathbf{w}$, where $\mathbf{w}$ is nonzero, has strictly larger length. □

Example D.7. Consider the following macroeconomic policy model

$$Y = C + I + G + X - M, \, C = 0.9Y_d, \, Y_d = Y - T,$$

$$T = 0.15Y, \, I = 0.25(Y - Y_{-1}) + 0.75G_I, \, G = G_C + G_I,$$

$$M = 0.02C + 0.08I + 0.06G_I + 0.03X,$$

$$N = 0.8Y, \, B = X - M, \, D = G - T,$$

where Y – GDP, Y_{-1} – GDP in the previous year, Y_d – Disposable Income, C – Private Consumption, T – Tax Revenues, I – Private Investment, M – Imports, N – Employment, B – Current Account of the Balance of Payments, D – Budget Deficit, X – Exports, G_C – Public Consumption Expenditures (instrument), G_I – Public Investment Expenditures (instrument).

All variables, except N, are measured in $ million. Employment (N) unit of measurement is 1,000 persons.

Suppose the following data is given: $Y_{-1} = \$1{,}200$ million, $X = \$320$ million.

The problem is the following.

i. Suppose that the government is interested in three targets: Employment (N), Balance of Payments (B) and Public Sector Deficit (D). Reduce the model into three equations by eliminating the 'irrelevant endogenous variables'.
ii. Does this model satisfy the Tinbergen's Theorem on the equality of the number of instruments and the number of targets? Why?
iii. Suppose that the government wants employment level (N) is 1,000. On the other hand the government is aware that the external borrowing limit of the country is 100 million, i.e., the country can not increase its current account deficit beyond this figure. Finally, government is eager to reduce public sector debt, and therefore aims at creating a budget surplus of $120 million. In other words, the government's targets are as follows:

$$N = 1{,}000,$$

$$B = -100,$$

$$D = -120.$$

How should government determine the values of its instruments, i.e., government investment and government consumption, under these circumstances?

Partial Solution

Tinbergen's approach requires 'targets' to be given. Starting from this data and using the model, the solution gives the necessary magnitudes of the instruments to achieve the desired target levels.

i. The *reduced form* of the model looks like

$$\begin{cases} N = \beta_{10} + \beta_{11}G_I + \beta_{12}G_C, \\ B = \beta_{20} + \beta_{21}G_I + \beta_{22}G_C, \\ D = \beta_{30} + \beta_{31}G_I + \beta_{32}G_C, \end{cases}$$

or in matrix terms

$$\begin{bmatrix} N \\ B \\ D \end{bmatrix} = \begin{bmatrix} \beta_{10} \\ \beta_{20} \\ \beta_{30} \end{bmatrix} + \begin{bmatrix} \beta_{11} & \beta_{12} \\ \beta_{21} & \beta_{22} \\ \beta_{31} & \beta_{32} \end{bmatrix} \begin{bmatrix} G_I \\ G_C \end{bmatrix}.$$

Using the above equations and through substitution the system can be reduced to the following form

$$\begin{cases} N = 1355.665 + 64.236G_I + 39.409G_C, \\ B = 274.581 - 2.954G_I - 1.739G_C, \\ D = -254.187 - 11.044G_I - 6.389G_C, \end{cases}$$

or in matrix terms

$$\begin{bmatrix} N \\ B \\ D \end{bmatrix} = \begin{bmatrix} 1355.665 \\ 274.581 \\ -254.187 \end{bmatrix} + \begin{bmatrix} -3.161 & -1.861 \\ 64.236 & 39.409 \\ -11.044 & -6.389 \end{bmatrix} \begin{bmatrix} G_I \\ G_C \end{bmatrix}.$$

This may be re-written as

$$\mathbf{y} = \mathbf{b} - \mathbf{Ax}$$

with

$$\mathbf{y} = \begin{bmatrix} N \\ B \\ D \end{bmatrix}, \mathbf{b} = \begin{bmatrix} 1355.665 \\ 274.581 \\ -254.187 \end{bmatrix}, \mathbf{x} = \begin{bmatrix} G_I \\ G_C \end{bmatrix},$$

or

$$\mathbf{Ax} = \mathbf{c},$$

where

$$\mathbf{c} = \mathbf{b} - \mathbf{y} = \begin{bmatrix} 1355.665 - N \\ 274.581 - B \\ -254.187 - D \end{bmatrix}.$$

The problem is then to find **y**, given **c**.

D.4 Problems

ii. No, it does not. The number of instruments (=2) is less than the number of targets (=3). The matrix of coefficients

$$A = \begin{bmatrix} -3.161 & -1.861 \\ 64.236 & 39.409 \\ -11.044 & -6.389 \end{bmatrix}$$

is an 3×2 matrix and rank $A = 2$. Therefore it does not have an inverse.

iii. In [31, pp. 37–42] the problem of inequality of targets and instruments is discussed. Tinbergen [31, pp. 39–40] correctly points out that when the number of targets exceed the number of instruments, an inconsistency problem arises.

In this case, one can calculate the pseudoinverse of the matrix A, which always exists. Notice that the rank of A is 2, i.e., A has full column rank. Then by Theorem D.2 the matrix A has the pseudoinverse of the following form

$$A^+ = (A^T A)^{-1} A^T = \begin{bmatrix} -0.261 & -0.288 & -1.532 \\ 0.451 & 0.470 & 2.498 \end{bmatrix}.$$

A least square approximate solution can be obtained by calculating

$$\mathbf{x}^* = A^+ \mathbf{c},$$

that is,

$$\mathbf{x}^* = \begin{bmatrix} -0.261 & -0.288 & -1.532 \\ 0.451 & 0.470 & 2.498 \end{bmatrix} \begin{bmatrix} 1355.665 - 1000 \\ 274.581 + 100 \\ -254.187 + 120 \end{bmatrix} = \begin{bmatrix} 4.701 \\ 1.231 \end{bmatrix}.$$

Thus $G_I = \$4.701$ million and $G_C = \$1.231$ million. These figures indicate that the government has to target extremely low levels for government expenditures.

D.4 Problems

1. Calculate $[1, 0]^+$.
2. Calculate

$$\begin{bmatrix} 0 \\ 1 \\ 2 \\ 3 \end{bmatrix}^+.$$

3. Calculate $[3, 2, 1, 0]^+$.

4. Calculate
$$\begin{bmatrix} 1 & 0 \\ -1 & 0 \\ 1 & 0 \\ 2 & 1 \end{bmatrix}^+.$$

5. Calculate
$$\begin{bmatrix} 1 & 2 & 3 \\ 0 & -1 & -2 \end{bmatrix}^+.$$

6.* Let A be an $m \times n$ matrix of rank r and let
$$K = \begin{bmatrix} G \\ 0 \end{bmatrix}$$

be its canonical form (see Definition 2.4), where G is an upper $r \times n$ submatrix without zero rows and 0 denotes a zero submatrix. Let $i_1, \ldots, i_r$ be the numbers of columns where the leading coefficients of the echelons appears, and let F be a submatrix of A formed by its columns $i_1, \ldots, i_r$. Prove that

$$A = FG$$

and this is a full rank factorization of A.

7. Find a full rank factorization of the matrix
$$A = \begin{bmatrix} 1 & 1 & 1 \\ 2 & 2 & 2 \\ 3 & 3 & 3 \\ 1 & 2 & 3 \end{bmatrix}.$$

8. Calculate
$$\begin{bmatrix} 1 & 0 & 0 \\ 1 & 1 & 1 \\ 0 & 1 & 1 \end{bmatrix}^+.$$

9. Let E_{ij} be an $n \times n$ matrix such that its element in i-th row and j-th column is unit and all other elements are zeroes. Find its full rank decomposition and pseudoinverse.

10. Prove the formulae:
    (a) $\operatorname{Im}(AA^+) = \operatorname{Im}(AA^T) = \operatorname{Im} A$.
    (b) $\operatorname{Ker}(AA^+) = \operatorname{Ker}(AA^T) = \operatorname{Ker} A^T$.
    (c) $\operatorname{Im} A^+ = \operatorname{Im} A^T$.
    (d) $\operatorname{Ker} A^+ = \operatorname{Ker} A^T$.
    (e) $\operatorname{Im} A^+ = (\operatorname{Ker} A)^\perp$.
    (f) $\operatorname{Ker} A^+ = (\operatorname{Im} A)^\perp$.

11. Find the solution of smallest length of the linear system

$$\begin{cases} 2x + 3y + 2z = 7, \\ 3x + 4y - z = 6. \end{cases}$$

12. Find the least-square solution of smallest length of the linear system

$$\begin{cases} x - 3y + t = -1, \\ 2y - 3z = -1, \\ x - 2y + z + t = 0, \\ x - 2z + t = 8. \end{cases}$$

Answers and Solutions

Chapter 1

1. (a) $\sqrt{137}$; (b) $\sqrt{113}$; (c) $2\sqrt{5}$; **4.** (a) $y = -2.25x + 1$; (b) $y = (8/7)x$; (c) $x = \sqrt{2}$. **8.** $x = 10/7$. **10.** (a) $x = 40/13, y = -15/13$; (b) $y = 0.6x - 3$; (c) no solutions. **11.** $z_1 = (a_{11}b_{11} + a_{12}b_{21})x_1 + (a_{11}b_{12} + a_{12}b_{22})x_2, z_2 = (a_{21}b_{11} + a_{22}b_{21})x_1 + (a_{21}b_{12} + a_{22}b_{22})x_2$.

Chapter 2

1. (a) $\mathbf{x} = (3, -1, -3, -3)$; (b) $\mathbf{x} = (-0.2, -0.4, -1)$. **4.** Yes. **7.** (a) 2; (b) 3. **11.** 2^n. **12.** For example, $A = B = \begin{bmatrix} 0 & 1 \\ 0 & 0 \end{bmatrix}$ and $C = 0$. **13.** For example, $A = \begin{bmatrix} 1 & 1 \\ 1 & 1 \end{bmatrix}, B = \begin{bmatrix} 0 & 0 \\ 0 & 0 \end{bmatrix}$ and $C = \begin{bmatrix} 1 & 1 \\ -1 & -1 \end{bmatrix}$. **17.** *Hint.* Use the property (1-b) of matrix multiplication. **18.** For example, $A = \begin{bmatrix} 1 & 0 & 0 \\ 0 & 1 & 0 \end{bmatrix}$ and $B = \begin{bmatrix} 1 & 0 \\ 0 & 1 \\ 1 & 1 \end{bmatrix}$.

19. (a) $x = \begin{bmatrix} a \\ a \end{bmatrix}$ for $a \in \mathbb{R}$; (b) $y = \begin{bmatrix} 3b & b \end{bmatrix}$ for $b \in \mathbb{R}$. **22.** $\begin{bmatrix} 1 & 16 & 0 \\ 0 & 10 & 0 \\ 0 & -180 & 5 \end{bmatrix}$.

23. (a) In AB, the i-th and j-th rows are interchanged as well; (b) c times j-th row of AB will be added to i-th row of AB; (c) i-th and j-th columns of AB are interchanged as well; (d) c times j-th column of AB will be added to i-th column of AB.
24. *Hint.* Use Problem 23. **27.** $\begin{bmatrix} a & b \\ c & -a \end{bmatrix}$, where $bc = -a^2$. **28.** Either $A = \begin{bmatrix} a & b \\ c & a \end{bmatrix}$, where $a^2 + bc = 1$, or $A = \pm I_2$.

29. $\begin{bmatrix} 2 & 1 & -4 & 5 \\ 0 & 0 & 1 & -3 \\ 0 & 0 & 0 & 0 \end{bmatrix}$, rank = 2. **30.** $\begin{bmatrix} 1 & 3 & 0 & 0 \\ 0 & 0 & 1 & 0 \\ 0 & 0 & 0 & 1 \end{bmatrix}$.

Chapter 3

1. (a) -2; (b) 0; (c) -1; (d) $-2b^3$; (e) $\sin(\alpha - \beta)$; (f) 0. **2.** (a) 2; (b) 30; (c) $abc + 2x^3 - (a+b+c)x^2$; (d) $\alpha^2 + \beta^2 + \zeta^2 + 1$; (e) $\sin(\beta - \zeta) + \sin(\zeta - \alpha) + \sin(\alpha - \beta)$.
3. $\begin{bmatrix} x' \\ y' \end{bmatrix} = \begin{bmatrix} \frac{1}{\sqrt{2}} & \frac{1}{\sqrt{2}} \\ -\frac{1}{\sqrt{2}} & \frac{1}{\sqrt{2}} \end{bmatrix} \begin{bmatrix} x \\ y \end{bmatrix} - \begin{bmatrix} \sqrt{2} - \frac{3}{\sqrt{2}} \\ \sqrt{2} - \frac{3}{\sqrt{2}} \end{bmatrix}$ **4.** $f(A) = 0$. **7.** 1875. **8.** $(-1)^n d$.
9. (a) 1 for $n = 4k$ and $n = 4k+1$, -1 for $n = 4k+2$ and $n = 4k+3$, where k is an integer; (b) $n+1$; c) $1 + (-1)^{n-1} 2^n$. **10.** $\det X = 1$. *Solution.* We have

$$X^3 - I_n = (X - I_n)(X^2 + X + I_n) = (X - I_n)\mathbf{0} = \mathbf{0},$$

hence $X^3 = I_n$. Then $\det(X^3) = \det I_n = 1$, that is, $(\det X)^3 = 1$ and $\det X = 1$.

Chapter 4

1. (a) $\begin{bmatrix} -2 & 1 \\ 1.5 & -0.5 \end{bmatrix}$; (b) $\begin{bmatrix} \cos\alpha & \sin\alpha \\ -\sin\alpha & \cos\alpha \end{bmatrix}$; (c) $\begin{bmatrix} -7/3 & 2 & -1/3 \\ 5/3 & -1 & -1/3 \\ -2 & 1 & 1 \end{bmatrix}$;

(d) $\frac{1}{9} \cdot \begin{bmatrix} 1 & 2 & 2 \\ 2 & 1 & -2 \\ 2 & -2 & 1 \end{bmatrix}$; (e) $\begin{bmatrix} -1 & 1 & 16 & -9 \\ -8 & 7 & 125 & -70 \\ -10 & 9 & 160 & -90 \\ -1 & 1 & 18 & -10 \end{bmatrix}$. **2.** (a) $\begin{bmatrix} 1 & -1 & 0 & \ldots & 0 \\ 0 & 1 & -1 & \ldots & 0 \\ 0 & 0 & 1 & \ldots & 0 \\ \vdots & \vdots & \vdots & \ddots & \vdots \\ 0 & 0 & 0 & \ldots & 1 \end{bmatrix}$;

(b) $\begin{bmatrix} 2-n & 1 & 1 & \ldots & 1 \\ 1 & -1 & 0 & \ldots & 0 \\ 1 & 0 & -1 & \ldots & 0 \\ \vdots & \vdots & \vdots & \ddots & \vdots \\ 1 & 0 & 0 & \ldots & -1 \end{bmatrix}$; (c) $\frac{1}{n-1} \begin{bmatrix} 2-n & 1 & 1 & \ldots & 1 \\ 1 & 2-n & 1 & \ldots & 1 \\ 1 & 1 & 2-n & \ldots & 1 \\ \vdots & \vdots & \vdots & \ddots & \vdots \\ 1 & 1 & 1 & \ldots & 2-n \end{bmatrix}$.

3. (a) $\begin{bmatrix} -1 & a-10 \\ 2 & 7.5 - 0.5a \end{bmatrix}$; (b) $\begin{bmatrix} 10.5 & 6.5 \\ 11 & -7 \end{bmatrix}$; (c) $\begin{bmatrix} 1 & 2 & 3 \\ 4 & 5 & 6 \\ 7 & 8 & 9 \end{bmatrix}$; (d) $\begin{bmatrix} 1 & 1 & 1 \\ 1 & 2 & 3 \\ 2 & 3 & 1 \end{bmatrix}$.

7. (a) 3; (b) 3 for $\lambda = \pm 1$, 4 for $\lambda \neq \pm 1$. **9.** rank $\begin{bmatrix} x_2 - x_1 & x_3 - x_1 \\ y_2 - y_1 & y_3 - y_1 \end{bmatrix} \leq 1$.

E Answers and Solutions

10. rank $\begin{bmatrix} a_1 & b_1 \\ a_2 & b_2 \\ a_3 & b_3 \end{bmatrix}$ = rank $\begin{bmatrix} a_1 & b_1 & c_1 \\ a_2 & b_2 & c_2 \\ a_3 & b_3 & c_3 \end{bmatrix}$ = 2.

Chapter 5

1. (a) $(x_1, x_2, x_3, x_4) = (1, 1, -1, -1)$. (b) $x_1 = (x_3 - 9x_4 - 2)/11$, $x_2 = (-5x_3 + x_4 + 10)/11$, $x_3, x_4 \in \mathbb{R}$. **2.** $x_1 = 1 - x_2 - x_3$ for $\lambda = 1$, $x_1 = x_2 = x_3 = 1/(\lambda + 2)$ for $\lambda \neq 1$. **3.** $x_1 = -53/208$, $x_2 = 59/16$, $x_3 = -29/16$, $x_4 = 27/13$. **4.** No solution.
5. $(x_1, x_2, x_3, x_4) = (3, 0, -5, 11)$. **6.** $(x_1, x_2, x_3, x_4, x_5) = (3, -5, 4, -2, 1)$. **7.** $(x_1, x_2, x_3, x_4, x_5) = (1/2, -2, 3, 2/3, -1/5)$. **8.** $x_1 = (-15x_2 + x_4 - 6)/10$, $x_3 = (4x_4 + 1)/5$, $x_2, x_4 \in \mathbb{R}$. **9.** $f(x) = x^2 - 5x + 3$. **10.** $f(x) = 2x^3 - 5x^2 + 7$.
12. No solution if $\alpha_1 + \beta_1 = 0$ and $\alpha_0 + \beta_0 \neq 0$; unique solution $p_i = \frac{\alpha_1 \beta_0 - \alpha_0 \beta_1}{\alpha_1 + \beta_1}$, $q_i^d = q_i^s = \frac{\alpha_0 + \beta_0}{\alpha_1 + \beta_1}$ (can be obtained by Kramer's rule); infinitely many solutions $q_i^d = q_i^s = \alpha_0 - \alpha_1 p_i$, p_i is arbitrary if $\alpha_1 + \beta_1 = \alpha_0 + \beta_0 = 0$.
13. i. For 'normal' goods, one expects demand to increase (decline) as its price falls (increases). This explains the negativity of the coefficients of own prices of all three goods. If the coefficient of the price of another good is positive (negative) in its demand function, this implies these two goods are *substitutes* (*complements*)[1].

ii. Equate supply and demand for each good, and rearrange the equations and express the linear equation system in matrix form as

$$\begin{bmatrix} -0,25 & -0,02 & 0,01 \\ 0,01 & -0,34 & 0,01 \\ -0,03 & 0,02 & -0,16 \end{bmatrix} \begin{bmatrix} p_1 \\ p_2 \\ p_3 \end{bmatrix} = \begin{bmatrix} -40 \\ -54 \\ -35 \end{bmatrix}$$

The solution is

$$p_1 \approx 154.87$$
$$p_2 \approx 169.58$$
$$p_3 \approx 210.91$$

(The values are rounded)

iii. $p_1 \approx 169.44$, $p_2 \approx 193.92$, $p_3 \approx 223.72$ (the values are rounded).

14. i. Y, C, I, G, M, N and B are endogenous, G_C, G_I, Y_{-1}, X are exogenous.
ii. Yes, both are endogenous variables.
iii. No, because it is an endogenous variable. It can not be controlled by the policy maker.

---

[1] For a further discussion of substitutes and complements, see, for example, [29, pp. 57–58] or [32, pp. 111–112].

iv. No, Tinbergen's theorem asserts that the number of instruments should be equal to the number of targets. See (vi).

v. Through substitution one gets

$$B = -0.375G_I - 0.268G_C + 317,$$
$$N = 6.789G_I + 5.829G_C - 1874.$$

vi. Notice that the equations given in (v) can be expressed as

$$\mathbf{y} = \mathbf{b} + A\mathbf{x}$$

where $\mathbf{y}$ is the vector of target variables, $\mathbf{x}$ is the vector of instruments, A is the coefficients matrix and $\mathbf{b}$ is the vector of intercept terms that are fixed. A unique solution to the above system can be obtained, if A^{-1} exists. The matrix A has an inverse if it is a square matrix, i.e. the number of rows (which is equal to the number of targets) should be equal to the number of columns (which is equal to number of instruments). This condition is Tinbergen's theorem. Secondly, A should have full rank, i.e. target variables, as well as instruments, should not be linearly dependent, i.e. they must be different.

For the given values of target variables the solution of the above system is obtained by calculating

$$\mathbf{x} = A^{-1}(\mathbf{y} - \mathbf{b})$$

which gives $G_I \approx \$325.64$ billion and $G_C \approx \$131.02$ billion.

Chapter 6

1. No. **3.** $(\xi_1, \xi_2, \ldots, \xi_n)$. **4.** $(\xi_1, \xi_2 - \xi_1, \xi_3 - \xi_2, \ldots, \xi_n - \xi_{n-1})$. **5.** $\dim M_n = n^2$. All matrices with all zero entries but one entry equal to 1 ("matrix units") form a basis of M_n. **7.** Yes. **8.** Yes. **9.** E. g., $f : \begin{bmatrix} a & b \\ c & d \end{bmatrix} \mapsto ax^3 + bx^2 + cx + d$.
10. $n(n+1)/2$. **15.** $(1/3, 11/6, 7/6, 11/6)$. **16.** $(-7, 11, 3)$. **17.** $(-2, -1, -2, -7)$.
18. (a) No. (b) No. (c) Yes for $c = 0$, no for all other c. **19.** Point (dim $= 0$), line (dim $= 1$), plane (dim $= 2$), the space $\mathbb{R}^3$ itself (dim $= 3$). **20.** The $n - 1$ vectors $(1, 0, \ldots 0, 1), (0, 1, 0, \ldots, 0), \ldots, (0, 0, \ldots, 1, 0)$ form a basis of $\mathcal{L}'$, so that $\dim \mathcal{L}' = n - 1$. **21.** The dimension is equal to 3; some possible bases are $\{\mathbf{a}_1, \mathbf{a}_2, \mathbf{a}_5\}$, $\{\mathbf{a}_1, \mathbf{a}_3, \mathbf{a}_5\}$ and $\{\mathbf{a}_1, \mathbf{a}_4, \mathbf{a}_5\}$.

Chapter 7

1. (a) Yes. (b) No. **2.** (a) No. (b) No. **3.** $\mathbf{e}_1 = \mathbf{f}_1$, $\mathbf{e}_2 = \left(\frac{1-n}{n}, \ldots, \frac{1}{n}, \frac{1}{n}\right)$, $\mathbf{e}_3 = \left(0, \frac{2-n}{n-1}, \frac{1}{n-1}, \ldots, \frac{1}{n}\right), \ldots,$ $\mathbf{e}_n = \left(0, \ldots, 0, -\frac{1}{2}, \frac{1}{2}\right)$. **6.** (a) E.g., $\mathbf{f}_3 = (2, 2, 1, 0)$, $\mathbf{f}_4 = (-2, 5, 6, 1)$. (b) E.g., $\mathbf{f}_3 = (1, -2, 1, 0)$, $\mathbf{f}_4 = (-25, -4, 17, 6)$. **7.** All 4 angles

are equal to $\pi/3$. **10.** $\mathbf{e}_1 = \left(1/\sqrt{15}\right)(2,1,3,1)$, $\mathbf{e}_2 = \left(1/\sqrt{23}\right)(3,2,-3,-1)$, $\mathbf{e}_3 = \left(1/\sqrt{127}\right)(1,5,1,10)$. **11.** $2\sqrt{7}$. **12.** $\mathbf{y} = (2,1,1,3)$, $\mathbf{x}-\mathbf{y} = (5,-5,-2,-1)$.
13. (a) 5; (b) 2. **14.** $[1/3, 11/3]$. **15.** $p(x) = 10.13 + 0.091x$, $p(27) = 12.6$.

Chapter 8

2. (a) 34. (b) 4. **3.** (a) Yes. (b) Yes. (c) No. (d) No. (e) No. **5.** $\begin{bmatrix} 1 & 0 & 0 \\ 0 & 0 & 0 \\ 0 & 0 & 0 \end{bmatrix}$. **6.** $\begin{bmatrix} 1 & 0 & 0 \\ 0 & 1 & 0 \\ 0 & 0 & 0 \end{bmatrix}$.

7. (a) $\begin{bmatrix} 0 & 1 & 1 \\ 2 & 0 & 1 \\ 3 & -1 & 1 \end{bmatrix}$. (b) Non-linear. (c) Non-linear. (d) $\begin{bmatrix} 1 & -1 & 1 \\ 0 & 0 & 1 \\ 0 & 1 & 0 \end{bmatrix}$. **8.** $\begin{bmatrix} 2 & -11 & 6 \\ 1 & -7 & 4 \\ 2 & -1 & 0 \end{bmatrix}$.

9. In the canonical basis: $\begin{bmatrix} 1 & 2 & 3 \\ 2 & 4 & 6 \\ 3 & 6 & 9 \end{bmatrix}$, in the given basis: $\begin{bmatrix} 20/3 & -5/3 & 5 \\ -16/3 & 4/3 & -4 \\ 8 & -2 & 6 \end{bmatrix}$.

10. (a) $\begin{bmatrix} 0 & -1 & 2 & 3 \\ 5 & 3 & 1 & 2 \\ 2 & 1 & 3 & 1 \\ 2 & 0 & 1 & 1 \end{bmatrix}$; (b) $\begin{bmatrix} -2 & 0 & 1 & 0 \\ 1 & -4 & -8 & -7 \\ 1 & 4 & 6 & 4 \\ 1 & 3 & 4 & 7 \end{bmatrix}$. **11.** $\begin{bmatrix} 1 & 0 & 6 \\ 0 & 2 & -2 \\ 0 & 0 & 3 \end{bmatrix}$. **12.** $\begin{bmatrix} 16 & 47 & -88 \\ 18 & 44 & -92 \\ 12 & 27 & -59 \end{bmatrix}$.

13. (a) $\begin{bmatrix} -70 & 17 \\ -243 & 59 \end{bmatrix}$; (b) $\begin{bmatrix} 23 & -29 \\ 27 & -34 \end{bmatrix}$; (c) $\begin{bmatrix} -10 & -9 \\ -1 & -1 \end{bmatrix}$.

Chapter 9

1. Basis: $\mathbf{e}_1 = (1,0)$. **2.** Basis: $\mathbf{e}_1 = (1,0,0)$, $\mathbf{e}_2 = (0,0,1)$. **3.** (a) $\lambda_1 = 1$, $\mathbf{e}_1 = (1,0,0)$. (b) $\lambda_1 = 1$, $\mathbf{e}_1 = (1,0,0)$. **4.** $\lambda_1 = -2i$, $\mathbf{e}_1 = c(1,-i)$ and $\lambda_1 = 2i$, $\mathbf{e}_1 = c(1,i)$, where $c \in \mathbb{C}$. **8.** (a) $\lambda_1 = 1$, $\lambda_2 = 3$; (b) $\lambda_1 = \frac{1-\sqrt{5}}{2}$, $\lambda_2 = \frac{1+\sqrt{5}}{2}$. **9.** *Hint.*

Put $B = \begin{bmatrix} \sqrt{\lambda_1} & 0 & \cdots & 0 \\ 0 & \sqrt{\lambda_2} & \cdots & 0 \\ \vdots & \vdots & \ddots & \vdots \\ 0 & 0 & \cdots & \sqrt{\lambda_n} \end{bmatrix}$. **10.** (a) $\begin{bmatrix} 1 & 0 & 0 \\ 0 & 2 & 0 \\ 0 & 0 & 2 \end{bmatrix}$. (b) $\begin{bmatrix} -2 & 0 & 0 & 0 \\ 0 & 2 & 0 & 0 \\ 0 & 0 & 2 & 0 \\ 0 & 0 & 0 & 2 \end{bmatrix}$. **11.** The only eigenvalue is $\lambda = 0$; the eigenvectors are constant polynomials. **14.** (a) $\begin{bmatrix} 2 & 0 \\ -4 & 2 \end{bmatrix}$;

(b) $2x_1^2 - 4x_1x_2 + 2x_2^2$, $\begin{bmatrix} 2 & -2 \\ -2 & 2 \end{bmatrix}$. **15.** Positive definite if $a > 0, b > 1/a, c > 0$; negative definite if $a < 0, b < 1/a, c < 0$.

Chapter 10

1. i. The total sales of firms of a sector to another firms of the same sector;

ii. $d_1 = 600, d_2 = 230, d_3 = 250$; iii. $A = \begin{bmatrix} 0.2 & 0.05 & 0.15 \\ 1/11 & 4/11 & 7/55 \\ 1/4 & 1/6 & 1/6 \end{bmatrix} \approx$

$\begin{bmatrix} 0.2 & 0.05 & 0.15 \\ 0.091 & 0.364 & 0.127 \\ 0.25 & 0.167 & 0.167 \end{bmatrix}$, $(I - A)^{-1} = \begin{bmatrix} 1.355 & 0.177 & 0.271 \\ 0.286 & 1.674 & 0.307 \\ 0.464 & 0.388 & 1.343 \end{bmatrix}$; iv. Yes (by Theorem 10.9, because $(I - A)^{-1} > 0$).

2. i. $A = \begin{bmatrix} \frac{20}{150} & \frac{40}{480} & \frac{10}{300} \\ \frac{30}{150} & \frac{200}{480} & \frac{100}{300} \\ \frac{20}{150} & \frac{60}{480} & \frac{50}{300} \end{bmatrix} \approx \begin{bmatrix} 0.133 & 0.083 & 0.033 \\ 0.2 & 0.417 & 0.333 \\ 0.133 & 0.125 & 0.167 \end{bmatrix}$, $(I - A)^{-1} \approx$

$\begin{bmatrix} 1.220 & 0.202 & 0.130 \\ 0.580 & 1.971 & 0.812 \\ 0.282 & 0.328 & 1.342 \end{bmatrix}$.

ii. By (10.26), $\mathbf{p}^T = \mathbf{v}^T(I - A)^{-1} = (80/150, 180/480, 140/300)(I - A)^{-1} = [1, 1, 1]$.

iii. From the information given in flow of funds table, the price equation can be written by reading the columns of the table as

$$\text{Outlay} = \text{payments made inputs} + \text{wage} + \text{profit}.$$

Since outlay is price times quantity, by dividing each side by corresponding output levels we can get the price equation for the Leontief model as

$$\mathbf{p} = \mathbf{p}A + \mathbf{w} + \pi, \tag{E.1}$$

where $\mathbf{p}$ is the prices vector as in (ii), $\mathbf{w}$ is the row vector of wage payments per unit of output, π is the row vector of profits per unit of output, and A is the input coefficients matrix as in (i). From (E.1) one gets

$$\mathbf{p}^T = (\mathbf{w} + \pi)(I - A)^{-1}. \tag{E.2}$$

In (iii) the question is to find the effect of a change in wage payments, on relative prices. We know that only $\mathbf{w}$ changes. So the new wage cost vector is $\mathbf{w}'$, substituting it to (E.2) we get

$$\mathbf{p}'^T = (\mathbf{w}' + \pi)(I - A)^{-1}. \tag{E.3}$$

From (E.2) and (E.3) we get

$$\mathbf{p}'^T - \mathbf{p} = (\mathbf{w}' - \mathbf{w})(I - A)^{-1}.$$

We have here $\mathbf{w}' = 1.2\mathbf{w}$, so that $\mathbf{p}'^T = \mathbf{p}^T + 0.2\mathbf{w}(I - A)^{-1} \approx [1.078, 1.085, 1.077]$. We see that the relative prices are changed.

E Answers and Solutions

3. i. From the original data the relative price vector (in terms of the price of the first commodity, p_j/p_1) can be computed as

$$\mathbf{p} = (1, 0.7754, 0.7537).$$

Notice that in this example $p_2/p_3 = 1.0288$.

When the technological progress takes place, all the coefficients in the second column of the above matrix declines by 10%.

Using the new matrix, we can calculate the relative price vector as

$$\mathbf{p} = (1, 0.7297, 0.7466).$$

Notice that in this example $p_2/p_3 = 0,9774$.

Notice that the technological progress led to a decline in the relative price of the manufacturing good, with respect to other two goods. On the other hand, relative price of the agricultural good increased with respect to other two goods.

ii. The Perron–Frobenius root of the matrix given in the table is 0.703 which gives the maximum rate of profit as 0.4225. After the technological change, the Perron–Frobenius root of the new matrix is 0.687, which corresponds to a higher rate of maximum profit 0.4556.

The finding indicates that, assuming competition which equalizes the rate of profit among sectors, an input saving technological progress in one sector, leads to an increase of the maximum rate of profit of the system as a whole.

4. (a) $\hat{\lambda} = 10, \hat{\mathbf{x}} = [1, 2]$; (b) $\hat{\lambda} = 0.961, \hat{\mathbf{x}} = [0.917, 0.398]$; (c) $\hat{\lambda} = 0.485, \hat{\mathbf{x}} = [0, 0.851, 0.526]$; (d) $\hat{\lambda} = 0.828, \hat{\mathbf{x}} = [0.247, 0.912, 0.448, 0.703]$.

5. productive matrices: b,c,d; irreducible matrices: a,b,c. **6.** (i) 18.9%; (ii) $[0.625, 0.539, 0.565]$; (iii) Let w be the wage rate, r the rate of profit, $\mathbf{a}_0$ the labor coefficients vector and $\mathbf{d}' = (1, 1, 1)$ the summation vector. Then the general expression for the wage rate can be derived as

$$w = \mathbf{d}'[I \quad (1+r)A]^{-1}\mathbf{a}_0$$

Substituting 10% for r one gets 5.379. **7.** Let $\mathbf{p} = (\mathbf{p}_1, \mathbf{p}_2)$ (price vector) and $\mathbf{a}_0 = (\mathbf{a}_{01}, \mathbf{a}_{02})$ (labor coefficients vector) with dimensions defined in accordance with the partition of the matrix A. Then it can be shown that (show this!) that

$$\begin{cases} \mathbf{p}_1 = (1+r)\mathbf{p}_1 A_{11} + w\mathbf{a}_{01}, \\ \mathbf{p}_2 = (1+r)[\mathbf{p}_1 A_{12} + \mathbf{p}_2 A_{22}] + w\mathbf{a}_{02}. \end{cases}$$

Here the first equation is sufficient to determine the maximum rate of profit. Therefore it is independent from the production technology of the second group of sectors (in technical terms, non-basics).

Chapter 11

6. (a) Square with vertices $(0, 2), (-2, 0), (0, -2), (2, 0)$; (b) square with vertices $(2, 2), (-2, 2), (-2, -2), (2, -2)$. **7.** (a) Yes; (b) Yes; (c) Yes. **8.** (a) $H^{[+]}$; (b) $H^{[-]}$; (c) H. **10.** Not necessary. For example, the union of two lines $l_1 : y = 0$ and $l_2 : x = 0$ in $\mathbb{R}^2$ is not convex. **11.** *Hint.* The problem can be posed in the following way:

$$\max_{x_1, x_2} 75x_1 + 55x_2$$

under the conditions

$$\begin{cases} x_1 + x_2 \leq 100, \\ 7x_1 + 5x_2 \leq 1000, \\ 2x_1 + 3x_2 \leq 150, \\ x_1, x_2 \geq 0, \end{cases}$$

where x_1 acres of land should be allocated to wheat and x_2 acres of land should be allocated to barley.
Answer. $x_1 = 50, x_2 = 0$. Net revenue will be $450.

12. *Hint.* Let X_1 be the amount invested in government bonds, X_2 the amount invested in auto company A, X_3 the amount invested in auto company B, X_4 the amount invested in textile company C, and X_5 the amount invested in textile company D.
Then the objective function is

$$\max 0.035X_1 + 0.055X_2 + 0.065X_3 + 0.06X_4 + 0.09X_5$$

with the constrains

$$\begin{cases} X_1 + X_2 + X_3 + X_4 + X_5 = 1000000, \\ X_2 + X_3 \leq 500000, \\ X_4 + X_5 \leq 500000, \\ X_1 - 0.35(X_2 + X_3) \geq 0, \\ X_5 - 0.65(X_4 + X_5) \leq 0 \end{cases}$$

and

$$X_1, X_2, X_3, X_4, X_5 \geq 0.$$

Answer. Projected return (value of the objective function): $68,361, $X_1 = 129,630$, $X_2 = 0$, $X_3 = 370,370$, $X_4 = 175,000$, $X_5 = 325,000$. Thus the LP solution suggests that the portfolio manager should not invest in auto company A.

13. *Hint.* The problem can be formulated as follows.
The objective is to minimize risk under the conditions given above.
Objective function: $\min 4x + 9y$.
Fund availability constraint: $50x + 150y \leq 10{,}000{,}000$.

E Answers and Solutions

Required revenue constraint: Since the return in money market is 4%, one money market certificate will earns $0.04 \cdot 50 = 2$. The same calculation leads to $0.1 \cdot 150 = 15$ for the stock market certificate. Therefore the constraint can be written as

$$2x + 15y \geq 400{,}000.$$

Liquidity Constraint: $x \geq 90{,}000$.

The problem therefore can be formulated as

$$\min 4x + 9y$$

under the constrains

$$\begin{cases} 50x + 150y \leq 10{,}000{,}000, \\ 2x + 15y \geq 400{,}000, \\ x \geq 90{,}000, \\ x, y \geq 0. \end{cases}$$

Answer. The solution is $x = 90{,}000$, $y = 14{,}667$. The value of the objective function is 492,000.

14. *Hint.* This is a linear programming problem. It can be formulated as

$$\max W = C_1 + 1.1 C_2$$

subject to

$$\begin{cases} X_1 \geq 0.1 X_1 - 0.2 X_2 - C_1, \\ X_2 \geq -0.3 X_1 + 0.15 X_2 - C_2, \\ 0.05 X_1 + 0.07 X_1 \leq 150, \\ C_1 \geq 1000, \\ C_2 \geq 440, \end{cases}$$

where C_i is the consumption of the output produced by sector i ($i = 1, 2$), X_i is the output of sector i ($i = 1, 2$).

Answer. $W = 1643.53$, $C1 = 11159.53$, $C2 = 440$, $X1 = 1522.83$, $X2 = 1055.12$.

15. *Hint.* All functions are linear. Therefore the problem can be formulated as a linear programming problem (using the given constraints and the assumption that all variables are non-negative). After making substitutions the problem can be formulated as

$$\max Y$$

subject to

$$\begin{cases} -4Y \leq -1200, \\ 0.05C + 0.6Y \leq 300, \\ 0.95C + 2.4Y \leq 1020, \\ -0.95C - 2.4Y \leq -920, \\ 3.8Y \leq 1220, \\ -3.8Y \leq -1200, \\ C + 3Y \leq 1220, \\ -C - 3Y \leq 1200, \\ Y, C \geq 0, \end{cases}$$

where, in particular, $I = 4Y - 1200$.
Answer. $Y = 321.053, C = 157.341, I = 4Y - 1200 = 84.211$.

Appendix A

2. $1 - 1/n$. **5.** $n = 1, n \geq 5$. **8.** (a) $\begin{bmatrix} \lambda^n & n\lambda^{n-1} \\ 0 & \lambda^n \end{bmatrix}$; (b) $\begin{bmatrix} \cos(n\alpha) & -\sin(n\alpha) \\ \sin(n\alpha) & \cos(n\alpha) \end{bmatrix}$.

Appendix B

3. 0. **4.** $x = 0, 1, \ldots, n - 1$. **5.** $8a + 15b + 12c - 19d$. **6.** $abcd$. **7.** $(-1)^n$. **8.** $2n + 1$ **9.** $2n + 1$. **10.** $(a_0 + a_1 + \cdots + a_n)x^n$. **11.** $a_0(x - a_1) \ldots (x - a_n)$. **12.** $(a - b + c + x)(a + b + c - x)(a + b - c + x)(a - b - c - x)$. **13.** $a_1 \ldots a_n - a_3 \ldots a_n - \cdots - a_2 \ldots a_{i-1} a_{i+1} \ldots a_n - \cdots - a_2 \ldots a_{n-1}$. **14.** $n + 1$. **15.** $2^{n+1} - 1$.
16. $\sum_{j=1}^{n} a_j \prod_{i \neq j} (x_i - a_i) + \sum_{j=1}^{n} (x_j - a_j)$

Appendix C

1. i. **2.** $-(14/13)i$. **3.** -0.5. **4.** $x = 1.72 - 0.04i, y = -2/6 + 1.6i$. **5.** $x = 3, y = 4$. **6.** $z = -1 \pm 6i$. **7.** $z_1 = -4 - 4i, z_2 = 1 + 2i$. **8.** $z = -3 + 2i$. **9.** $z_1 = 1 + i, z_2 = 4 - 5i$. **10.** 1. **11.** 2^{4k} for $n = 8k, 2^{4k}(1-i)$ for $n = 8k+1, -2^{4k+1}i$ for $n = 8k+2$, $-2^{4k+1}(1+i)$ for $n = 8k+3, -2^{4k+2}$ for $n = 8k+4, 2^{4k+2}(i-1)$ for $n = 8k+5$, $2^{4k+3}i$ for $n = 8k+6, 2^{4k+3}(1+i)$ for $n = 8k+7$. **12.** $z = \cos\left(\pi \frac{4k+1}{12}\right) + i\sin\left(\pi \frac{4k+1}{12}\right)$ for $k = 0, 1, \ldots, 5$. **13.** $z = 4\left(\cos\left(\pi \frac{3k+1}{6}\right) + i\sin\left(\pi \frac{3k+1}{6}\right)\right)$ for $k = 0, 1, 2, 3$, that is, either $z = \pm 2\left(\sqrt{3} + i\right)$ or $z = \pm 2\left(-1 + \sqrt{3}i\right)$. **15.** $z = \sqrt[5]{5}e^i = \sqrt[5]{5}(\cos(1) + i\sin(1))$. **16.** 3. **17.** 2. **20.** ± 1. **23.** $1 + 2\pi ki, k \in \mathbb{Z}$. **24.** $\begin{bmatrix} -5 + i & -6 \\ 4 & 5 + i \end{bmatrix}$. **25.** $2n$.

Appendix D

1. $\begin{bmatrix} 1 \\ 0 \end{bmatrix}$. **2.** $\begin{bmatrix} 0 & \frac{1}{14} & \frac{1}{7} & \frac{3}{14} \end{bmatrix}$. **3.** $\begin{bmatrix} \frac{3}{14} \\ \frac{1}{7} \\ \frac{1}{14} \\ 0 \end{bmatrix}$. **4.** $\begin{bmatrix} \frac{1}{3} & -\frac{1}{3} & -\frac{1}{3} & 0 \\ -\frac{2}{3} & \frac{2}{3} & \frac{2}{3} & 1 \end{bmatrix}$. **5.** $\begin{bmatrix} \frac{5}{6} & \frac{4}{3} \\ \frac{1}{3} & \frac{1}{3} \\ -\frac{1}{6} & -\frac{2}{3} \end{bmatrix}$.

6. $\begin{bmatrix} 1 & 1 \\ 2 & 2 \\ 3 & 3 \\ 1 & 2 \end{bmatrix} \begin{bmatrix} 1 & 0 & -1 \\ 0 & 1 & 2 \end{bmatrix}$. **8.** $\begin{bmatrix} \frac{2}{3} & \frac{1}{3} & -\frac{1}{3} \\ -\frac{1}{6} & \frac{1}{6} & \frac{1}{3} \\ -\frac{1}{6} & \frac{1}{6} & \frac{1}{3} \end{bmatrix}$ **9.** $\begin{bmatrix} 0 \\ \vdots \\ 1_i \\ \vdots \\ 0 \end{bmatrix} \begin{bmatrix} 0 & \ldots & 1_j & \ldots & 0 \end{bmatrix}$. **11.** $\begin{bmatrix} \frac{71}{93}, \frac{109}{93}, \frac{91}{93} \end{bmatrix} \approx$ $[0.763, 1.172, 0.978]$. **12.** $[3.7, 2.8, 1.2, 3.7]$.

References

1. Abadir, K.M., Magnus, J.R.: Matrix Algebra, Econometric Exercises Series, No. 1. Cambridge University Press, Cambridge (2005)
2. Ben-Israel, A., Greville, T.: Generalized Inverses. Theory and Applications, 2nd edn. Springer, New York (2003)
3. Bapat, R.B., Raghavan, T.E.S.: Nonnegative Matrices And Applications. Cambridge University Press, Cambridge (1997)
4. Braverman, E.M.: Mathematical Models of Planning and Control in Economic Systems. Nauka, Moscow [in Russian] (1976)
5. Campbell, S.L., Meyer, C.D., Jr.: Generalized Inverses of Linear Transformations. Dover, New York (1991)
6. Dasgupta, D.: Using the correct economic interpretation to prove the Hawkins-Simon-Nikaido theorem: One more now. J. Macroecon. **14**(4), 755–761 (1992)
7. Debreu, G., Herstein, I.N.: Nonnegative square matrices. Econometrica **21**, 597–607 (1953) (see also in [20, pp. 57–67])
8. Dhrymes, P.J.: Mathematics for Econometrics, 3rd edn. Springer, New York (2000)
9. Diewert, E.: Index numbers. In: Durlauf, S.N., Blume, L.E. (eds.) The New Palgrave Dictionary of Economics, 2nd edn. Palgrave Macmillan, New York (2008)
10. Gantmacher, F.R.: Theory of Matrices, Vols. 1, 2. Chelsea Publishing, New York (1959)
11. Gel'fand, I.M.: Lectures on Linear Algebra (with the collaboration of Z. Ya. Shapiro), Dover Books on Advanced Mathematics. Dover Publications, New York (1989)
12. Hawkins, D., Simon, H.A.: Note: some conditions of macroeconomic stability. Econometrica **17**, 245–248 (1949)
13. Hawkins, Th.: Continued fractions and the origins of the Perron–Frobenius theorem. Archive History Exact Sci. **62**(6), 655–717 (2008)
14. Judge, G.G., Griffiths, W.E., Hill, R.C., Lütkepohl, H., Lee, T.-C.: The Theory and Practice of Econometrics, 2nd edn. Wiley, New York (1985)
15. Krattenthaler, C.: Advanced Determinant Calculus, The Andrews Festschrift (Maratea, 1998). Sém. Lothar. Combin. 42, Art. B42q, 67 pp. (electronic) (1999)
16. Kurz, M.D., Salvatori, N.: Theory of Production. Cambridge University Press, Cambridge (1995)
17. Kurz, M.D., Salvadori, N.: "Classical" roots of input–output analysis: a short account of its long history. Econ. Syst. Res. *XII*, 153–179 (2000)
18. Kurz, M.D., Salvadori, N.: Input–output analysis from a wider perspective: a comparison of the early works of Leontief and Sraffa. Econ. Syst. Res. *18*(4), 373–390 (2006)
19. Lang, S.: Linear Algebra, Undergraduate Texts in Mathematics. Springer, New York (1989)

20. Newman, P.: Readings in Mathematical Economics-I, Value Theory. The Johns Hopkins Press, Baltimore (1968)
21. O'Neill, M.J., Wood, R.J.: An Alternative Proof of the Hawkins-Simon Condition. Asia-Pacific J. Oper. Res. *16*(2), 173–184 (1999)
22. Pasinetti, L.: Lectures on the Theory of Production. Columbia University Press, New York (1977)
23. Persky, J.: Price indexes and general exchange values. J. Econ. Perspect. *12*(1), 197–205 (1998)
24. Prasolov, V.V.: Problems and Theorems in Linear algebra, Translations of Mathematical Monographs, vol. 134. American Mathematical Society, Providence, RI (1994)
25. Proskuryakov I.V.: Problems in Linear Algebra. Mir, Moscow (1978)
26. Searle, S.R., Willett, L.S.: Matrix Algebra for Applied Economics. Wiley, New York (2001)
27. Shilov, G.E.: Linear Algebra. Dover Publications, New York (1977)
28. Sraffa, P.: Production of Commodities by Means of Commodities-Prelude to a Critique of Economic Theory. Cambridge University Press, Cambridge (1960)
29. Stiglitz, J.E., Walsh, C.E.: Economics, 4th edn. W.W. Norton, New York (2006)
30. Takayama, A.: Mathematical Economics, 2nd edn. Cambridge University Press, Cambridge (1985)
31. Tinbergen, J.: On The Theory of Economic Policy. North Holland, Amsterdam (1952)
32. Varian, H.R.: Intermediate Microeconomics – A Modern Approach, 6th edn. W.W. Norton, New York (2003)
33. Vinberg, E.B.: A Course in Algebra, Graduate Studies in Mathematics, vol. 56. American Mathematical Society, Providence, RI (2003)
34. Yaari, M.E.: Linear Algebra for Social Sciences. Prentice-Hall, Englewood Cliffs, New Jersey (1971)

Index

Adjoint matrix, 67
Argument of a complex number, 239

Basic commodity, 187, 189
Basis, 95
    canonical for $\mathbb{R}^n$, 95
    orthogonal, 109
    orthonormal, 109
Basis transformation matrix, 100
Bilinear form, 157
    symmetric, 157

Canonical basis for $\mathbb{R}^n$, 95
Canonical form (of a matrix), 42
Cauchy inequality, 26
Characteristic polynomial, 145
Closed half-space, 204
Cofactor, 55
Collinear vectors, 95
Commodity
    basic, 187, 189
Complex multiplication, 237
Complex number, 237
Complex vector space, 244
Complexification, 245
Component of a vector, 17
Condition
    Hawkins–Simon, 176
Constant returns to scale technology, 104
Convex set, 200
coordinates, 96

Decreasing returns to scale, 104
Determinant, 55

Dimension of a linear space, 92
Direct product of matrices, 34
Distance
    between a vector and a subspace, 115
    between two vectors, 108
Dot product, 23, 107

Eigenvalue, 142
    Perron–Frobenius, 185
Eigenvector, 142
    Perron–Frobenius, 185
Elementary operations of matrices, 38
Elementwise product of matrices, 35
Exponent of a complex number, 241

Full image, 123
Full rank, 71
Full rank factorization, 253
Fundamental theorem of algebra, 242

Gaussian elimination procedure, 42, 43
Gram–Schmidt orthogonalization process, 112

Hadamard product of matrices, 35
Half-space
    closed, 204
    open, 204
Hawkins–Simon condition, 176
Hawkins–Simon theorem, 176
Homogeneous system of linear equations, 37
Homomorphism, 123
Hypercube, 206
Hyperplane, 100, 202

Identical operator, 125
Identity matrix, 31, 51
Image, 123, 133
    full, 123
Imaginary axis, 237
Imaginary part of a complex number, 237
Imaginary unit, 237
Inconsistent system of linear equations, 81
Increasing returns to scale, 104
Indecomposable matrix, 178
Induction, 217
Induction assumption, 219
Induction principle, 219
Induction variable, 219
Inner product, 23, 107
Invariant subspace, 141
Inverse of a mapping, 123
Inverse of a matrix, 65
Inverse of an operator, 132
Irreducible matrix, 178
Isomorphic linear spaces, 97

Kernel, 134
Kronecker delta function, 51
Kronecker product of matrices, 34
Kronecker–Capelli theorem, 76

Leading coefficient (of a row of a matrix), 42
Leontief model, 172
Linear combination, 92
Linear mapping, 123
Linear operator, 124
Linear programming problem, 195
Linear space, 91
    dimension of, 92
Linear spaces
    isomorphic, 97
Linear transformation, 124
Linearly dependent vectors, 20, 92
Linearly independent vectors, 21, 92
Lower triangular matrix, 56

Mapping, 123
Mathematical induction, 219
Matrices
    similar, 136
Matrix, 30
    symmetric, 36
    transpose of, 35
    adjoint, 67
    basis transformation, 100

    canonical form of, 42
    elementary operations, 38
    full rank factorization of, 253
    identity, 31, 51
    indecomposable, 178
    inverse, 65
    irreducible, 178
    lower triangular, 56
    modal, 155
    nilpotent, 73
    non-negative, 31
    null, 31
    of a full rank, 71
    of a linear transformation, 126
    order, 31
    polynomial of, 52
    positive, 31
    power of, 52
    productive, 172
    rank, 38
    reducible, 177
    row echelon form of, 42
    singular, 67
    upper triangular, 56
Minor, 55
Modal matrix, 155
Multiplication
    complex, 237
Multiplicity of a root, 243

Natural numbers, 217
Nilpotent matrix, 73
Non-collinear vectors, 95
Non-negative matrix, 31
Non-negative vector, 18
Non-singular orthogonal operator, 153
Null matrix, 31
Null operator, 125
Null vector, 18
Numéraire, 182
Number
    complex, 237
    pure imaginary, 237

Open half-space, 204
Operator, 124
    identical, 125
    linear, 124
    null, 125
    orthogonal, 153
    orthogonal non-singular, 153
    orthogonal singular, 153
    self-adjoint, 150

Index

Order of the matrix, 31
Orthogonal basis, 109
Orthogonal operator, 153
Orthonormal basis, 109
Oversized enterprise, 104

Peano axioms, 218
Perron–Frobenius eigenvalue, 185
Perron–Frobenius eigenvector, 185
Polynomial of a matrix, 52
Polytope, 204
Positive matrix, 31
Positive vector, 18
Power of a matrix, 52
Product of matrices, 32
    direct, 34
    elementwise, 35
    Frobenius, 46
    Hadamard, 35
    Kronecker, 34
    Shur, 35
    tensor, 34
Productive matrix, 172
Productive system, 172
Projection of a vector to a subspace, 114
Projection of a vector to a unit vector, 113
Pseudoinverse, 250
Pure imaginary number, 237

Quadratic form, 157

Rank of a matrix, 38
Real axis, 237
Real part of a complex number, 237
Real space, 17
Real vector space, 244
Reducible matrix, 177
Root
    simple, 243
Rotation matrix, 33
Row echelon form (of a matrix), 42

Scalar, 17
Scalar product, 23, 107
Self-adjoint operator, 150
Set
    convex, 200
Shur product of matrices, 35

Similar matrices, 136
Simple root, 243
Singular matrix, 67
Singular orthogonal operator, 153
Span, 99
Sraffa Model, 183
Strongest Induction Principle, 222
Subspace, 98
    invariant, 141
    spanned by given vectors, 99
Sylvester criterion, 159
Symmetric matrix, 36
System
    productive, 172
System of linear equations
    homogeneous, 37
    inconsistent, 81

Tensor product of matrices, 34
Theorem
    Hawkins–Simon, 176
Trace of a matrix, 35
Transformation, 124
    linear, 124
Translation, 124
Transpose of a matrix, 35

Upper triangular matrix, 56

Vandermonde determinant, 232
Vector, 17, 91
    components, 17
    non-negative, 18
    positive, 18
    projection of, 113, 114
Vector space, 91
Vector space over complex numbers, 244
Vector space over real numbers, 244
Vectors
    collinear, 95
    linearly dependent, 20, 92
    linearly independent, 21, 92
    non-collinear, 95
    orthogonal, 25
    parallel, 27
Vertex of a polytope, 205

Zero vector, 18

Made in the USA
Las Vegas, NV
09 February 2024